NGSS VERSION

nutty brown press

NEW EDITION

ARGUMENT-DRIVEN INQUIRY©

THREE-DIMENSIONAL INVESTIGATIONS FOR BIOLOGY

VICTOR SAMPSON, PH.D.

nutty brown press

Argument-Driven Inquiry© Three-Dimensional Investigations for Biology, NGSS Version

Author: Victor Sampson, Ph.D.

ISBN: 978-1-963684-04-9

Nutty Brown Press
12707 Nutty Brown Rd, Suite G
Austin, TX 78737
www.nuttybrownpress.com

Permissions

Disclaimers

Library of Congress Cataloging-in-Publication Data
Nutty Brown Press
Argument-Driven Inquiry© Three-Dimensional Investigations for Biology, NGSS Version
LCCN: 2024918174 (print)
ISBN: 978-1-963684-04-9

There are several good reasons for teaching students about science. First, science courses give students a strong foundation in the basics to prepare them for what they will be expected to know or do after they graduate from high school. Second, teaching students about science can get them interested in the subject so they are more likely to go into a science or science-related career. Science courses can also help satisfy students' natural curiosity about how the world works. We, however, think that one of the most important reasons for teaching students about science is that it will be useful for them in everyday life.

Science is useful because it, along with engineering, mathematics, and the technologies that are made possible by the other three fields, affects almost every aspect of modern life. For example, people need to understand science to think meaningfully about policy issues that affect their communities or to make informed decisions about what food to eat, what medicine to take, or what products to use. People can use their understanding of science to help evaluate the acceptability of different ideas or convince others about the best course of action when faced with a wide range of options. In addition, understanding how science works and all the new scientific findings reported each year in the media can be interesting, relevant, and meaningful on a personal level and can open doors to exciting new professional opportunities. The more a person understands science—which includes the theories, models, and laws that scientists have developed over time to explain how and why things happen and how these ideas are developed and refined based on evidence—the easier it is for that person to have a productive and fulfilling life in our technology-based and information-rich society. Thus, science is useful to everyone, not just to future scientists.

A Framework for K-12 Science Education,[1] which was released in 2012, outlined a broad set of expectations for students to help ensure that all students, rather than just a select few, are able to engage in public discussions on science-related issues, be careful consumers of scientific and technical information, and enter the careers of their choice by the time they graduate from high school. These expectations were then used to guide the development of the *Next Generation Science Standards* (NGSS) and, subsequently, revisions to curriculum, instruction, assessment, and professional development for educators across the United States. *A Framework for K-12 Science Education* is organized around three dimensions of science that convey what is important in terms of what students should be able to do at the end of the fifth, eighth, and twelfth grades. The three dimensions included in *A Framework for K-12 Science Education* are (1) science and engineering practices (SEPs), (2) crosscutting concepts (CCs), and (3) disciplinary core ideas (DCIs). SEPs are the different activities scientists engage in as they attempt to generate new concepts, models, theories, or laws that are both valid and reliable. CCs are ideas that are used across disciplines. These concepts provide a framework or a lens that people can use to explore a wide range of natural phenomena; thus, these concepts often influence what people focus on or pay attention to when they attempt to understand how something works or why something happens. Finally, the DCIs are a set of key organizing principles with broad explanatory power within a discipline. All three

[1] National Research Council, *A Framework for K-12 Science Education: Practices, Crosscutting Concepts, and Core Ideas* (Washington, DC: National Academies Press, 2012).

of these dimensions of science are equally important. Students, as a result, not only need to know about the SEPs, CCs, and DCIs but must also be able to use all three dimensions at the same time to figure things out or to solve problems.

Science teachers will need to use new instructional approaches to help students learn how to use SEPs, CCs, and DCIs to figure things out or solve problems. These instructional approaches must give students an opportunity to do science. To help Biology teachers make this instructional shift, we have developed a way of teaching called Argument-Driven Inquiry (ADI). ADI is a research-developed and classroom-tested instructional model that is designed to give students an opportunity to use SEPs, CCs, and DCIs to construct and critique claims about how things work or why things happen. As part of this process, students must talk, listen, read, and write to obtain, evaluate, and communicate information. As a result, ADI creates a rich learning environment that enables students to learn science in a deep and meaningful way and develop essential language and literacy skills simultaneously.

This book describes how ADI works and why it is important. It also includes eight investigations that Biology teachers can use to help students reach the performance expectations found in the NGSS. The eight investigations described in this book will also enable students to develop many of the essential reading, writing, speaking, and listening skills outlined in the Common Core State Standards for ELA[2] because ADI gives students an opportunity to read scientific texts, make presentations, respond to questions and critiques, and then write and revise reports as part of each investigation. Finally, and perhaps most importantly, ADI can help emerging multilingual students meet WIDA's English Language Development Standards (ELDS)[3] standards because it provides a language-rich context where students can use receptive and productive language to communicate and negotiate meaning with others. Biology teachers, as a result, can use these investigations to not only help students learn what they need to learn to meet academic expectations but also help them learn how to actually use (rather than recite) ideas and practices from the fields of science, ELA, and language development in a meaningful and useful way.

---

[2] National Governors Association Center for Best Practices & Council of Chief School Officers, *Common Core Standards for English language arts and literacy in history/social studies, science, and technical subjects* (Washington, DC: 2010).

[3] WIDA, *WIDA English Language Development Standards Framework, 2020 Edition Kindergarten-Grade 12* (Wisconsin: 2020).

About the Author

Victor Sampson is an Associate Professor of STEM (science, technology, engineering, and mathematics) education at The University of Texas at Austin (UT-Austin). He received a BA in zoology from the University of Washington, an MIT from Seattle University, and a Ph.D. in curriculum and instruction with a specialization in science education from Arizona State University. Victor also taught high school biology and chemistry for nine years in public schools. He is an expert in argumentation and three-dimensional instruction in science education, teacher learning, and assessment. He is also an expert in the design and development of new educational apps. Victor has published over 50 research articles and more than 10 books in the field of STEM education. While at Florida State University (2007-2014) and University of Texas (2014-current), he served as the PI or a Co-PI on six large research projects that were funded by either the National Science Foundation (NSF) or the Institute of Education Sciences (IES). Victor has received several awards for his scholarship, including the 2008 National Association for Research in Science Teaching (NARST) Outstanding Dissertation award and the 2012 NARST Early Career award. He is also a National Science Teaching Association (NSTA) Fellow.

Table of Contents

Introduction 9

A Vision for Science Education in the United States 9

The Need for New Ways of Teaching Science in Biology 12

Notes for "Future You" 15

The Argument-Driven Inquiry (ADI) Instructional Model 16

Development of ADI 16

An Overview of the Stages of an ADI Investigation 17

Stage 1: TASK 18

Stage 2: IDEAS 20

Stage 3: PLAN 22

Stage 4: DO 25

Stage 5: SHARE 27

Stage 6: REFLECT 37

Stage 7: REPORT 41

Quick Reference Guide for the Stages of an ADI Investigation 46

Notes for "Future You" 48

Ways to Differentiate Instruction When Using the ADI instructional Model 49

Meaningful, Rigorous, and Equitable by Design 49

The Problem with Making Modifications to 3D Instructional Materials 50

Accommodations Already Embedded into the ADI Instructional Model 52

Additional Differentiation Based on Need 52

Notes for "Future You" 55

Support for Emerging Multilingual Students When Using the ADI Instructional Model 56

Principles and Best Practices in the Context of Science Instruction 56

Additional Support for Students Learning Academic English and Science 57

Notes for "Future You" 59

Ways to Create a Safer Learning Environment Using the ADI Instructional Model 60

The Physical Environment 60

The Proper Use of Personal Protective Equipment 61

The Creation of Classroom Rules and Procedures 61

Cleaning Up After an Investigation or Design Challenge 62

Notes for "Future You" 63

Assessment When Using the ADI Instructional Model 64

Purposes of Assessment 64

Assessment as Making Thinking Visible 64

Formative Assessments 65

Embedded Performance Tasks 65

Exit Tickets 65

Q&As without Scores 66

Educative Assessments 67

Embedded Peer Review 67

Embedded Reflective Discussions 67

Q&As with Feedback 67

Summative Assessments 68

Embedded Investigation Reports 68

Q&As with Scores 68

Notes for "Future You" 69

Resources for Teachers 70

Lesson Plans 70

Student Handout Copy Masters 71

The ADI Learning Hub 71

Investigation Kits 72

Classroom Materials 72

Notes for "Future You" 73

Investigations 75
10.1 Swimming in the Dead Sea - Lesson Plan 77
10.1 Swimming in the Dead Sea - Student Handout 121
10.2 Tissue Repair - Lesson Plan 135
10.2 Tissue Repair - Student Handout 181
11.1 Elk Population Decline in Colorado - Lesson Plan 195
11.1 Elk Population Decline in Colorado - Student Handout 243
11.2 Bark Beetles and Climate Change in Washington - Lesson Plan 259
11.2 Bark Beetles and Climate Change in Washington - Student Handout 305
12.1 Chromosome Reduction in Reproductive Cells - Lesson Plan 321
12.1 Chromosome Reduction in Reproductive Cells - Student Handout 365
12.2 Fruit Fly Body Color and Wing Shape - Lesson Plan 379
12.2 Fruit Fly Body Color and Wing Shape - Student Handout 425
13.1 Antibiotic Resistance in Bacteria - Lesson Plan 443
13.1 Antibiotic Resistance in Bacteria - Student Handout 491
13.2 Cichlids of Lake Malawi - Lesson Plan 507
13.2 Cichlids of Lake Malawi - Student Handout 553

Other Useful Stuff 567
Exit Ticket 567
ADI Learning Hub Subscription 568
Complete Hands-On Materials List 569
Safety Rules Acknowledgment Form 572
Peer Review Guide and Teacher Scoring Rubric (PRG/TSR) 573

A Vision for Science Education in the United States

The current aim of science education in the United States is for all students to become proficient in science by the time they finish high school. Science proficiency consists of four interrelated aspects.[4] First, it requires an individual to know important scientific explanations about the natural world, to be able to use those explanations to solve problems, and to be able to understand new explanations when they are introduced. Second, it requires an individual to be able to generate and evaluate scientific explanations and scientific arguments. Third, it requires an individual to understand the nature of scientific knowledge and how scientific knowledge develops over time. Finally, and perhaps most importantly, an individual who is proficient in science should be able to participate in scientific practices (such as planning and carrying out investigations, analyzing and interpreting data, and arguing from evidence) and communicate in a manner that is consistent with the norms of the scientific community. These four aspects of science proficiency include the knowledge and skills that all people need to be able to pursue a degree in science, be prepared for a science-related career, and participate in a democracy as an informed citizen.

This view of science proficiency served as the foundation for the development of *A Framework for K-12 Science Education*.[5] *A Framework for K-12 Science* was then used to guide the creation of *Next Generation Science Standards (NGSS)*.[6] The *NGSS*, in turn, outlines what all students should be able to do by the time they complete grades 2, 5, 8, and 12 as they progress toward the goal of science proficiency. The *NGSS*, like *A Framework for K-12 Science Education*, are organized around three important dimensions of science. These dimensions are science and engineering practices (SEPs), crosscutting concepts (CCs), and disciplinary core ideas (DCIs). Figure 1 shows the three dimensions of *A Framework for K-12 Science Education* and the *NGSS* and how they overlap.

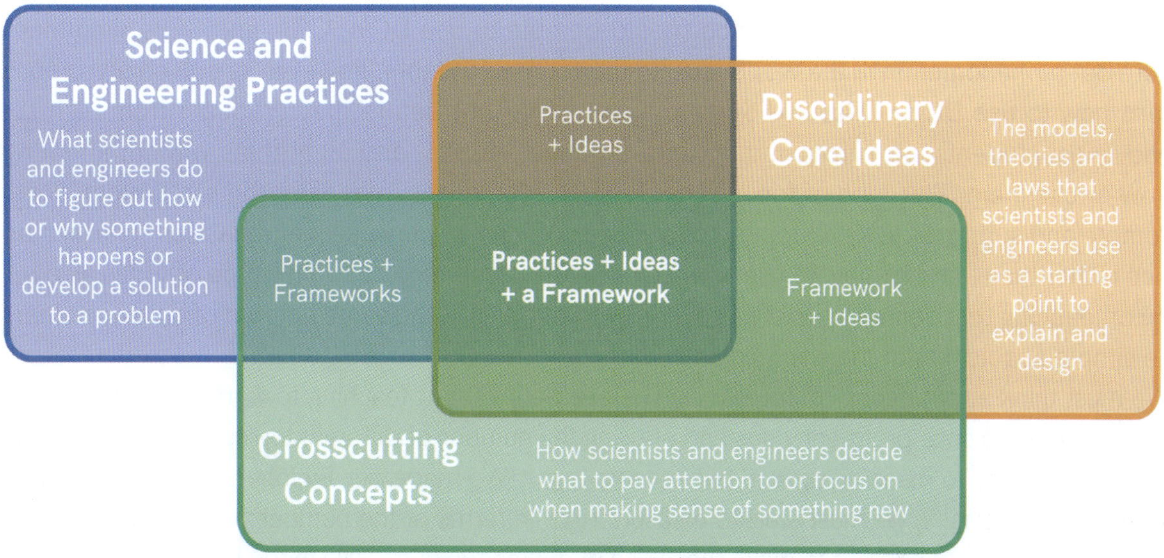

Figure 1. The three dimensions of science in *A Framework for K-12 Science Education* and the *NGSS*

4 Richard A. Duschl, Heidi A. Schweingruber, and Andrew W. Shouse,Taking science to school: Learning and teaching science in grades K-8 (Washington, DC: National Academies Press, 2007).

5 National Research Council, *A Framework for K-12 Science Education*.

6 NGSS Lead States, *Next Generation Science Standards: For States, By States* (Washington, DC: National Academies Press, 2013).

A DCI is central to understanding a variety of natural phenomena. An example of a DCI in physical science is forces and motion. This DCI can be used to help explain not only why a stationary ball starts to move after it has been hit by a rolling ball but also why it takes several people to make a large car move and why the rope in a tug-o-war will sometimes not move at all even when many people are pulling on it at the same time. An example of a DCI in life science is natural selection. This DCI can be used to not only explain why animals and plants are so well adapted to where they live but also why specific traits become more common in population, why there are so many different species on Earth, and how farmers can create crops with a desired trait through selective breeding. Finally, an example of a DCI in Earth-Space Science is plate tectonics. This DCI can be used to explain such things as the locations and sizes of mountain ranges, chains of islands in the middle of the ocean, volcanoes, earthquakes, and the locations of different types of fossils.

The CCs, in contrast, are principles that are important across the disciplines of science. The CCs help scientists think about what to focus on or pay attention to during an investigation or when developing a solution to a problem. For example, one of the CCs found in the NGSS is cause-and-effect relationships. This CC is important in many different fields of Biology, including ecology, biochemistry, anatomy, and cell biology. It is equally important in physics and chemistry. Physicists use this CC when they study how things move, why things change temperature, and how circuits behave. Chemists use this CC when they study the energetics and dynamics of chemical processes.

The SEPs describe what scientists do as they attempt to make sense of the natural world (or, in the case of engineers, design solutions to problems). Some of the SEPs include familiar activities that we typically associate with "doing" science, such as (a) asking questions, (b) planning and carrying out investigations, and (c) analyzing and interpreting data. The rest of the SEPs, however, focus on activities that are not always associated with "doing science". These SEPs include (a) developing and using models, (b) using mathematics and computational thinking, (c) constructing explanations and designing solutions, (d) engaging in argument from evidence, and (e) obtaining, evaluating, and communicating information. All these SEPs are important because scientists engage in different practices at different times and in different orders depending on what they are studying and what they are trying to do at that point in time.

Students who are proficient in science should be able to use DCIs, CCs, and SEPs simultaneously to figure out a phenomenon in the world around them. Few students, however, are given an opportunity to learn how to use DCIs, CCs, and SEPs to figure things out before they graduate from high school. Instead, most students are introduced to facts, concepts, terms, and algorithms without a real reason to know or use them. This type of focus on facts, concepts, terms, and algorithms in Biology does little to promote and support the development of science proficiency because it emphasizes "learning about" science rather than learning how to use science to "figure something out". This type of focus also reflects a view of teaching that defines "rigor" as covering more topics in the same amount of time and "learning" as the simple acquisition of more information.

We must think about rigor in different ways before we can start teaching toward the goal of helping all students develop science proficiency. Instead of using the number of different topics covered in a particular grade level as a way to measure rigor in our schools (e.g., "We made our Biology course more challenging by adding more topics."), we must start to measure rigor in terms of the number of opportunities students have to use DCIs, CCs, and SEPs to make sense of a phenomenon (e.g., "We made Biology more challenging because students have to figure out how body color and wing shape traits are inherited in fruit flies."). A rigorous class, in other words, is one where students are expected to do science, not just learn about science. From this perspective, our goal as teachers should be to help our students learn how to use DCIs and CCs as tools to plan and carry out investigations, construct and evaluate explanations, and question how we know what we know instead of just ensuring that we "cover" all the different DCIs and CCs that are found in the NGSS by the end of the school year.

To better promote and support the development of science proficiency, we must also revise and expand our definition of learning and how it happens. A narrow definition of learning views it as an individual process where students accumulate more and more information over time. This view of learning often defines conceptual understanding as developing abstract representations from concrete examples. Rather than using this narrow definition, we define learning as both a social and individual process that involves being exposed to new ideas and ways of doing things, trying out these new ideas and practices under the guidance of more experienced people, and then adopting the ideas and practices that are found to be useful for understanding the world.[7] This more expanded definition of learning is often referred to as sense-making. Sense-making is the process by which a learner uses their own ideas and experiences along with the ideas (DCIs and CCs) and practices of science to develop new and more robust explanations for how and why things happen in the world around them. Sense-making includes using new ideas and practices to build a deeper understanding of an aspect of the world the student is already familiar with and the process by which students try to understand unfamiliar phenomena.

Finally, and perhaps most important to supporting the development of science proficiency, we must also revise and expand our definition of conceptual understanding. Instead of defining it as "the ability to make an abstraction from several concrete examples", conceptual understanding should be defined as "the ability to transfer ideas and practices learned in one context to other contexts and knowing which ideas and practices are the most appropriate to use to predict and explain a novel phenomenon within a given context". This expanded definition of conceptual understanding is necessarily intertwined with the definition of sense-making because a student who has a conceptual understanding of an idea will then be able to transfer that knowledge to make sense of a novel situation.

From this perspective on learning, sense-making, and conceptual understanding, students "do science" while in school not because it is fun or interesting (which is true for many), but because it gives students a reason to use the ideas and practices of the disciplines. When students are given repeated opportunities to use DCIs, CCs, and SEPs to make sense of the world, they will begin to see why these ideas and practices are valuable and where or when they are most useful. Over time, students will adopt these ideas and practices and start using them independently. We, therefore, must give our students opportunities to experience how scientists figure things out and share ideas, so they can learn and adopt the practices of the scientific community with its specific goals, methods for collecting, analyzing, and interpreting data, ways of explaining, and ways of determining if a new idea is valid, acceptable, and useful or not.

It is important to keep in mind that helping students learn how to use the ideas and practices of science to figure things out is not a "hands-off" approach to teaching. The process requires constant input and guidance about "what counts" from teachers familiar with the goals, norms, and the ways things are done in science. Thus, learning how to use DCIs, CCs, and SEPs to figure things out depends on supportive and informative interactions with teachers. This is important because students must have a supportive and educative learning environment to try out new ideas and practices, make mistakes, and refine what they know and how they do things before they are able to adopt the ideas and practices of science as their own.

---

[7] Sarah Michaels, Andrew W. Shouse, and Heidi A. Schweingruber, *Ready, Set, Science!: Putting Research to Work in K-8 Science Classrooms* (Washington, DC: National Academies Press, 2008); National Academies of Science, Engineering, and Medicine, *How People Learn II: Learners, Contexts, and Cultures* (Washington, DC: National Academies Press, 2018); National Research Council, *A Framework for K-12 Science Education.*

The Need for New Ways of Teaching Science in Biology

Biology has historically been taught through a combination of direct instruction and hands-on activities. A typical lesson often begins with the teacher introducing students to a new concept through direct instruction. Next, the teacher illustrates the concept by showing a demonstration or asking students to complete a hands-on activity to provide them with a memorable experience with the concept. If the memorable experience is a hands-on activity, the teacher will often provide their students with a step-by-step procedure to help ensure that no one in the class "gets lost" or "does the wrong thing" and that everyone "gets the right results". The teacher will usually assign a set of questions for the students to answer independently or in groups after the activity to ensure that everyone in the class "reaches the right conclusion". The lesson then ends with the teacher reviewing the concept and all related terms to make sure that everyone in the class learned what they were "supposed to have learned". The teacher often accomplishes this last step by leading a whole-class discussion, assigning a worksheet to complete, or having the students play an educational game. This approach can be described as a "content-activity-practice" lesson structure.

However, a great deal of research suggests that the content-activity-practice lesson structure does little to help students learn key concepts.[8] This finding is troubling because, as noted earlier, one of the main goals of this type of lesson is to help students understand an important concept by giving them a memorable experience. In addition, this type of lesson does little to help students learn how to plan and carry out investigations or analyze and interpret data, because students have no voice or choice during the activity. Students are expected to follow a set of directions rather than think about what data they will collect, how they will collect it, and what they will need to do to analyze it once they have it. Because of the emphasis on following a procedure and getting the "right" results, these activities can also lead to misunderstandings about the nature of scientific knowledge and how this knowledge is developed over time. As a result, these hands-on activities do not reflect how science is done.

In the early 2000s, many instructional materials started to incorporate inquiry-based lesson structures, such as the 5Es, to help address the shortcomings of the content-activity-practice approach. Instructional materials that use an inquiry-based lesson structure tend to introduce students to new ideas at the end of a lesson instead of at the beginning. Inquiry-based lessons, in other words, give students a firsthand experience with a concept before introducing terms and vocabulary.[9] Inquiry-based lessons are thus often described as an "activity before content" approach to teaching science.[10] The focus of these "activity before content" lessons, as the name implies, is to help students understand important science content. Inquiry-based lessons give students more opportunities to learn how to plan and carry out investigations, analyze and interpret data, and develop explanations. An inquiry-based lesson structure also provides students with more agency than the content-activity-practice lesson structure, which in turn helps students see themselves as "knowers and doers" of science.

Although classroom-based research has indicated that inquiry-based lessons are effective at helping students understand science content and give students more agency than content-activity-practice lessons, they do not do as much as they could to help students develop all four aspects of science proficiency.[11,12]

[8] Michaels, Shouse, and Schweingruber, *Ready, Set, Science!*

[9] National Research Council, *A Framework for K-12 Science Education.*

[10] Sean Cavanagh, "Science Labs: Beyond Isolationism," *Education Week* 26, no. 18 (2007): 24-26.

[11] Michaels, Shouse, and Schweingruber, *Ready, Set, Science!*

[12] National Research Council, *A Framework for K-12 Science Education.*

For example, inquiry-based lessons are usually not designed in a way to encourage students to learn how to use DCI, CCs, and SEPs at the same time, because the DCI or CC are not introduced until the end of the lesson when these important concepts are "discovered" by the students. These lessons also do not give students an opportunity to participate in the full range of SEPs because they tend to be designed in a way that gives students few opportunities to learn how to participate in practices that focus on how new ideas are developed, shared, refined, and eventually validated within the scientific community. These important sensemaking practices include developing and using models, constructing and evaluating experimental designs, and communicating explanations individually and collaboratively. Inquiry-based lessons that do not focus on sensemaking also do not create a need for students to read, write, or speak in science or engineering terms because they focus on introducing students to new ideas and then how to design and carry out investigations rather than how to develop, share, critique, and revise ideas. As a result, these inquiry-based lessons are often not used to help them develop essential literacy skills. To help address this problem, teachers need to start using instructional materials that give students more opportunities to "figure things out" in a scientific way instead of just "learning about things" in science.

This emphasis on "figuring things out" instead of "learning about things" represents a big change in how we teach science. To figure out how things work or why things happen in a way that is consistent with how science or engineering is done, students must have opportunities to use DCI, CCs, and SEPs simultaneously to make sense of the world around them or to design solutions to problems. The focus on using these elements together during a lesson is called three-dimensional (3D) instruction because students have an opportunity to use these three dimensions of science or engineering at the same time to understand how something works, to explain why something happens, or to develop a novel solution to a problem. The image below shows the difference between 3D instruction, 2D instruction, and 1D instruction in terms of classroom focus.

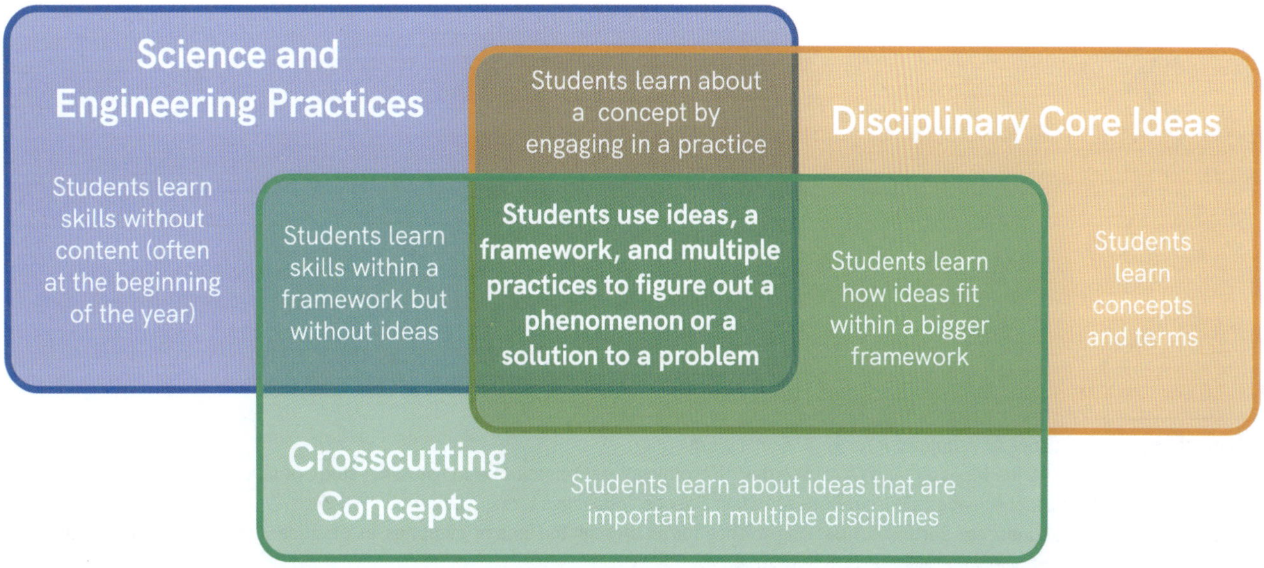

Figure 2. The difference between 1D, 2D, and 3D instruction

When teachers create learning experiences using 3D instruction, they encourage their students to develop or use conceptual models, develop explanations, share and critique ideas, and use evidence to support or challenge claims, all of which allow students to develop the knowledge and skills they need to be proficient in science. A large body of research suggests that all students benefit from 3D instruction because it makes the learning process more active, equitable, and inclusive.[13]

One of the best ways to integrate 3D instruction into Biology is through investigations that focus on making sense of how the world works. Well-designed investigations can provide opportunities for students to not only use DCIs to understand how something works or to explain why something happens but also to use several different CCs and SEPs during the same lesson. A teacher, for example, can give their students an opportunity to figure out why drinking water from the Dead Sea is hazardous. The teacher can then encourage students to use what they know about the structure and function in multicellular organisms (a DCI) and their understanding of the movement of matter into, within, and out of systems (a CC), to plan and conduct an investigation with the goal of developing an explanation (two SEPs) for what happens to a person's body when concentrated salt water enters the digestive system. In addition to planning and carrying out an investigation and developing an explanation, students must also ask questions, analyze and interpret data, argue from evidence, and obtain, evaluate, and communicate information (an additional four SEPs).

Using DCIs, CCs, and SEPs concurrently is also important because it creates a classroom experience that parallels how science is done. This, in turn, gives all students who participate in the investigation an opportunity to deepen their understanding of what it means to do science and to develop a positive science-related identity.[14,15,16] When students develop such a positive identity in school, they are more likely to describe themselves as someone who enjoys learning about science and is "good at" science. A positive science-related identity is also associated with an increased likelihood of pursuing a science degree in college and choosing a science-related career later in life.[17] In the next section of this book, we will describe how to foster the development of science proficiency through 3D instruction using an instructional model called Argument-Driven Inquiry (ADI).

---

[13] National Research Council, *A Framework for K-12 Science Education*.

[14] Heidi B. Carlone, Catherine M. Scott, and Cassi Lowder, "Becoming (less) scientific: A longitudinal study of students' identity work from elementary to middle school science," *Journal of Research in Science Teaching* 51, no. 7 (2014): 836-869.

[15] Edna Tan and Angela Calabrese Barton, "Unpacking science for all through the lens of identities-in-practice: the stories of Amelia and Ginny," *Culture Studies of Science Education* 3, no. 1 (2008): 43-71.

[16] Edna Tan and Angela Calabrese Barton, "Transforming Science Learning and Student Participation in Sixth Grade Science: A Case Study of a Low-Income, Urban, Racial Minority Classroom," *Equity & Excellence in Education* 43, no. 1 (2010): 38-55.

[17] National Research Council, *A Framework for K-12 Science Education*.

Notes for "Future You"

Think about everything you have read in this chapter. Then, write down some things (including the definitions of important terms) that you think will be helpful for you to remember when you use an ADI investigation in class. We all forget things, so do what you can now to help "future you".

Some things I want to make sure that I remember

The Argument-Driven Inquiry (ADI) Instructional Model

Development of ADI

The ADI instructional model was first introduced in 2009.[18,19] The original version of this instructional model included eight stages of activities and was designed to reflect the most up-to-date research-based recommendations about how to promote and support learning in science that was available at that time.[20,21,22] The 2009 version of the ADI instructional model was then tested and improved through design-based research that was carried out in a wide range of classrooms over the next ten years. It was also refined over this period as new recommendations or "best practice" guidelines about how to foster student learning in science were published.[23,24,25] This development process not only resulted in numerous peer-reviewed journal articles that document how the use of this instructional model in science supports the development of all four aspects of science proficiency,[26,27,28,29,30,31,32,33] but also a series of books for teachers to use in grades 3-12.[34] The ADI instructional model was transformed into

---

18 Victor Sampson and Leeanne Gleim, "Argument-Driven Inquiry to Promote the Understanding of Important Concepts & Practices in Biology," *The American Biology Teacher* 71, no. 8 (2009): 465-472.

19 Victor Sampson, Jonathon Grooms, and Walker Joi, "Argument-Driven Inquiry," *The Science Teacher* 76, no. 8 (2009): 42-47.

20 J. D. Bransford, A. L. Brown, and R. R. Cocking, *How People Learn: Brain, Mind, Experience, and School: Expanded Edition* (Washington, DC: National Academies Press, 2000).

21 Michaels, Shouse, and Schweingruber, *Ready, Set, Science!*

22 National Research Council, *America's Lab Report: Investigations in High School Science* (Washington, DC: National Academies Press, 2006).

23 National Research Council, *A Framework for K-12 Science Education*.

24 National Academies of Science, Engineering, and Medicine, *How People Learn II*.

25 Brett Moulding, Nancy Songer, and Kerry Brenner, eds., *Science and Engineering for Grades 6 -12: Investigation and Design at the Center* (Washington, DC: National Academies Press, 2019).

26 Victor Sampson, Jonathon Grooms, and Joi Phelps Walker, "Argument-Driven Inquiry as a way to help students learn how to participate in scientific argumentation and craft written arguments: An exploratory study," *Science Education* 95, no. 2 (2010): 217-257.

27 Victor Sampson and Joi Phelps Walker, "Argument-Driven Inquiry as a Way to Help Undergraduate Students Write to Learn by Learning to Write in Chemistry," *International Journal of Science Education* 34, no. 10 (2012): 1443-1485.

28 Joi Phelps Walker and Victor Sampson, "Learning to Argue and Arguing to Learn: Argument-Driven Inquiry as a Way to Help Undergraduate Chemistry Students Learn How to Construct Arguments and Engage in Argumentation During a Laboratory Course," *Journal of Research in Science Teaching* 50, no. 5 (2013): 561-596.

29 Tuba Demircioglu and Sedat Ucar, "Investigating the Effect of Argument-Driven Inquiry in Laboratory Instruction," *Educational Sciences: Theory and Practice* 15, no. 1 (2015): 267-283.

30 Hsiang-Ting Chen et al., "Using a modified argument-driven inquiry to promote elementary school students' engagement in learning science and argumentation," *International Journal of Science Education* 38, no. 2 (2016): 170-191.

31 Pınar Seda Çetin and Gülüzar Eymur, "Developing Students' Scientific Writing and Presentation Skills through Argument Driven Inquiry: An Exploratory Study," *Journal of Chemical Education* 94, no. 7 (2017): 837-843.

32 Anna M. Strimaitis et al., "Promoting Equitable Biology Lab Instruction by Engaging All Students in a Broad Range of Science Practices: An Exploratory Study," *School Science and Mathematics* 117, no. 3-4 (2017): 92-103.

33 Pinar Seda Cetin et al., "Exploring the effectiveness of engagement in a broad range of disciplinary practices on learning of Turkish high-school chemistry students," *International Journal of Science Education* 40, no. 5 (2018): 473-497.

34 Victor Sampson et al., *Argument-Driven Inquiry in Biology: Lab Investigations for Grades 9-12* (Arlington, VA: NSTA Press, 2014); Victor Sampson et al., *Argument-Driven Inquiry in Chemistry: Lab Investigations for Grades 9-12* (Arlington, VA: NSTA Press, 2014).

the current seven-stage version in 2020. This book's focus is the most current, seven-stage iteration of the ADI instructional model.

The ADI instructional model is intended to serve as a guide or a template for creating meaningful, rigorous, and equitable 3D science investigations. When teachers use these investigations inside their classrooms, their students not only have an opportunity to learn new DCIs, CCs, and SEPs that are found in NGSS as they figure out how or why something happens, but they are also encouraged to use DCIs, CCs, and SEPs that they learned during prior investigations (or grades) as part of the process. The instructional model also ensures that each investigation includes the following features:

1. Students make sense of a phenomenon (an observable event), so they figure out why or how things happen in the world instead of just "learning about" terms.
2. Students have agency to make decision about how they go about figuring out a phenomenon.
3. Students are encouraged to value other people's ideas and participation as helpful for sense-making.
4. Students interact with the content in multiple ways, have access to the supports they need, are pushed to think critically, and have numerous opportunities to learn from their mistakes.
5. Investigations create a language-rich learning environment where students must read, write, and talk to obtain, evaluate, and communicate information in the service of sense-making.
6. Student thinking is made visible throughout the investigation, which provides ongoing opportunities for educative, formative, and summative assessments by teachers.

An Overview of the Stages of an ADI Investigation

The ADI instructional model includes seven stages of classroom activity (see Figure 3). These seven stages are included in the model to provide a structure that supports students as they make sense of a phenomenon, create an explanation for that phenomenon, share arguments to validate these explanations, and then refine their explanations and arguments based on feedback. ADI also provides an authentic context for students to practice essential literacy skills. Finally, and perhaps most importantly, these investigations create a language-rich learning environment that enables emerging multilingual students to learn a new language as they learn science.

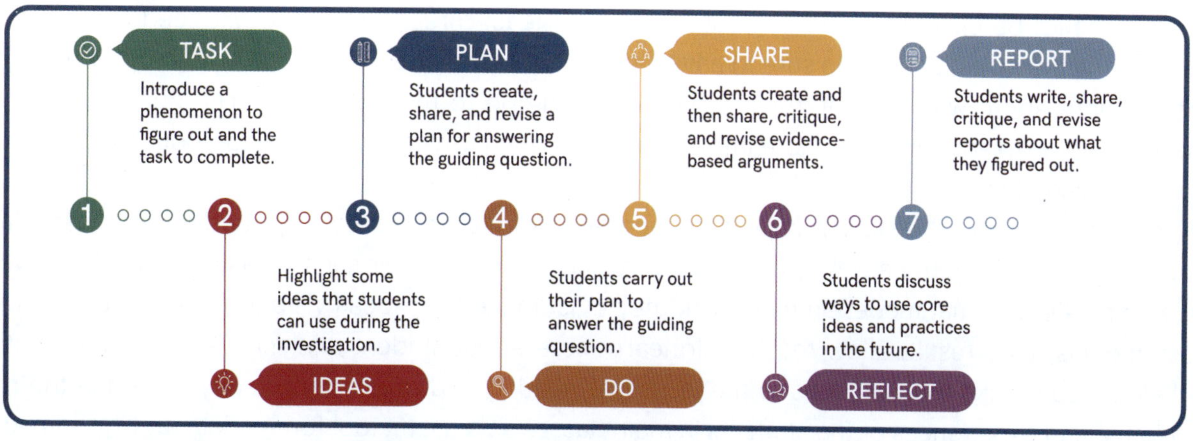

Figure 3. The ADI Instructional Model

In this section, we will explain what happens during each stage of an ADI investigation. These stages are the same across investigations. As a result, students quickly learn what is expected of them during each stage of an investigation and can focus their attention on learning the DCIs, CCs, and SEPs needed to develop science proficiency. We will also highlight some important things to "keep in mind" when guiding students through each stage of an investigation to maximize student learning and minimize wasted time.

Stage 1: TASK

The Task stage includes three activities:

1. An introduction of a phenomenon and students sharing what they notice about it.
2. An introduction to the guiding question of the investigation (what students need to figure out).
3. Students sharing their ideas about how or why the phenomenon or problem happens with their group and identifying things they need to learn more about.

The first activity of this stage of an investigation begins with the introduction of a phenomenon. This introduction at the beginning of the investigation creates a need for students to figure something out. The phenomenon will usually be presented to students in video format. Students should be encouraged to record what they notice about the phenomenon on the Task stage handout as they watch the video introduction. They students should then be given an opportunity to share their observations with the rest of the class. At this point, students will be interested and would want to know more about the phenomenon.

The Task (what students need to do) and the guiding question (what students need to figure out) of the investigation are introduced to the students during the second activity of this stage. The guiding question is intentionally designed to align with DCI and one or more CCs found in the NGSS.

During the third activity of this stage, students are given an opportunity to share what they already know about the phenomenon. Students begin by creating an initial model on the Task stage handout to illustrate what they know. They should also be encouraged to use words to help explain their ideas or thinking as part of the model. Once all the students have created a basic conceptual model of their thinking, they should be given time to share their ideas with the other students in their group. As they share their ideas, they will start asking more questions about the phenomenon and begin to think about what else they will need to know before they can answer the guiding question. Finally, each group of students should be encouraged to generate a list of at least three things they will need to learn more about during the investigation. Be sure to have each group share these questions or the topics they want to learn more about with the entire class before moving on to the next stage. The questions provide students with a "need" to obtain more information and will motivate them to read about DCIs and CCs during the next stage.

The third activity of this stage is important because students' prior knowledge and experiences related to the phenomenon should always be used as a starting point for their sense-making. In addition, when students create an initial model on their own, they must make their ideas visible as part of the process. These models, as a result, allow teachers to learn more about student thinking. With this information, the teacher can make modifications to the lesson as needed and, more importantly, leverage the prior knowledge and experiences of the students in their classes as a useful tool for figuring out why something happens in the world around them or to develop a solution to the problem. Generating a list of "questions we have now" also helps students identify gaps in their understanding and creates a desire to learn.

The three activities that make up the Task stage also provide multiple opportunities to practice, develop, and demonstrate mastery of the following SEP expectations for high school:

1. Ask questions that arise from careful observation of phenomena to clarify and/or seek additional information (*Asking Questions*).
2. Ask questions that arise from examining models to clarify and/or seek additional information and relationships (*Asking Questions*).
3. Use a model to provide a mechanistic account of a phenomenon (*Developing and Using Models*).
4. Evaluate the merits and limitations of two different models of the same phenomenon (*Developing and Using Models*).

Things to keep in mind during the Task stage:

1. Be sure that students are in small groups for this stage. They will need to be able to work on their own and share their ideas with others.
2. Emerging multilingual students should be allowed to use English, their native language, or some combination of the two during this stage of the investigation.
3. Move around as students work and ask questions such as:
    - What are you trying to figure out?
    - Can you explain your thinking to me?
    - Why do you think that happens like that?
4. Be sure to frame the goal of the investigation as needing to explain a phenomenon; this will
    - create a need to learn,
    - provide an end goal for instruction, and
    - support ongoing changes in student ideas.
5. Starting an investigation with a phenomenon also helps to
    - increase student interest (no more "Why do we need to know this?"),
    - motivate students, and
    - develop their identity as a knowers and doers of science.
6. Activities such as (a) introducing a phenomenon, (b) making the guiding question of the investigation explicit, and (c) eliciting students' current ideas about why or how the phenomenon occurs are designed to help students comprehend more of what they read during later stages of the investigation. These activities also encourage students to be active readers who engage with a text on a deeper level because they create a need to read to learn and provide a framework for making sense of the information found in the text.

Stage 2: IDEAS

The Ideas stage consists of three activities. Each activity requires students to complete a short reading about an important DCI, CC, or, in some cases, an SEP, which will be helpful for them to use during the investigation. The short readings are purposefully designed to:

1. Clearly and accurately present a DCI, CC, or SEP in a way that promotes and supports meaningful sense-making through reading.
2. Provide students with "pieces of the puzzle" that they can use to help figure out an answer to the guiding question of the investigation.
3. Provide students with opportunities to engage with appropriate scientific texts to gather evidence and develop an understanding of concepts.

Each reading highlights a different DCI, CC, or SEP. Figure 4 shows two examples of the texts that are provided to students during the Ideas stage of an investigation. Notice that these texts include

1. images to illustrate key ideas found in the text,
2. key concepts in bold, and
3. a space for students to identify the ideas in the text that they think will be the most helpful for figuring out an answer to the guiding question.

It is important to note that the DCI, CC, or SEP found in the text do *not* describe or explain what the students need to figure out during the investigation; instead, the ideas are intended to be tools for students to use during an investigation as they make sense of the phenomenon. The goal of the teachers at this stage of the investigation is to *put science ideas on the table* so students can work with them and learn more about them as they use the SEPs to figure out how or why the phenomenon happens.

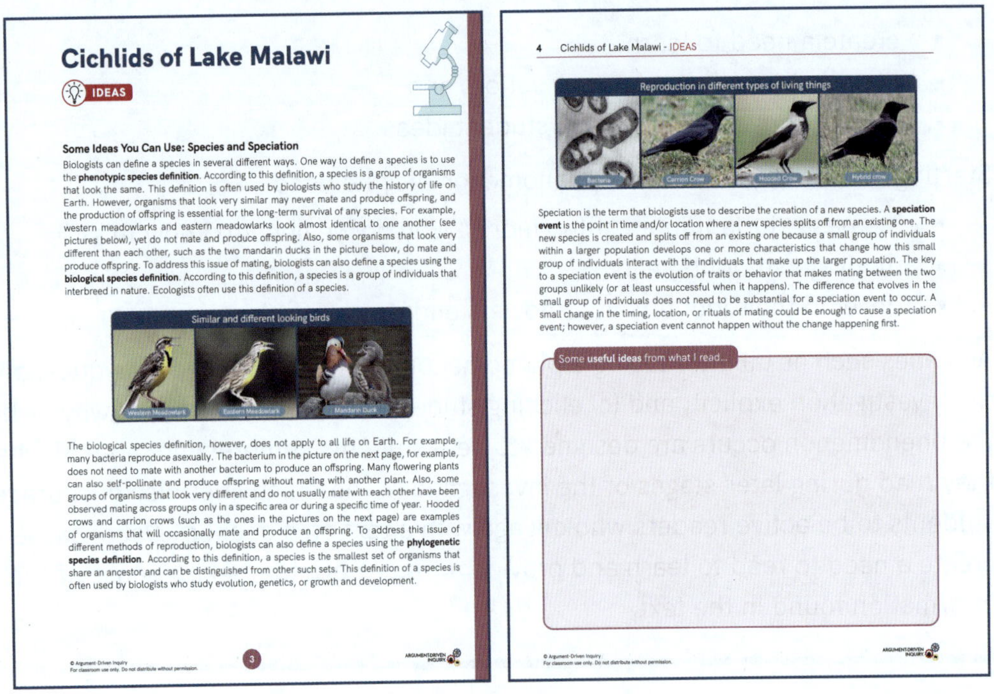
Cichlids of Lake Malawi

IDEAS

Some Ideas You Can Use: Species and Speciation

Biologists can define a species in several different ways. One way to define a species is to use the **phenotypic species definition**. According to this definition, a species is a group of organisms that look the same. This definition is often used by biologists who study the history of life on Earth. However, organisms that look very similar may never mate and produce offspring, and the production of offspring is essential for the long-term survival of any species. For example, western meadowlarks and eastern meadowlarks look almost identical to one another (see pictures below), yet do not mate and produce offspring. Also, some organisms that look very different than each other, such as the two mandarin ducks in the picture below, do mate and produce offspring. To address this issue of mating, biologists can also define a species using the **biological species definition**. According to this definition, a species is a group of individuals that interbreed in nature. Ecologists often use this definition of a species.

Similar and different looking birds

The biological species definition, however, does not apply to all life on Earth. For example, many bacteria reproduce asexually. The bacterium in the picture on the next page, for example, does not need to mate with another bacterium to produce an offspring. Many flowering plants can also self-pollinate and produce offspring without mating with another plant. Also, some groups of organisms that look very different and do not usually mate with each other have been observed mating across groups only in a specific area or during a specific time of year. Hooded crows and carrion crows (such as the ones in the pictures on the next page) are examples of organisms that will occasionally mate and produce an offspring. To address this issue of different methods of reproduction, biologists can also define a species using the **phylogenetic species definition**. According to this definition, a species is the smallest set of organisms that share an ancestor and can be distinguished from other such sets. This definition of a species is often used by biologists who study evolution, genetics, or growth and development.

© Argument-Driven Inquiry
For classroom use only. Do not distribute without permission.
3
ARGUMENT-DRIVEN INQUIRY

4 Cichlids of Lake Malawi - IDEAS

Reproduction in different types of living things

Speciation is the term that biologists use to describe the creation of a new species. A **speciation event** is the point in time and/or location where a new species splits off from an existing one. The new species is created and splits off from an existing one because a small group of individuals within a larger population develops one or more characteristics that change how this small group of individuals interact with the individuals that make up the larger population. The key to a speciation event is the evolution of traits or behavior that makes mating between the two groups unlikely (or at least unsuccessful when it happens). The difference that evolves in the small group of individuals does not need to be substantial for a speciation event to occur. A small change in the timing, location, or rituals of mating could be enough to cause a speciation event; however, a speciation event cannot happen without the change happening first.

Some **useful ideas** from what I read...

© Argument-Driven Inquiry
For classroom use only. Do not distribute without permission.
ARGUMENT-DRIVEN INQUIRY

Figure 4. Examples of text that students read during an investigation

Each activity in this stage should begin with students reading (or listening to the teacher read aloud) the scientific text. The students should then identify 2-3 key ideas found in the text that they think will be useful for figuring out an answer to the guiding question by talking with the other students in their small groups. Be sure to give each group of students an opportunity to share with the rest of the class what they think are the most important ideas. Teachers can write these ideas on a whiteboard (or have the students write them) to create a "class record" of helpful ideas to use during the investigation. The teacher can also differentiate the Ideas stage by presenting the scientific texts in four ways:

1. The teacher reads the text aloud and helps the class annotate the important points.
2. The teacher reads the text aloud and the students annotate.
3. The students read the text independently and discuss what to annotate as a whole group.
4. The students read and annotate the text independently.

The three activities that make up the Ideas stage also provide multiple opportunities to practice, develop, and demonstrate mastery of the following SEP expectations for high school:

1. Critically read scientific literature adapted for classroom use to determine the central ideas or conclusions and/or to obtain scientific and/or technical information to summarize complex evidence, concepts, processes, or information presented in a text by paraphrasing them in simpler but still accurate terms (*Obtaining, Evaluating, and Communicating Information*).
2. Compare, integrate and evaluate sources of information presented in different media or formats (e.g., visually or quantitatively) as well as in words to address a scientific question (*Obtaining, Evaluating, and Communicating Information*).
3. Gather, read, and evaluate scientific and/or technical information from multiple authoritative sources, assessing the evidence and usefulness of each source (*Obtaining, Evaluating, and Communicating Information*).

Things to keep in mind during the Ideas stage:

1. Be sure that students are in small groups for this stage. They will need to be able to share their ideas with others.
2. Many supports for helping students comprehend what they read are already embedded into this stage (e.g., activating prior knowledge, providing a shared experience, making connections, synthesizing, and talking with peers).
3. Do not front-load vocabulary or tell students what they "should learn" by the end of the investigation during this stage, because that will remove an opportunity for students to construct, build, and develop knowledge through authentic application and performance of SEPs, CCs, and DCIs. The later Reflect stage will provide an opportunity to focus on the meaning of key terms and important vocabulary.
4. Let students know that they should write down or highlight important ideas because they will revisit the texts later in the lesson.
5. Students will likely use some or most of the information that they include in the "Things we know from what we read..." box to help justify their evidence in their arguments later.

Stage 3: PLAN

The Plan stage includes three activities:

1. Students explore the materials that will be used to help answer the guiding question.
2. Students work together in a teams to create a plan for collecting the information that they will need and how they will make sense of it.
3. Students share their plans with other students or the teacher to get feedback that they will use to revise their plans and make them better.

Small groups of students develop plans for collecting the data they need to answer the guiding question during the Plan stage of ADI. The overall intent of this stage is to provide students with an opportunity to use DCIs and one or more CCs to plan an investigation (an SEP). It also provides a context for students to learn how to use appropriate data collection techniques (such as controlling variables, making multiple observations, and quantifying observations) and how to use different types of data collection tools (such as rulers to measure length, scales to measure mass, and graduated cylinders to measure volume). Finally, and perhaps most importantly, this stage of an ADI investigation gives students a chance to see why some tools and approaches to data collection work better than others and how the method used during a scientific investigation is based on the nature of the question and the phenomenon under investigation. Students even begin to learn how to deal with the uncertainties that are associated with all empirical work as they discover the importance of attending to precision when they take measurements and attempt to eliminate factors that may change the results of their tests.

The first activity of this stage is called a "tool talk", which is simply an opportunity for students to explore and learn more about the available materials and equipment. The tool talk is important because high school students are often unfamiliar with the materials and specialized science equipment they will need to use during an investigation. Even if the students are familiar with the specialized equipment, they may use one or more tools incorrectly or in an unsafe manner unless they are reminded of proper use. Teachers should, therefore, review specific safety protocols and precautions as part of the tool talk. Including a tool talk during this stage is useful because students often find it difficult to design a method to collect the data needed to answer the guiding question (which is the goal of the Plan stage) when they do not understand what they can and cannot do with the available materials and equipment.

The second activity in the Plan stage is for each group of students to develop a plan for collecting and making sense of data. To facilitate this process, teachers should direct the students to fill out an "Investigation Plan" during this activity. The investigation plan is a simple graphic organizer that is designed to encourage students to think about what type of data they will need to collect, how to collect it, and how to analyze it (see Figure 5 for an example). Note that there is a space for students to identify the observations or measurements they plan to collect, a space to write out a procedure for collecting observations or measurements, and a space to draw a picture of how they plan to set up their available equipment. There is also space for the students to explain how they plan to make sense of their observations or measurements once they have them. The investigation plan is designed to not only encourage students to think through the task they need to complete but to also make their thinking about planning an investigation to answer a specific question visible to others (which, in turn, makes it easier to give them useful feedback because they have described what they plan to do before they begin).

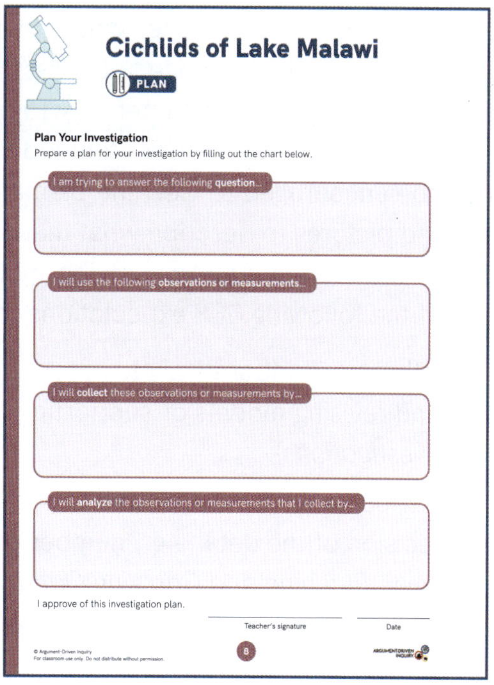
Cichlids of Lake Malawi

PLAN

Plan Your Investigation

Prepare a plan for your investigation by filling out the chart below.

I am trying to answer the following **question**...

I will use the following **observations or measurements**...

I will **collect** these observations or measurements by...

I will **analyze** the observations or measurements that I collect by...

I approve of this investigation plan.

Teacher's signature

Date

© Argument-Driven Inquiry
For classroom use only. Do not distribute without permission.

8

Figure 5. Example of an investigation plan

Be sure to check in with each group of students as they work during this activity. One way to do a quick check-in with a group is to use the "two questions, a nudge, and a goal" technique:

1. Sit down with a group of students and ask, "Would one of you tell me a little about what you all have been doing or thinking about?" (Question 1)
2. A student answers on behalf of the group.
3. Ask, "Would anyone else like to add anything to that?" (Question 2)
4. A different student provides additional information on behalf of the group.
5. Ask, "Have you thought about [a hint or something you want to highlight]?" (A nudge)
6. Say, "Okay, I'm going to check on the other groups. When I get back, I want to see [something that you want the group to accomplish]." (A goal)

If students are struggling to develop a plan, there are several ways to help them without telling them exactly what to do. One option is to have a student from each group share what they are doing and anything they are unsure about. Students from other groups can then offer advice. A second option is to have all the students in a group number off, and then have students with the same assigned number meet together in a new group (e.g., all the 1s meet in this part of the room, all the 2s in this part of the room, etc.), so students from different groups can share what they have planned and what they are stuck on and can offer advice. The students can then return to their original groups to share ideas. Both techniques are a great way to help spread good ideas. However, keep in mind that the student-designed investigations should not all be the same. The students will learn more about how to do science during the later stages of the instructional model when the groups use different methods to collect and analyze data.

Students are given feedback on their plan from the teacher or other groups during the last activity of this stage, which they will use to revise their plan. Once any revisions are made, the small group should be able to start working on the task. Do not require students to "be as detailed as possible" or write full paragraphs, use complete sentences, or incorporate proper writing conventions as they fill out one of the graphic organizers. Their original plans will likely change as they start collecting data and realize that they "forgot about something important" or discover "a better way to do it" (which is what happens all the time in professional science). Therefore, it is not important for students to spend a lot of time making a perfect plan. Instead, they should be encouraged to view the plan as a rough draft that will change, and use numbered or bulleted lists and pictures to describe what that will do.

The three activities that make up the Plan stage also provide multiple opportunities to practice, develop, and demonstrate mastery of the following SEP expectations for high school:

1. Plan an investigation individually and collaboratively to produce data to serve as the basis for evidence as part of building and revising models or supporting explanations for phenomena (*Planning and Carrying Out Investigations*).
2. Decide on types, quantity, and accuracy of data needed to produce reliable measurements, consider limitations on the precision of the data (e.g., number of trials, cost, risk, and time), and refine the design accordingly (*Planning and Carrying Out Investigations*).
3. Consider possible variables or effects and evaluate the confounding investigation's design to ensure variables are controlled (*Planning and Carrying Out Investigations*).
4. Select appropriate tools to collect, record, analyze, and evaluate data (*Planning and Carrying Out Investigations*).

Things to keep in mind during the Plan stage:

1. Students should be in small groups for this stage.
2. Encourage students to think about how to use the CCs as a tool to help them decide what observations or measurements to collect and how to best make sense of the collected data. You can encourage them to think about the CCs by asking question such as:
    a. What types of patterns might we look for to help answer the guiding question?
    b. What information do we need to find a relationship between a cause and an effect?
    c. What measurement scale and units might we use as we collect data?
    d. How might energy transfer into, within, or out of the system?
    e. What are the components of the system we are studying and how do they interact?
    f. How might the structure of what you are studying relate to its function?
3. Not all investigations in science are experiments. Scientists use different methods to answer different questions.
4. The investigation plan is a graphic organizer designed to help students think about how to answer the guiding question of the investigation, what data they need to collect, and how they will need to analyze it. It is not a scientific method; it is a scaffold.
5. The graphic organizer makes student thinking about investigation design visible so teachers can use it as an embedded formative assessment. Students will be able to use the information that they include in the graphic organizer to help write their investigation reports later in the lesson.
6. For those groups who may need a more challenging learning experience, you can use a different graphic organizer that requires them to provide more details about their plan (see Appendix), or not give them any graphic organizer.

Stage 4: DO

The Do stage of the investigation or design challenge gives students a chance to collect data and test their ideas. This stage includes three activities:

1. Students carry out the plan that they previously created for collecting and analyzing data.
2. Students make sense of the data by creating tables or graphs.
3. Students develop a tentative answer to the guiding question based on evidence.

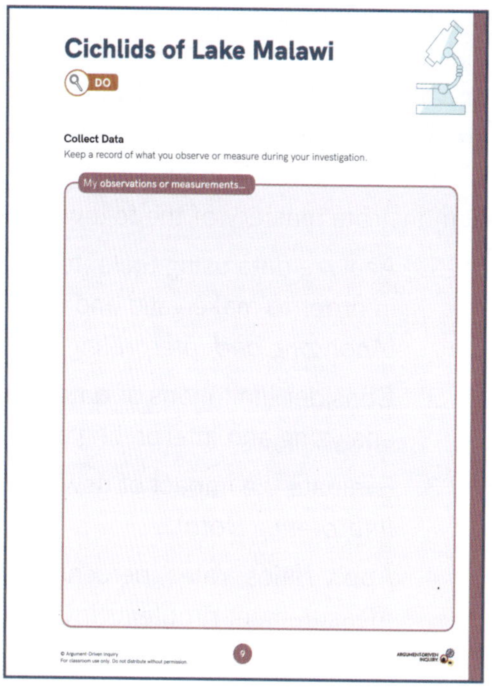

Figure 6. Handout for collecting observations and measurements

The first activity of the Do stage gives students an opportunity to put their plans into practice. It is a good idea to revisit the guiding question as a whole class to refocus the smaller groups on the question they are trying to answer. Allow students to ask any final questions about what they need to do, then give them 20–30 minutes to carry out their data collection plans and record their observations and measurements on their handouts (see Figure 6). Students use different tools, materials, and techniques depending on the nature of the phenomenon. As students work in small groups to collect data, be sure to move from group to group to answer questions and offer advice.

The first activity of the Do stage also provides multiple opportunities to practice, develop, and demonstrate mastery of the following SEP expectations for high school:

1. Conduct an investigation individually and collaboratively to produce data to serve as the basis for evidence (*Planning and Carrying Out Investigations*).
2. Conduct an investigation in a safe and ethical manner (*Planning and Carrying Out Investigations*).
3. Refine the design of an investigation as needed during an investigation to improve the accuracy and precision of the data (*Planning and Carrying Out Investigations*).

Figure 7. Handout for analyzing observations and measurements

The intent of the second activity of the Do stage is to give students a chance to make sense of the data they collected during the first activity of this stage. Therefore, students will need to analyze the data they collected to look for trends over time, differences between conditions, relationships between variables, and other patterns. Figure 7 shows the handout the students use to make sense of their data.

If students are struggling to make sense of their data, there are several ways to help without telling them exactly what to do. One option is to have a student from each group share what they are doing and anything they are unsure about. Students from other groups can then offer advice. A second option is to have all the students in a group number off, and then have students with the same assigned number meet in a new group so students from different groups can share what they have planned, are stuck on, or can offer advice. The students can then return to their original groups to share ideas. Both techniques are a great way to help spread good ideas.

The second activity of the Do stage also provides multiple opportunities to practice, develop, and demonstrate mastery of the following SEP expectations for high school.

1. Analyze data using tools, technologies, and/or models (e.g., computational and mathematical) in order to make valid and reliable scientific claims or determine an optimal design solution (*Analyzing and Interpreting Data*).
2. Consider limitations of data analysis (e.g., measurement error, sample selection) when analyzing and interpreting data (*Analyzing and Interpreting Data*).
3. Evaluate the impact of new data on a working explanation and model (*Analyzing and Interpreting Data*).
4. Apply ratios, rates, percentages, and unit conversions in the context of complicated measurement problems involving quantities with derived or compound units (*Using Mathematics and Computation Thinking*).

Students develop a tentative answer to the guiding question in their small groups during the third (and last) activity of this stage. Often, but not always, developing an adequate answer to the guiding question requires the students to construct a mechanistic explanation. Other times, an answer to the guiding question consists of a more descriptive relationship between independent and dependent variables. The third activity of the Do stage, as a result, provides multiple opportunities for students to practice, develop, and demonstrate mastery of the following SEP expectations for high school:

1. Make a quantitative and/or qualitative claim regarding the relationship between dependent and independent variables (*Constructing Explanations*).
2. Apply scientific ideas, principles, and/or evidence to provide an explanation of phenomena (*Constructing Explanations*).
3. Construct and revise an explanation based on valid and reliable evidence obtained from a variety of sources (including students' own investigations, models, theories, simulations, and peer review) and the assumption that theories and laws that describe the natural world operate today as they did in the past and will continue to do so in the future (*Constructing Explanations*).

This final activity in the Do stage can be challenging for students because they are rarely asked to construct explanations based on their own interpretation of the results of an analysis of data. It is important for teachers to actively work to support student sense-making during this activity. In this stage, teachers should circulate from group to group to act as a resource for the students, asking questions that prompt them to think about what they are doing, and why. To help students remember the goal of the activity and to encourage students to think about whether their analysis is informative or not, teachers can ask questions such as "What are you trying to figure out?", "Why is that information important?", and "Why is that analysis useful?".

Things to keep in mind during the Do stage:

1. Students should be in small groups for this stage.
2. One way to do a quick check-in with a group during this activity is to use the "two questions, a nudge, and a goal" technique (see description of this technique in the Plan stage).
3. Be sure to remind students to keep a record of any observations or measurements they make during this stage. They will need this information later in the lesson.
4. For groups that finish collecting and analyzing data early, you can use the questions posted to the Wonder Wall you created during the Task stage as an extension activity. Students can choose a second question to investigate while their classmates are still working on the guiding question of the investigation.

Stage 5: SHARE

The Share stage of the investigation gives students a chance to communicate to their classmates what they have learned and to revise their thinking in response to feedback and other ideas that are shared with them. This stage has three activities:

1. Students work together in groups to create an evidence-based argument to share what they have figured out.
2. Students share their own arguments and critique the arguments made by students in the other groups.
3. Students revise and improve their arguments based on their learning by interacting with other students and their ideas.

Each group of students creates a draft argument to share with the class what they have learned during the first activity of the Do stage. The intent of having students craft arguments is to encourage them to focus not only on "what they figured out" but also on "how they know what they know". Figure 8 illustrates the ADI argument framework. In this framework, an argument has three components: a claim, evidence, and justification of the evidence. The *claim* is the students' answer to a guiding question and thus is often an explanation for how or why something happens. The *evidence* consists of an analysis of the data they collected during the investigation and an interpretation of that analysis to make sense of their measurements or observations. The evidence often includes a graph that shows a difference between groups, a trend over time, or a relationship between variables; it also includes a statement or two that explains what the analysis means (but not why it matters). The *justification of the evidence* is a statement that explains why the evidence matters. In other words, the justification is used to defend their results by making explicit the DCI, CC, assumptions underlying the data collection, analysis of the data, and interpretation of the analysis so other people can understand why the evidence is important and relevant.

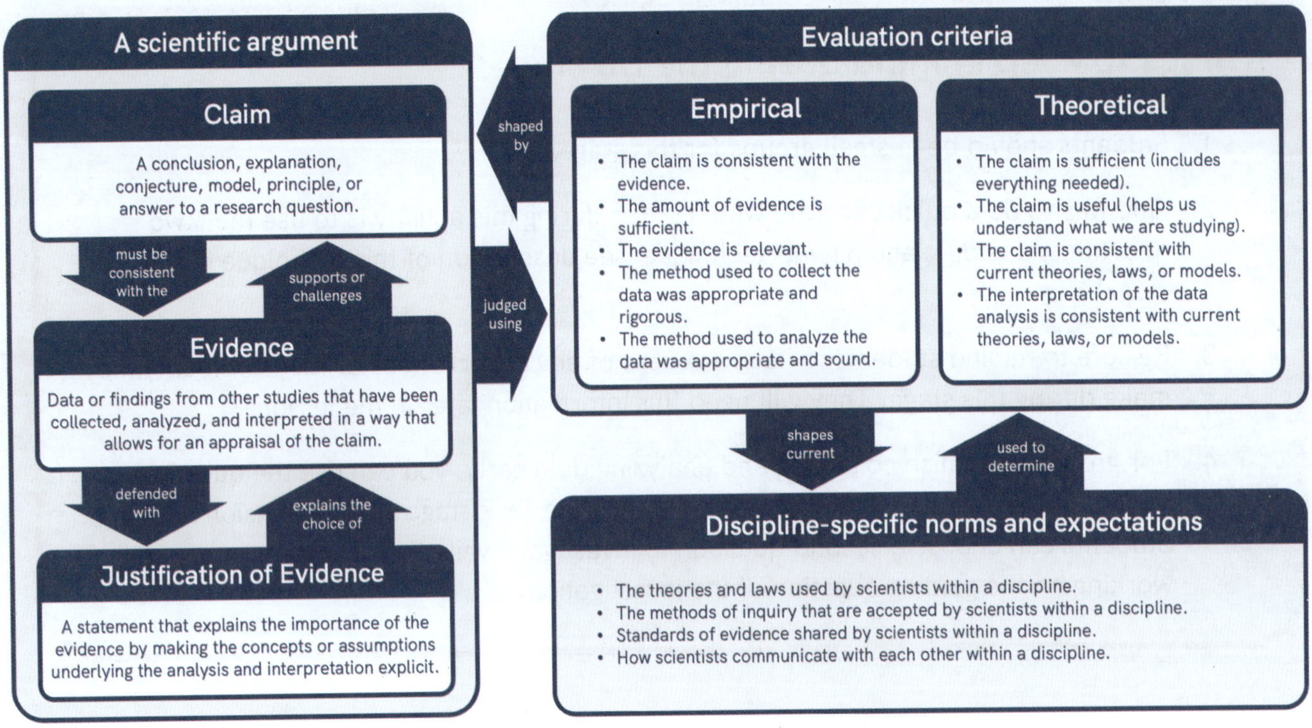

Figure 8. The ADI Argument Framework

It is not enough for students to be able to include all three components of an argument when they have an opportunity to argue from evidence. It is also important for students to understand that in science, some arguments are better than others. Therefore, an important aspect of arguing from evidence in science involves the evaluation of various components of arguments put forward by others. Figure 8, therefore, also highlights three types of criteria that students can and should be encouraged to use to evaluate an argument in science: empirical, theoretical, and discipline-specific norms and expectations.

Empirical criteria include

- how well the claim fits with all available evidence,
- the sufficiency of the evidence,
- the relevance of the evidence,
- the appropriateness and rigor of the method used to collect the data, and
- the appropriateness and reliability of the method used to analyze the data.

Theoretical criteria include

- the sufficiency of the claim (e.g., Does it include everything needed?),
- the usefulness of the claim (e.g., Does it help us understand the phenomenon?),

- how consistent the claim is with accepted theories, laws, or models (e.g., Does it fit with our current understanding of motion and foaces?), and
- how consistent the interpretation of the results of the analysis is with accepted theories, laws, or models (e.g., Is the interpretation based on what we know about the relationship between structure and function?).

Finally, what counts as "quality" in terms of these different criteria varies from discipline to discipline (e.g., biology, geology, and chemistry) and within the specific fields of each discipline (e.g., the fields of ecology, botany, and zoology within the discipline of biology). This variation is due to differences in the types of phenomena investigated by scientists within these disciplines or fields (e.g., how the surface of Earth changes over time, how a trait is inherited in a plant, or how a change in the forces acting on an object influences the motion of that object), the types of methods they use (e.g., descriptive studies, experimentation, or computer simulations), and the different DCIs they use to figure things out. It is, therefore, important to keep in mind that what counts as a quality argument in science depends on the discipline and field.

At this point, it is important to compare this definition of a scientific argument with the claim-evidence-reasoning (CER) framework that many science educators have been using for years to help students develop scientific explanations. The ADI scientific argument framework contains all three components of the CER framework but adds a *justification of the evidence*. It also has a more restrictive definition for what does and does not count as evidence in science. Under the CER framework, evidence is data (observations, measurements, or other information) that support the claim. Reasoning is then defined as an explanation of how or why the evidence supports a claim. From this perspective, evidence and reasoning are related but independent components. In other words, evidence can support a claim without any explanation from the author of the argument on *how* the evidence supports the claim. The reasoning component is then used to fill in that gap. The ADI definition of evidence, in contrast, defines evidence as an analysis of data along with an interpretation of that analysis. From this perspective, it is not enough to support a claim with measurements or observations (that would be using data to support a claim). The measurements or observations must be first transformed into a representation that shows, through analysis, a difference over time, a difference between groups, or a relationship between variables. This analysis then must be interpreted for an audience by explaining how it provides support for a claim. The analysis of data and the interpretation of the analysis is what makes evidence and data different. Without an analysis and interpretation of the data, the claim is not being supported by evidence; it is just being supported by data. The justification of the evidence is then used to explain why the authors of the argument decided to use the evidence in the first place (or do a specific type of analysis and provide that specific interpretation of the analysis). To provide a justification that goes beyond, "that made the most sense to us" or "because that is what we were told to do", students must use one or more overarching scientific ideas, which includes DCIs and/or CCs, to explain why the evidence matters in this situation or why the evidence is important to focus on in this specific case.

Each group of students should create their draft argument in a medium that can easily be viewed by the other groups. This is important because each group will share their draft argument with the other students in the class during the next activity of this stage. Figure 9 shows one way for students to present their draft arguments. We recommend that students use dry-erase markers to create their draft argument on a 2″×3″ whiteboard during this stage. Students can also create their draft arguments using presentation software such as Microsoft PowerPoint, Apple Keynote, or Google Slides, devoting one slide to each component of their argument. They can then share their arguments with others using a tablet or a laptop. The choice of medium is not important as long as students are able to easily modify the content of their argument as they work, and other students can easily see each component of one another's arguments.

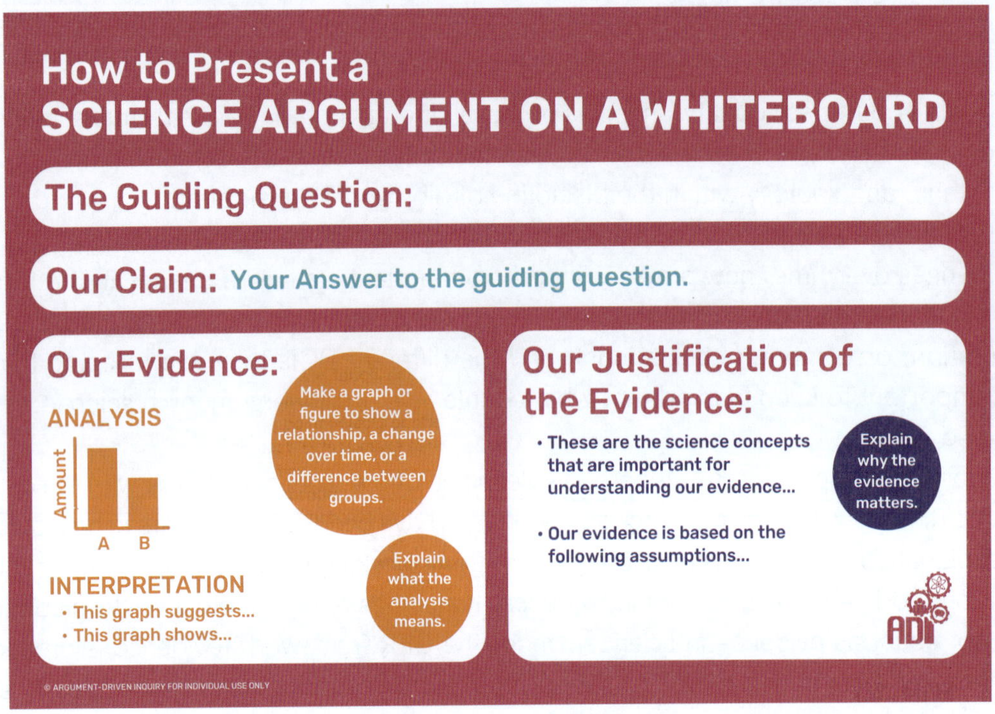

Figure 9. Presentation of a scientific argument

Figure 10 shows an example of an argument crafted by students. This argument is written in English; however, it can also be written in another language, such as Spanish, or a combination of several languages to help support emerging multilingual students in learning science and English at the same time. It is important to allow students to communicate their ideas in their first language, English, or a combination of languages because (a) it allows students to use their home language and culture as a resource for making sense of the phenomenon,[35,36,37] (b) students learn new languages through meaningful use and interaction,[38,39,40] and (c) students' academic language development in their native language facilitates their academic language development in English.[41,42,43]

---

[35] Kathy Escamilla and Susan Hopewell, "Transitions to biliteracy: Creating positive academic trajectories for emerging bilinguals in the United States," in *International Perspectives on Bilingual Education: Policy, Practice, and Controversy*, ed. John E. Petrovic (Charlotte, NC: Information Age Publishing, Inc., 2010): 69-94.

[36] Claude Goldenberg and Rhoda Coleman, *Promoting Academic Achievement Among English Learners: A Guide to the Research* (California: Corwin, 2010).

[37] Norma González, Luis C. Moll, and Cathy Amanti, eds., *Funds of Knowledge: Theorizing Practices in Households, Communities, and Classrooms* (New Jersey: Lawrence Erlbaum Associates, 2005).

[38] H. Douglas Brown, *Principles of Language Learning and Teaching*, 5th ed. (New York: Pearson Education, 2007).

[39] Eugene E. García, *Teaching and Learning in Two Languages: Bilingualism & Schooling in the United States* (New York: Teachers College Press, 2005).

[40] Else Hamayan, "What is the role of culture in language learning?," in *English Language Learners at School: A Guide for Administrators*, 2nd ed., ed. Else Hamayan and Rebecca Freeman Field (Philadelphia: Caslon Publishing, 2012), 61-64.

[41] Escamilla and Hopewell, *"Transitions to biliteracy: Creating positive academic trajectories for emerging bilinguals in the United States,"* 69-94.

[42] Ofelia Garcia and Jo Anne Kleifgen, *Educating Emergent Bilinguals: Policies, Programs, and Practices for English Language Learners* (New York: Teachers College Press, 2010).

[43] Margo Gottlieb, Anne Katz, and Gisela Ernst-Slavit, *Paper to Practice: Using the TESOL English Language Proficiency Standards in PreK-12 Classrooms* (Alexandria, VA: Teachers of English to Speakers of Other Languages, Inc., 2009).

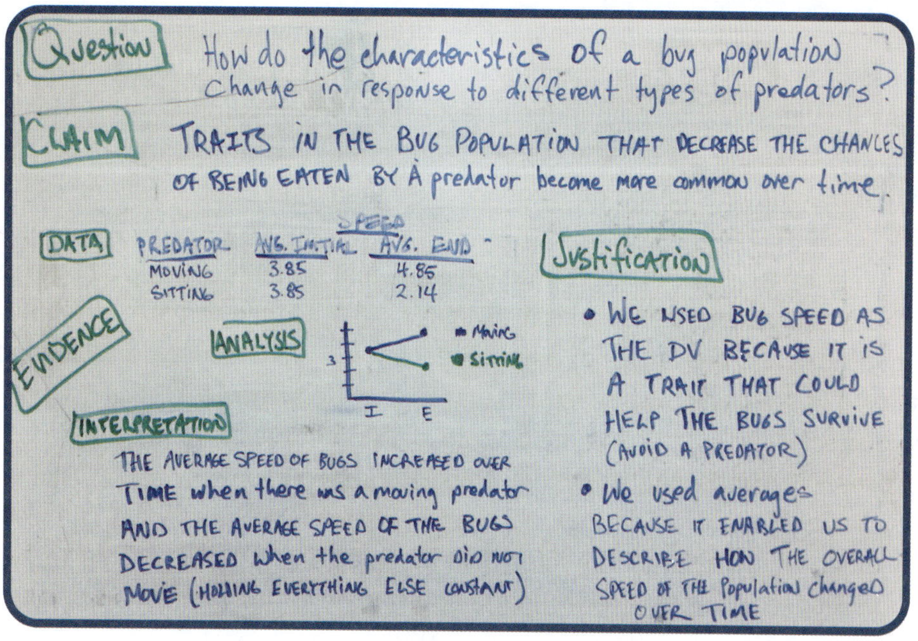

Figure 10. An example argument written by students

It is important to remember that at the beginning of the school year, students will struggle to develop strong evidence-based arguments and will often rely on inappropriate support for their claims, such as plausibility (e.g., "That sounds good to me.") or fit with personal experience (e.g., "But that is what I saw on TV once.") as they attempt to make sense of phenomena. However, as students learn why it is useful to use evidence in an argument, what makes evidence valid or acceptable from a scientific or engineering perspective, and the importance of providing a justification for their evidence through repeated practice, they will improve their ability to support their ideas with sufficient evidence and adequate justification thereof.[44,45,46,47] This activity provides opportunities for students to construct developmentally appropriate written and verbal arguments that justify explanations of phenomena and solutions to problems using evidence acquired from the learning experience.

In summary, the first activity of the Share stage provides multiple opportunities to practice, develop, and demonstrate mastery of the following SEP expectations for high school:

1. Apply scientific reasoning, theory, and/or models to link evidence to the claims to assess the extent to which the reasoning and data support the explanation or conclusion (*Constructing Explanations*).
2. Construct a written argument based on data and evidence (*Engaging in Argument from Evidence*).
3. Make and defend a claim based on evidence about the natural world that reflects scientific knowledge and student-generated evidence (*Engaging in Argument from Evidence*).

---

[44] Çetin and Eymur, "Developing Students' Scientific Writing and Presentation Skills through Argument Driven Inquiry: An Exploratory Study," 837-843.

[45] Chen et al., "Using a modified argument-driven inquiry to promote elementary school students' engagement in learning science and argumentation," 170-191.

[46] Strimaitis et al., "Promoting Equitable Biology Lab Instruction by Engaging All Students in a Broad Range of Science Practices: An Exploratory Study," 92-103.

[47] Hsiang-Ting Chen et al., "Bridging the gender gap of children's engagement in learning science and argumentation through a modified argument-driven inquiry," *International Journal of Science and Mathematics Education* 17, no. 1 (2019): 635-655.

Each student shares, evaluates, and critiques the draft arguments during the second activity of this stage. The process of sharing their arguments with their classmates requires students to communicate their explanations for the phenomenon under investigation and to support their ideas with evidence. Other students in the class are expected to listen to presentation of these arguments, ask questions as needed, evaluate the arguments based on evidence, and then offer critiques along with suggestions for improvement.

This activity is important because research has indicated that students develop a better understanding of DCIs, CCs, and SEPs and acquire better critical-thinking skills when they are exposed to alternative ideas, respond to the questions and challenges of other students, and are encouraged to evaluate the merits of competing ideas.[48,49,50] Research has also suggested that students learn how to distinguish between alternative ideas using rigorous scientific criteria and are able to develop scientific habits of mind (such as treating ideas with initial skepticism, insisting that the reasoning and assumptions be made explicit and that claims be supported by valid evidence) when they have an opportunity to participate in these argumentation sessions.[51,52,53] In other words, this activity integrates argumentation and discourse to support students' development of knowledge and skills in a way that is appropriate for the concept and the grade level.

Teachers can choose from three formats to give students an opportunity to share and critique one another's arguments during this stage. One is the *whole-class presentation format*, in which each group gives a presentation to the whole class. This format is often useful when students are first learning how to propose, support, evaluate, challenge, and refine ideas in the context of science, because the teacher can provide more support as students interact with one another. This format begins with the teacher asking a group to present their argument to the class. The teacher can then encourage students in the class to ask questions, offer critiques, and give the presenters suggestions about ways to improve. The teacher can also ask questions as needed.

Another format for an argumentation session is called the *gallery walk*, in which each student group displays their argument at their workstation. The entire group then moves to each group's workstation in turn. Students are expected to read the argument at that workstation (which was created by a different group) and offer critiques and suggestions for improvement (see Figure 11). The students should move to a different workstation after a few minutes so they can read and critique the arguments written by different groups. They should also be encouraged to keep a record of good ideas or potential ways to improve their own arguments as they move from workstation to workstation. After repeating this rotational process in the gallery walk 3-4 times, each group will have had an opportunity to read and critique three or four different arguments and receive feedback from three or four different groups. Students often use sticky notes to provide feedback to one another during a

---

48 Chen et al., "Using a modified argument-driven inquiry to promote elementary school students' engagement in learning science and argumentation," 170-191.

49 Strimaitis et al., "Promoting Equitable Biology Lab Instruction by Engaging All Students in a Broad Range of Science Practices: An Exploratory Study," 92-103.

50 Cetin et al., "Exploring the effectiveness of engagement in a broad range of disciplinary practices on learning of Turkish high-school chemistry students," 473-497.

51 Kathryn N. Hosbein, Meghan A. Lower, and Joi P. Walker, "Tracking Student Argumentation Skills across General Chemistry through Argument-Driven Inquiry Using the Assessment of Scientific Argumentation in the Classroom Observation Protocol," *Journal of Chemical Education* 98, no.6 (2021): 1875-1887.

52 Fina Fakhriyah, Ani Rusilowati, and Endang Susilaningsih, "Argument-Driven Inquiry Learning Model: A Systematic Review," *International Journal of Research in Education and Science* 7, no. 3 (2021): 767-784.

53 Niklas Gericke, Per Högström, and Johan Wallin, "A systematic review of research on laboratory work in secondary school." *Studies in Science Education* 59, no. 2 (2023): 245-285.

gallery walk argumentation session. This type of format is useful because everyone in the classroom will be actively engaged during the argumentation session and will have a chance to see different arguments. This format, however, does not give students a chance to support, critique, and challenge the ideas, explanations, and arguments of other groups through talk, because no two groups are at the same workstation at the same time.

Figure 11. An example of a gallery walk

A third format for the argumentation session is called the *modified gallery walk* (see Figure 12). When a teacher chooses to use this format, one or two members of a group stay at their workstation to share their group's ideas (we call these students the *presenters*) while the other group members move to different groups' workstations one at a time to listen to and critique the arguments developed by their classmates (we call the students who go to different groups *travelers*). The travelers should visit at least three different workstations during an argumentation session that is organized using the modified gallery walk format. We recommend that teachers use the modified gallery walk format rather than the other formats whenever possible because it provides more opportunities for student-to-student talk. This is especially important for helping students develop speaking and listening skills. It also provides a context for emerging multilingual students to use both productive (speaking and writing) and receptive (listening and reading) language and to learn science through meaningful interactions with others.[54,55,56]

---

[54] Brown, *Principles of Language Learning and Teaching*.

[55] Rebecca M. Callahan, Victor Sampson, and Stephanie Rivale, "Activating Bilingual English Language Learners' Strengths in Science: The Pedagogy of Argument Driven Inquiry (ADI)," in *Teaching the Content Areas to English Language Learners in Secondary Schools. English Language Education*, vol 17, ed. Luciana C. de Oliveira et al. (New York: Springer Cham, 2019).

[56] Garcia and Kleifgen, *Educating Emergent Bilinguals*.

Figure 12. An example of a modified gallery walk

We also recommend that the presenters keep a record of their classmates' critiques and any suggestions for improvement (see Figure 13). The travelers should also be encouraged to keep a record of good ideas or potential ways to improve their own arguments as they travel from group to group. The presenters and the travelers can record this information in the Share section of their handouts.

Figure 13. Students keeping track of good ideas during a modified gallery walk

It is important to note that supporting and promoting productive interactions between students inside the classroom can be difficult because for many students this will be the first time that they are expected to participate in SEPs such as engaging in argument from evidence and obtaining, evaluating, and communicating information. These are not typical expectations of students in classrooms where science is taught using a content-activity-practice lesson structure. To encourage more productive interactions between students, materials, and ideas, the ADI instructional model requires students to generate their arguments in a medium that can be seen by others. By looking at the ideas that are presented on whiteboards, tablets, or paper, students tend to focus their attention on evaluating the claims, evidence, and justifications as well as how the various components of the argument align with the DCIs and CC(s) that they are using to figure things out—rather than attacking the source of the ideas. This strategy often creates an opportunity to share, critique, and give feedback more productively, and makes it easier for students to identify and discard faulty ideas. It is also important for the students to view these interactions between themselves and ideas as opportunities to learn. The teacher, therefore, should describe the goal of this activity as an opportunity for students to collaborate with their peers and as a chance to give one another feedback so the quality of all the arguments in the classroom can be improved, rather than as an opportunity to determine who is "right" or "wrong". This is why we ask students to keep track of what they see and hear during this activity: so they can use the ideas of others to improve their own arguments.

Just like in the earlier stages of ADI, it is important for the classroom teacher to be involved in (without leading) the discussions that take place between students as they share, critique, and refine their ideas. Once again, teachers should move from group to group or from student to student to keep everyone involved and model what it means to use evidence to support or challenge a claim. Teachers can ask the presenters' questions such as "Why did you decide to analyze your data like that?" or "Was there any data that did not fit with your claim?" to encourage students to use criteria such as the fit with evidence to evaluate the quality of the arguments. Teachers can also ask the presenters to explain how the claim they are presenting fits with a DCI or a CC or to explain why the evidence they used is important. In addition, teachers can ask the students who are listening to a presentation questions such as "Do you think their analysis is accurate?" or "Do you think their interpretation makes sense?" or even (in the case of an investigation related to matter and its interactions) "Do you think their claim fits with what we know about the properties of matter?". These questions can serve to remind students to use scientific criteria to evaluate an argument during the discussions. Overall, at this stage of the lesson, it is the goal of the teacher to encourage students to think about how they know what they know and why some claims are more valid or acceptable in science. This stage of the model, however, is not the time to tell students that they are right or wrong.

In summary, the first activity of the Share stage provides multiple opportunities for students to practice, develop, and demonstrate mastery of the following SEP expectations for high school:

1. Make and defend a claim based on evidence about the natural world that reflects scientific knowledge and student-generated evidence (*Engaging in Argument from Evidence*).
2. Present an oral and written argument based on data and evidence (*Engaging in Argument from Evidence*).
3. Evaluate the claims, evidence, and/or reasoning used to support an explanation to determine the merit of an argument (*Engaging in Argument from Evidence*).
4. Respectfully provide and/or receive critiques on scientific arguments by probing reasoning and evidence and challenging ideas and conclusions, responding thoughtfully to diverse perspectives, and determining what additional information is required to resolve contradictions (*Engaging in Argument from Evidence*).
5. Compare, integrate, and evaluate sources of information presented in different media or formats (e.g., visually and quantitatively) as well as in words to address a scientific question (*Obtaining, Evaluating, and Communicating Information*).
6. Communicate scientific and/or technical information or ideas in multiple formats including orally, graphically, textually, and mathematically (*Obtaining, Evaluating, and Communicating Information*).
7. Evaluate scientific and/or technical information from multiple sources, assessing the evidence and usefulness of each source (*Obtaining, Evaluating, and Communicating Information*).

The last activity in the Share stage provides time for original groups to reconvene to discuss what they learned by interacting with other groups so they can revise their draft arguments. This process can begin with the presenters sharing the critiques and suggestions for improvement that they heard during the argumentation session. The students who visited the other groups during the argumentation can then share their ideas for making the arguments better based on what they observed and discussed at other workstations. At this point in the process, students often realize that way they collected or analyzed data was flawed in some way. Teachers should encourage students to collect new data or reanalyze the data they collected as needed and give students time to conduct additional tests of ideas or claims. At the end of this stage, each group should have a final argument that is much better than their draft argument.

The third activity of the Share stage, as a result, provides students with multiple opportunities to practice, develop, and demonstrate mastery of the following SEP expectations for high school:

1. Revise an explanation based on valid and reliable evidence obtained from a variety of sources (including students' own investigations, models, theories, simulations, peer review) and the assumption that theories and laws that describe the natural world operate today as they did in the past and will continue to do so in the future (*Constructing Explanations*).
2. Make a claim based on evidence about the natural world that reflects scientific knowledge and student-generated evidence (*Engaging in Argument from Evidence*).
3. Construct a written argument based on data and evidence (*Engaging in Argument from Evidence*).

Things to keep in mind during the Share stage:

1. Be sure to have students revisit the notes they took during the Ideas stage and reread the scientific texts before they start developing their evidence-based arguments. The DCI and CCs found in these texts will not only make more sense to them at this point in the investigation, but will also help them provide an appropriate justification of their evidence by enabling them to explain why their evidence matters in this context.
2. All the arguments should not be the same (or perfect) during the first activity of this stage. The sharing and critiquing of arguments during the second activity will be more interesting (and students will learn more about how to do science) if each group has a different claim, evidence, and justification of the evidence so there is something to discuss.
3. One of the best ways to ensure that important ideas spread through the class is to help one or two groups develop a very strong component of their argument. Other groups will see these examples and add a version of them to their own arguments during the argumentation session. For example, help one group make a perfect graph, another group write out a perfect interpretation of their analysis, and a third group include a core idea that the other groups are missing in their justification.
4. Make sure to stress that the goal of sharing and critiquing arguments during this stage is to make thinking visible and to help each other out. The goal of sharing and critiquing arguments should not be to determine which group has the best one. Students need to see value in the ideas of others and view their classmates as a resource. This does not happen when students are trying to show others that they have the best argument.

Stage 6: REFLECT

The Reflect stage gives students a chance to think about what they have learned during the investigation, set goals for the future, and make connections to other learning experiences. This stage includes two activities. Each activity is a whole class discussion about one of the following topics:

1. A DCI and/or CC that the students used during the investigation (Activity 1).
2. How to improve the use of scientific and engineering practices during future investigations (Activity 2).

The intent of the first activity is to give students an opportunity to discuss a DCI and/or CC that they used to make sense of the phenomenon during the investigation. To support student sense-making, the discussion asks students to use a DCI and/or CC from the investigation to explain an image of a similar phenomenon or an image that was created to illustrate the DCI and/or CC. As an example, the image in Figure 14 is used in the investigation "Tissue Repair" to provide a starting point for students to discuss their ideas about how new cells are made from existing cells without losing any genetic information in the process. To begin the discussion, show the image and have students discuss it in their small groups. Be sure to encourage them to focus their small-group discussion on how they can use a DCI and/or CC to explain what is shown in the image. Then, as a whole class, invite students to share the ideas they discussed within their small groups and to add to, challenge, or refine the ideas shared by students from other groups. Keep asking questions and encouraging students to reach a consensus about how best to explain the image. Be sure to have a plan for the discussion before it begins. For example, teachers should identify the ideas related to the DCI and/or CC that the students should discuss during this activity and decide how to steer or guide the topic of the discussion toward the target idea if the students do not bring it up on their own.

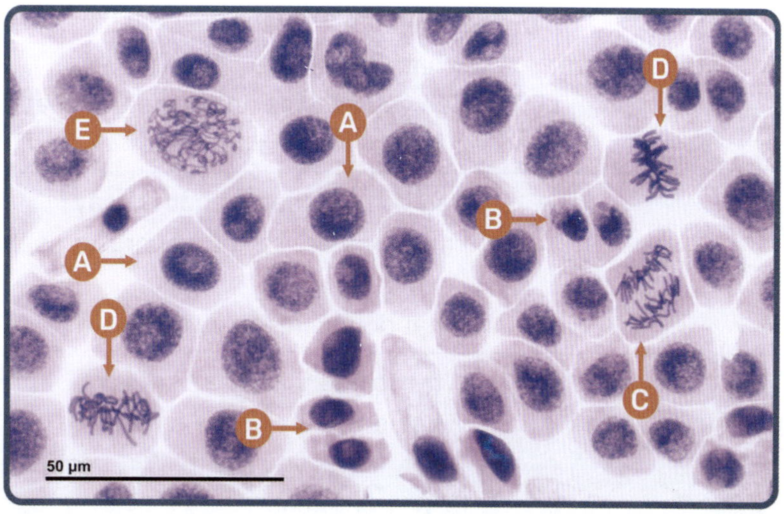

Figure 14. An example of an image used to start a discussion about a DCI or a CC

It is important to emphasize at this point that the whole-class discussion that takes place during this activity (as well as during the other two activities in this stage) is not intended to be a lecture or a time to ask a series of review questions related to the topic; instead, these discussions should be an opportunity for students to share what they are thinking and what they do not understand. The more students talk about their ideas during this stage, the more meaningful the experience will be for them and the more the teacher will be able to learn about their thinking. The teacher can then use what they learn about student thinking to decide if a DCI and/or CC has been mastered or not. If the DCI and/or CC has not been mastered, the teacher can then decide if the students need a simple clarification of an important idea or term, an extensive review of an entire concept, more opportunities for practice and reinforcement during the unit, or an additional lesson to address major gaps in their understanding.

Table 1 lists some useful talk goals for students (e.g., encourage students to support an idea or encourage them to think with others) and some talk moves (e.g., press for support or invite contributions) that teachers can use to make progress toward these goals during a whole-class discussion. Table 1 also includes examples of questions that teachers can ask students as part of the discussion. These questions are organized by talk goal and move. For example, if a teacher has a student talk goal of "encourage students to share, expand on, or clarify an idea" at some point during the reflect discussion, the teacher can use the "share an idea" talk move by asking the class a question such as "How might we explain that?" or the "expand on an idea" talk move by asking "Can you tell me more about that?". These talk goals and moves are only suggestions; however, many teachers find them useful when leading a whole-class discussion.

Table 1. Examples of talk goals and moves that teachers can use during a whole-class discussion

Talk goal	Talk move	Examples
Encourage students to share, expand on, or clarify an idea	Share an idea	What do you see going on here? What do you think is happening? How might we explain that? What are you thinking about?
	Expand on an idea	Can you tell me more about that? What do you mean by that? Can you give me an example of that? Is that all you wanted to say?
	Clarify an idea	Can you say that in another way? I think you are saying [rephrase]. Is that what you mean? What do you mean by that?
Encourage students to support an idea	Press for support	Why do you think that? What is your evidence for that? What makes you say that? How do you know?

(*Continued*)

Encourage students to think with others	Invite contributions	Would anyone like to add to that idea? Can anyone else give me some examples of that? Can anyone take that idea and push it a little further? Does anyone want to respond to that idea?
	Revoice an idea	Who can explain that to me in another way? Can anyone share what [name] said again for those who might have missed it?
	Critique an idea	Do you agree or disagree with that idea, and why? Does that fit with what we have figured out so far? Does it always work that way?
Encourage students to think about what they know and what they do not know yet	Assess progress	I'm confused now. What have we figured out so far? What do we need to figure out now? So, what do we agree on right now and what are we still trying to figure out? What are we sure and not so sure about? What questions do we still have?

This first activity of the Reflect stage provides an opportunity not only for students to demonstrate their mastery of DCIs and/or CCs included in the NGSS but also for the teacher to review and reinforce the DCIs and/or CCs as needed throughout a school year. Research has suggested that this more contextualized approach to reviewing (as well as practicing using DCIs and CCs to make sense of new phenomena), which intentionally spirals key concepts and ideas throughout the year to support mastery and retrieval practices, supports greater conceptual understanding and retention over time when compared to more traditional, decontextualized approaches that tend to favor additional lectures, readings, and practice problems after students do poorly on an assessment.[57,58,59]

The second activity of the Reflect stage gives students an opportunity to think about what they learned about the practices of science and what they should do the next time they engage in these practices. This activity is important because students are expected to use the SEPs during every ADI investigation. The SEPs are complex, and students cannot be expected to master them without being given opportunities to try, fail, and then learn from their mistakes. To encourage students to learn from their mistakes during an investigation, they must have an opportunity to reflect on what went well and what went wrong. Teachers should therefore encourage the students to think about what they did during the investigation, how well what they did worked, and what they think they could have done better. Figure 15 provides a set of questions that can be used by teachers to facilitate a discussion about how students engaged in or used the practices of science during the current investigation.

[57] Bransford, Brown, and Cocking, *How People Learn.*

[58] National Academies of Science, Engineering, and Medicine, *How People Learn II.*

[59] National Research Council, *A Framework for K-12 Science Education.*

1. What things did you do as you carried out your investigation to increase your confidence in the quality of the evidence you relied on to make, support, or evaluate a claim?

2. What would you have done differently during this investigation to improve the quality of the evidence that you had available?

3. What should we do in the future to plan and carry out the best investigations possible?

Figure 15. An example of questions that can be used to start a discussion about SEPs

Once the teacher elicits student ideas about an SEP, they can use those ideas to highlight and draw attention to what does and does not count as quality or rigor in science and then offer advice about ways to improve. Teachers can also use these ideas about what counts to generate class rules or "norms" that the entire class agrees to follow when they plan and carry out investigations or design solutions to problems in the future. Over time, students will gradually improve their skills and develop the abilities they need to participate in the SEPs in a grade-level appropriate way as they learn more about what works and what does not during an investigation, and why. This activity is an important component of this stage of the instructional model because it goes beyond providing multiple opportunities to practice the SEPs to also provide a mechanism for continuous improvement.

Things to keep in mind during the Reflect stage:

1. Make sure that students are doing most of the talking during this stage.
2. Your goal during the reflective discussion is to figure out what students are thinking and build on their ideas as needed.
3. Make sure that every student has an opportunity to contribute to the discussion and that they feel like their ideas and contributions are valued.

Stage 7: REPORT

The Report stage of the investigation gives students an opportunity not only to learn how to write in the context of science (an important literacy skill) but also to write to learn (to develop a better understanding of the DCIs and CCs of science through writing). This stage has three activities:

1. Students write draft reports to share what they figured out during the investigation.
2. Students participate in a peer review of the reports.
3. Students revise their draft reports based upon peer review suggestions before submitting a final version to the teacher for evaluation.

The first activity of the Report stage calls for each student to write a draft report that consists of three sections. Each section is intended to address one of the following fundamental questions:

1. What were you trying to figure out, and why?
2. What did you do to answer your question, and why?
3. What is your argument?

The format of the report is designed to emphasize the informative and persuasive nature of science writing and to help students learn how to communicate in multiple modes (words, figures, tables, and equations). The three-question format is intended to align with the sections of scientific research articles (i.e., introduction, method, results, and discussion) in a way that will help students understand the important role argument plays in science.

The student handout for this stage of the investigation includes a report template with section headings and sentence starters (see Figure 16). The sentence starters are intended to act as a guide for students as they learn to write in the context of science and should make the assignment less intimidating. We recommend using these sentence starters to encourage students to write in a clear and concise manner.

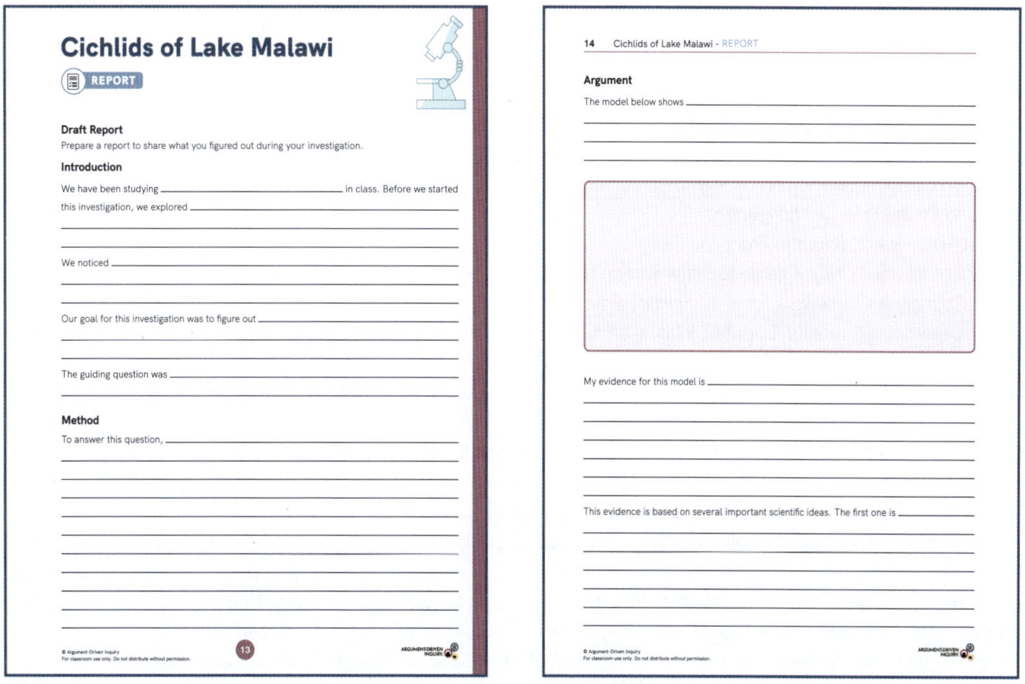
Cichlids of Lake Malawi

REPORT

Draft Report

Prepare a report to share what you figured out during your investigation.

Introduction

We have been studying ______ in class. Before we started this investigation, we explored ______

We noticed ______

Our goal for this investigation was to figure out ______

The guiding question was ______

Method

To answer this question, ______

13

14 Cichlids of Lake Malawi - REPORT

Argument

The model below shows ______

My evidence for this model is ______

This evidence is based on several important scientific ideas. The first one is ______

Figure 16. An example of a handout designed to help students write a draft report

ARGUMENT-DRIVEN INQUIRY

Over time, it may be decided that they are no longer needed. In addition, we recommend allowing emerging multilingual students to write and communicate what they have learned in their first language, English, or a combination of the two because this writing task will help these students learn science and English at the same time. However, it is important to scaffold this process for the emerging multilingual students based on their current level of English language proficiency because (a) students' development of academic language and academic content knowledge are interrelated processes[60,61] and (b) students' access to instructional tasks that require complex thinking is enhanced when linguistic complexity and instructional support match their levels of language proficiency.[62]

Students write a report because writing is an important part of doing science. Scientists must be able to share the results of their research through writing. They also must be able to read and understand the writing of others as well as evaluate its worth.[63,64] In addition, writing helps students learn how to articulate their thinking in a clear and concise manner, encourages metacognition, and improves student understanding of DCI, CCs, and SEPs.[65,66,67,68] It also enables students to master the disciplinary-based writing process and composition skills[69,70] that are included in the Common Core Standards for ELA[71] and the expressive language development expectations outlined in the WIDA English Language Development Standards.[72] Finally, and perhaps most importantly, writing makes each student's thinking visible to the teacher (which facilitates assessment) and enables the teacher to provide students with the educative feedback they need to improve.

The second activity of the Report stage gives students an opportunity to review the reports in pairs using the peer review guide and teacher scoring rubric (PRG/TSR; see Figure 17). The PRG/TSR is designed as an analytical rubric that explicitly defines the criteria for mastery of the task. It contains four sections. One section focuses on the introduction of the report, one section focuses on the method, one on the argument, and one on the overall writing mechanics. Each section contains specific criteria unique to that section of the report. These criteria are framed as questions that can be answered with a "yes" (meets expectations for that criterion), "somewhat" (approaches expectations for the criterion), or "no" (does not meet expectations for that criterion). The pair of students can simply answer each question as they evaluate the quality of each section of the report as well as the quality of the writing. There is also space for the reviewers to provide the author with feedback about how to improve the report.

---

60 Jana Echevarria, Deborah Short, and Kristin Powers, "School Reform and Standards-Based Education: A Model for English-Language Learners," *The Journal of Educational Research* 99, no. 4 (2006): 195-211.

61 García, *Teaching and Learning in Two Languages*.

62 Gottlieb, Katz, and Ernst-Slavit, *Paper to Practice*.

63 Robert Barrass, *Scientists must write: A guide to better writing for scientists, engineers and students,* 2nd ed. (London, Routledge, 2002).

64 National Research Council, *A Framework for K-12 Science Education*.

65 Çetin and Eymur, "Developing Students' Scientific Writing and Presentation Skills through Argument Driven Inquiry: An Exploratory Study," 837-843.

66 Strimaitis et al., "Promoting Equitable Biology Lab Instruction by Engaging All Students in a Broad Range of Science Practices: An Exploratory Study," 92-103.

67 Fakhriyah, Rusilowati, and Susilaningsih, "Argument-Driven Inquiry Learning Model: A Systematic Review," 767-784.

68 Sumeyye Erenier and Pinar Seda Cetin, "Utilizing Argument-Driven Inquiry to Develop Pre-Service Teachers' Metacognitive Awareness and Writing Skills," *International Journal of Research in Education and Science* 5, no. 2 (2019): 628-638.

69 Sampson and Walker, "Argument-Driven Inquiry as a Way to Help Undergraduate Students Write to Learn by Learning to Write in Chemistry," 1443-1485.

70 Victor Sampson et al., "Writing to Learn by Learning to Write During the School Science Laboratory: Helping Middle and High School Students Develop Argumentative Writing Skills as They Learn Core Ideas," Science Education, 97, no. 5 (2013): 643-670.

71 National Governors Association Center for Best Practices & Council of Chief School Officers, *Common Core Standards*.

72 WIDA, *WIDA English Language Development Standards Framework*.

Educational research has suggested that analytical rubrics are particularly effective for promoting growth throughout the year, as the norms for what counts as "quality" consistently evolve with students' increased knowledge and skill. This type of rubric stands in contrast to one measured against a static standard of performance. A rubric with a specific standard as the end goal cannot provide continued opportunities for growth once students meet the standard used to develop the rubric. Research has also shown that analytic rubrics are effective for focusing students' attention on important questions, such as the quality of evidence, alongside details such as font size, neatness, and word count. While these details are important, they tend to be the sole focus of rubrics normed against a static standard of performance.

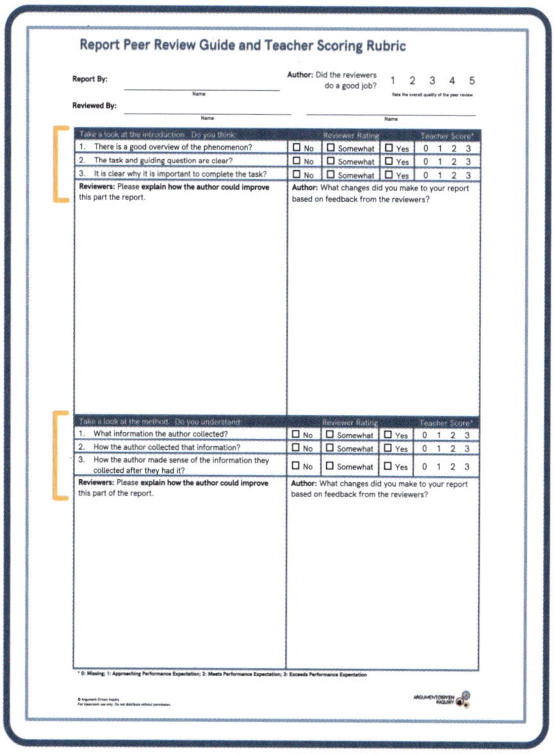

Report Peer Review Guide and Teacher Scoring Rubric

Report By: ______ (Name)

Reviewed By: ______ (Name) ______ (Name)

Author: Did the reviewers do a good job? 1 2 3 4 5
Rate the overall quality of the peer review

Take a look at the introduction. Do you think:	Reviewer Rating			Teacher Score*
1. There is a good overview of the phenomenon?	☐ No	☐ Somewhat	☐ Yes	0 1 2 3
2. The task and guiding question are clear?	☐ No	☐ Somewhat	☐ Yes	0 1 2 3
3. It is clear why it is important to complete the task?	☐ No	☐ Somewhat	☐ Yes	0 1 2 3
Reviewers: Please **explain how the author could improve** this part the report.	**Author:** What changes did you make to your report based on feedback from the reviewers?			

Take a look at the method. Do you understand:	Reviewer Rating			Teacher Score*
1. What information the author collected?	☐ No	☐ Somewhat	☐ Yes	0 1 2 3
2. How the author collected that information?	☐ No	☐ Somewhat	☐ Yes	0 1 2 3
3. How the author made sense of the information they collected after they had it?	☐ No	☐ Somewhat	☐ Yes	0 1 2 3
Reviewers: Please **explain how the author could improve** this part of the report.	**Author:** What changes did you make to your report based on feedback from the reviewers?			

* 0: Missing; 1: Approaching Performance Expectation; 2: Meets Performance Expectation; 3: Exceeds Performance Expectation

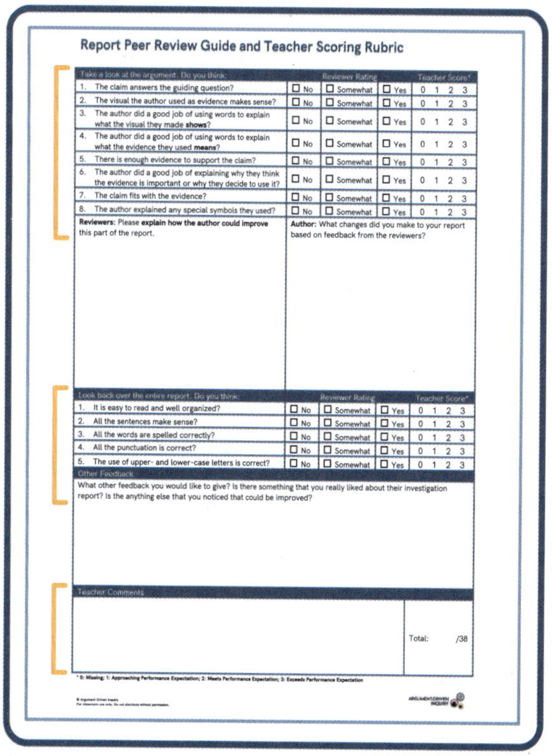

Report Peer Review Guide and Teacher Scoring Rubric

Take a look at the argument. Do you think:	Reviewer Rating			Teacher Score*
1. The claim answers the guiding question?	☐ No	☐ Somewhat	☐ Yes	0 1 2 3
2. The visual the author used as evidence makes sense?	☐ No	☐ Somewhat	☐ Yes	0 1 2 3
3. The author did a good job of using words to explain what the visual they made **shows**?	☐ No	☐ Somewhat	☐ Yes	0 1 2 3
4. The author did a good job of using words to explain what the evidence they used **means**?	☐ No	☐ Somewhat	☐ Yes	0 1 2 3
5. There is enough evidence to support the claim?	☐ No	☐ Somewhat	☐ Yes	0 1 2 3
6. The author did a good job of explaining why they think the evidence is important or why they decide to use it?	☐ No	☐ Somewhat	☐ Yes	0 1 2 3
7. The claim fits with the evidence?	☐ No	☐ Somewhat	☐ Yes	0 1 2 3
8. The author explained any special symbols they used?	☐ No	☐ Somewhat	☐ Yes	0 1 2 3
Reviewers: Please **explain how the author could improve** this part of the report.	**Author:** What changes did you make to your report based on feedback from the reviewers?			

Look back over the entire report. Do you think:	Reviewer Rating			Teacher Score*
1. It is easy to read and well organized?	☐ No	☐ Somewhat	☐ Yes	0 1 2 3
2. All the sentences make sense?	☐ No	☐ Somewhat	☐ Yes	0 1 2 3
3. All the words are spelled correctly?	☐ No	☐ Somewhat	☐ Yes	0 1 2 3
4. All the punctuation is correct?	☐ No	☐ Somewhat	☐ Yes	0 1 2 3
5. The use of upper- and lower-case letters is correct?	☐ No	☐ Somewhat	☐ Yes	0 1 2 3

Other Feedback

What other feedback you would like to give? Is there something that you really liked about their investigation report? Is the anything else that you noticed that could be improved?

Teacher Comments

Total: /38

* 0: Missing; 1: Approaching Performance Expectation; 2: Meets Performance Expectation; 3: Exceeds Performance Expectation

Figure 17. The Peer Review Guide/Teacher Scoring Rubric (PRG/TSR)

Once a pair of students finishes reviewing a report as a team, they are given another report to review. When students are grouped in pairs, they only need to review two different reports. Give students only 15 minutes to review each report (we recommend setting a timer). When students are grouped into pairs and given 15 minutes to complete each review, the entire peer review process can be completed in 30 minutes (2 reports × 15 minutes = 30 minutes).

Reviewing each report as a pair using the PRG/TSR is an important component of the peer review process because it provides students a forum to discuss what counts as high quality or acceptable and, in so doing, forces them to reach a consensus. This method also helps prevent students from checking off "yes" for each criterion on the PRG/TSR without thorough consideration of the merits of the report. It is also important for students to provide constructive and specific feedback to the author when areas of the report are not found to meet the standards established by the PRG/TSR. The peer review process provides students with an opportunity to read good and bad examples of reports. This helps them learn new ways to organize and present information, which in turn will help them write better subsequent reports. It also provides an opportunity and a mechanism for all students in the classroom, including emerging multilingual students, to develop shared norms for what counts as high-quality writing in the context of science.

This stage of the model is intended to give students opportunities to ask questions and obtain and evaluate information. This stage also gives students an opportunity to develop the reading skills that they need to be successful in science. Students must be able to determine the central ideas or conclusions of a scientific text and determine the meaning of symbols, key terms, and other domain-specific words. In addition, students must be able to assess the reasoning and evidence that an author includes in a text to support their claim and compare or contrast findings presented in a text with those from other sources. Students can develop all these skills (as well as other discipline-based reading skills found in the Common Core ELA standards) when they are required to read and critically review reports written by their classmates. This stage is also beneficial for emerging multilingual students because, as noted earlier, people learn productive and responsive forms of language through its meaningful use, such as reading and reviewing a report, and through authentic interactions with others.

The final activity of the Report stage is to revise the report. Each student is required to rewrite their report using the reviewers' comments and suggestions as a guideline. The author is also required to explan in the Author Response section of the PRG/TSR what they did to improve each section of the report in response to the reviewers' suggestions. This approach provides all students with a chance to improve their writing mechanics and develop their reasoning and understanding of the content. This process also offers students the added benefit of reducing academic pressure by providing support in obtaining the highest possible grade for their final product. Once the report is revised, it is submitted to the teacher for evaluation. The teacher can then provide a score on the PRG/TSR in the column labeled "Teacher Score" and use these ratings to assign an overall grade for the report.

The PRG/TSR, as noted earlier, is designed to be an analytical rubric rather than a holistic one. Analytical rubrics break down the characteristics of an assignment or products into parts, allowing the scorer to itemize and define exactly what aspects are strong and which ones need improvement. Holistic rubrics, in contrast, often list 35 levels of performance along with a broad description of the characteristics that define each level. The main advantage of an analytical rubric is that it provides targeted feedback to students. We take this one step further and identify the criteria that, when used together, indicate that a student can write a report to share what was figured out during an investigation using DCI, CCs, and SEPs. Thus, a score of 2 on each criterion (meets expectations) indicates that a student has reached a level of mastery. Each criterion is also phrased as a question (rather than a description) to help facilitate scoring and reduce bias. For example, two questions included in the PRG/TSR are, "Do you think the task and the guiding question are clear?" and "Do you think the analysis of the data is correct?". Answering "yes" to one of these questions indicates that the student meets expectations for that criterion and should be awarded a score of 2. Answering "somewhat" or "partially" indicates that a student is approaching expectations for that criterion and should be awarded a score of 1. Finally, answering "no" indicates that the student needs improvement or did not include that aspect of the report and should earn a score of 0. Research on this approach has not only shown that both students and teachers can review/score a report accurately using the PRG/TSR but has also shown that students' writing improves substantially over time because of what they learn from reviewing other reports and from the targeted feedback they receive on their own reports from their peers and their teachers.[73,74]

---

[71] Sampson et al., "Writing to Learn by Learning to Write During the School Science Laboratory: Helping Middle and High School Students Develop Argumentative Writing Skills as They Learn Core Ideas," 643-670.

[72] Sampson and Walker, "Argument-Driven Inquiry as a Way to Help Undergraduate Students Write to Learn by Learning to Write in Chemistry," 1443-1485.

The PRG/TSR is designed to be used with any ADI investigation, thus allowing teachers to use the same scoring rubric throughout the entire year. This is beneficial for several reasons. First, the criteria for what counts as a high-quality report do not change from investigation to investigation. Students quickly learn what is expected from them when they write a report, and teachers do not have to spend valuable class time explaining the various components of the PRG/TSR each time they assign a report. Second, the PRG/TSR makes it clear which components of a report need to be improved next time, because the grade is not based on a holistic evaluation of the report. Students, as a result, can see the aspects of their writing that is strong and the aspects that need improvement. Finally, and perhaps most importantly, the PRG/TSR provides teachers with a standardized measure of student performance that can be compared over multiple reports across semesters, thus allowing teachers to track improvement over time.

In summary, the writing, reading, and reviewing that students engage in during the Report stage is purposefully designed to:

1. Provide multiple opportunities for students to engage in various written and graphic modes of communication to support students in developing understanding of DCIs, CCs, and SEPs.
2. Promote and support students to act as scientists who can learn from engaging with phenomena, make sense of concepts, and productively struggle.
3. Strategically and systematically develop students' content knowledge by having them revisit and review DCIs and CCs that were introduced during an earlier stage of the investigation.
4. Help students develop and use science vocabulary in context.
5. Provide an opportunity for students to construct and present developmentally appropriate written and verbal arguments that justify explanations of phenomena and solutions to problems using evidence acquired from the learning experience.

Things to keep in mind during the Report stage:

1. There are many writing supports (e.g., sentence starters, pre-writing, and graphic organizers) already embedded into this stage. You might not need to provide much extra support for your students.
2. Do not worry if students do not produce a perfect report during the first activity. All the reports will be reviewed and revised before they are turned in to you. The reports should be viewed as a starting point. Encourage students to "just get something down on paper" so "we can work on it together".
3. Writing is an important component of this model. Do not skip this stage. When a student writes the report on their own, it not only helps them learn to write in the context of science (an important literacy skill) but also gives each student an opportunity to write to learn (develop a better understanding of the core ideas of science by writing about them).

(Continued)

4. Do not over-scaffold the writing process. Mistakes are opportunities to learn. Do not take these opportunities away from your students.
5. The goal of the peer review process is not to assign grades. Rather, it gives students an opportunity to give and receive feedback about their writing.
6. The peer review process is one of the best ways for students to learn how to write. When students read and review two reports, they have an opportunity to (1) see examples of texts written by others for the same purpose, (2) discuss what counts as quality and why some reports are stronger than others, (3) discuss ways to strengthen a report, and (4) pick up things that they can do to improve their own reports.
7. Teachers can review the writing of the emerging multilingual students depending on their English language proficiency, but make sure that these students participate in the peer review process.
8. Students don't always use the feedback they receive from others to improve the quality of their reports. Make sure you move from student to student and encourage them to make changes based on the feedback.
9. If students get stuck as they are writing their final reports, model how to use feedback. For example, place yourself in the student's shoes and say, "Okay, based on the peer review guide, I need to do more to describe what I did to collect data. I bet if I add information about ______________, readers will have a better idea of what I did."

Quick Reference Guide for the Stages of an ADI Investigation

Table 2 provides a one-page summary of the seven stages of an ADI investigation. This table is organized by stage because students and the teacher need to accomplish different things during each stage. The first column indicates the stage. The second column lists the activity or activities that happen during each stage. The activities are separated into rows to improve ease of use and to help provide teachers with guidance about what they and the students need to accomplish during each stage. The third column in the table indicates the activity format (whole class, small group, or independent). Notice that this instructional model incorporates three student grouping strategies over the course of an investigation to promote and support student learning. The fourth column provides the amount of time that is needed to complete each activity within a stage. This time is included to make it easier for a teacher to maintain the pace of an investigation and to facilitate lesson planning. Including the times for each activity also makes it possible to break an investigation over several instructional blocks during a day or over several different days of instruction.

Table 2. Summary of the activities that take place during each stage of the ADI Instructional model

Stage	Activities	Format	Time (Min)
Task	Introduce a phenomenon.	Whole class	10
	Introduce the task and the guiding question of the investigation.	Whole class	5
	Give students time to create a model of their initial ideas about the topic.	Individual	10
Ideas	Highlight some ideas that students can use during the investigation.	Small group	20
Plan	Tool talk.	Whole class	5
	Students develop a method to answer the guiding question.	Small group	15
Do	Students carry out their plan to collect data.	Small group	20
	Students make sense of the data they collected and then develop a tentative answer to the guiding question.	Small group	10
Share	Students develop draft evidence-based arguments to share what they figured out.	Small group	20
	Students share and critique the draft arguments.	Small group	20
	Students revise draft arguments based on feedback and other discussions.	Small group	10
Reflect	Discuss the DCI and/or a CC that was used during the investigation.	Whole class	10
	Discuss how SEPs were used and how to improve.	Whole class	10
Report	Students write reports to share what they were trying to do, what they did, what they figured out, and how they know.	Individual	30
	Students read and critique reports based on established criteria, see strong and weak examples, and develop shared norms for what counts as quality.	Small group	30
	Students revise draft reports based on feedback and other discussions.	Individual	30
	Students submit the final reports to the teacher for evaluation.	Individual	5

Notes for "Future You"

Think about everything you have read in this chapter. Then write down some things (including the definitions of important terms) that you think will be helpful for you to remember when you use an ADI investigation in class. We all forget things, so do what you can now to help "future you".

Some things I want to make sure that I remember

Ways to Differentiate Instruction When Using the ADI Instructional Model

Today, there are profound differences between specific demographic groups in terms of their educational achievements in science. The reasons for these differences are complex, and they cannot all be adequately addressed in this section of the book. However, it is important to focus on five key areas when considering different approaches or methods for differentiating instruction for students. The first key area concerns differences in opportunities to learn because of unequal access. The second key area considers how instructional materials can be made more inclusive and motivating for diverse groups of students. The third is how low learning expectations and biased stereotypical views about the interests or abilities of specific students or demographic groups contribute, in both overt and subtle ways, to different learning experiences and supports for students. A fourth key area to consider is the students' own motivations and interest in science, which is often shaped by how a subject is presented inside a classroom and can play a role in achievement and the students' decision to pursue science in secondary school and beyond. Thus, attention to factors that may motivate or fail to motivate students from specific demographic groups is important to keep in mind when designing instructional materials. Finally, students' preparation in other subjects, especially English language arts and mathematics, also affects their achievement in science. Teachers often report that they have difficulty helping students who have failed to become effective readers and writers by high school to make progress in science and other subject areas. These reported difficulties, in turn, reinforce lowered expectations and tracking of many students as they move through school, thereby significantly reducing their access to science pathways through K-12 and limiting their opportunities to pursue an advanced degree in science. To help address these five issues, 3D instructional materials need to be meaningful, rigorous, and equitable by design.

Meaningful, Rigorous, and Equitable by Design

It is important to define the terms meaningful, rigorous, and equitable given how these terms are used to mean different things for different educators. A learning experience is meaningful when it sparks a student's curiosity because it focuses on something that is interesting, important, or relevant to them. Meaningful learning experiences, as a result, are key factors in motivating students to learn. A rigorous learning experience invites students to figure out how or why something happens or to develop a solution to a meaningful problem. Rigor should not be viewed as covering more content in a shorter amount of time or assigning more practice problems or extra assignments for students to complete on their own; instead, rigor should be viewed as pressing students to go beyond what they currently know and can do. A rigorous learning experience, in other words, requires a student to learn how to use the DCI, CCs, and SEPs from science to make sense of the world around them, think in new ways, and learn from their mistakes. Finally, a learning experience is equitable when it provides all students, rather than a select few, with a fair opportunity to learn in this way. It also means doing everything possible to ensure that every student inside a classroom feels like their ideas and participation are valued. A meaningful, rigorous, and equitable learning experience makes it possible for every student to see themselves as a "knower and doer" of science. A knower and doer of science is someone who can use the DCI, CCs, and

SEPs, along with their own ideas, to make sense of how or why observable events in the world happen in a way that is consequential to them.

The 3D instructional materials that are found in this book are designed to make learning experiences meaningful, rigorous, and equitable for all students. The phenomenon that is used to anchor each investigation makes the content interesting, important, and relevant to students. The focus on using DCI, CCs, and SEPs to make sense of these phenomena provides students with multiple opportunities to learn how to use these principles from science to think in new ways. Finally, the focus on constructing, evaluating, and revising explanations and other ideas during an investigation makes a learning experience more equitable because students have more opportunities to receive feedback about what they can do to improve. This feedback allows them to learn from their mistakes. It also helps creates a classroom community that values the ideas and participation of all students in the process of sense-making. Finally, the ADI instructional model provides ample opportunities for students to become effective readers and writers because of the emphasis on reading and writing (as well as speaking) in science as a way for students to obtain, evaluate, and communicate information in the service of figuring out phenomena.

There are several studies in science education that have provided strong support for these claims. For example, a study conducted in 2017[75] compared the learning gains of students enrolled in the general sections and honors sections of a high school biology course and found that the students enrolled in the general sections made similar or larger gains than the students enrolled in the honors sections when both groups were taught using ADI (which, as noted earlier, is a 3D instructional approach). In fact, the students in the general sections of the course were able to plan and carry out an investigation, argue from evidence, and write a science-specific persuasive essay just as well as the students in the honors sections by the end of the academic year, thereby eliminating the achievement gaps that existed between the two groups at the beginning of the study. In another study that was conducted in 2016,[76] students from underrepresented minority groups, females, and students with low GPAs who were enrolled in a chemistry course that was taught using ADI outperformed their counterparts who were enrolled in traditional sections taught through a more structured inquiry approach on an assessment that required the use of DCI and SEP. In addition, students from underrepresented minority groups and females in the ADI sections performed as well as the students from majority groups and males in the ADI section of the course and outperformed the students from majority groups and males enrolled in the traditional sections. These studies, and others like them, indicate that all students can be successful and achievement gaps can be closed when learning experiences are designed to be meaningful, rigorous, and equitable from the start.

The Problem with Making Modifications to 3D Instructional Materials

Numerous studies in science education have shown that all students, with a small number of notable exceptions (such as individuals with severe cognitive impairments), can engage in and learn complex DCI, CCs, and SEPs when provided with supportive conditions and enough time to learn.[77,78] Although significant and persistent achievement gaps in science exist on national and state assessments for low-income students, minority students, and emerging multilingual learners, the differences in scores on these assessments should not be seen as an inability of these students to learn specific content or reach a

[75] Strimaitis et al., "Promoting Equitable Biology Lab Instruction by Engaging All Students in a Broad Range of Science Practices: An Exploratory Study," 92-103.

[76] Joi Phelps Walker et al., "Using the laboratory to engage all students in science practices," *Chemistry Education Research and Practice* 17, no. 4 (2016): 1098-1113.

[77] Michaels, Shouse, and Schweingruber, *Ready, Set, Science!*

[78] National Research Council, *A Framework for K-12 Science Education.*

specific level of mastery. Rather, these achievement gaps should be viewed as a consequence of unequal access to meaningful, rigorous, and equitable opportunities to learn. Any science instructional materials that are used in Biology must therefore include learning experiences that give students a fair opportunity to learn and master the DCI, CCs, and SEPs found in the NGSS.

Unfortunately, many of the features that make the 3D instructional materials included in this book more meaningful, rigorous, and equitable are often seen by educators as too complex or challenging for many students, especially those students who have IEPs or 504s and those students who are learning English. Teachers, as a result, often want to modify the instructional materials to "make them easier" for these students. One common modification that is intended to "make it easier for these students to learn" is to use a content-activity-practice lesson structure or confirmation labs and structured or guided inquiry instead of 3D instruction. This type of modification, however, can be problematic because it prevents the students who use these "simplified materials" from participating in the SEPs or to use DCI and CCs to make sense of phenomena. For example, the practice of planning and carrying out investigations in the NGSS for states that all students in high school should be able to do things such as (a) "plan an investigation to produce data to serve as the basis for evidence as part of building and revising models or supporting explanations for phenomena" and (b) "decide on types, how much, and accuracy of data needed to produce reliable measurements and consider limitations on the precision of the data (e.g., number of trials, cost, risk, and time), and refine the design accordingly". However, if teachers use instructional materials that never give students an opportunity to plan a descriptive, comparative, or experimental investigation because the instructional materials use a content-activity-practice or a structured lesson structure, then these students will never get an opportunity to learn how to plan out these different types of investigation. Another common modification for students who are labeled as low-achieving or "at risk" of failure is to limit the number of opportunities for them to participate in investigations so they can be given more time to learn about and review the DCIs. Once again, however, this type of modification is problematic because these students cannot be expected to master the SEPs and CCs if they never have a chance to use the SEPs and CCs to make sense of a wide range of phenomena. These types of modifications, although well intentioned, do not provide these students with a fair opportunity to learn.

Students who are labeled as "honors" or "AP" in high school are also often provided with instructional materials that have been modified by educators. However, instead of attempting to simplify the instructional materials, teachers change them to make them more "challenging". One common modification that is made to instructional materials to make them more challenging is adding more problems to solve as part of each lesson. Another common modification is to increase the amount or the complexity of the content. Unfortunately, to provide the instructional time that is needed to complete the extra problems or to teach the additional content, multiple opportunities for these students to engage in SEPs or use CCs are eliminated to help save time or because "they are not as important as the content". As a result, these students are exposed to a great deal of content, but often do not have a chance to learn how to use the SEPs and CCs to make sense of a wide range of phenomena. These types of modifications, as a result, also do not provide these "honors" or "AP" students with a fair opportunity to learn.

The investigations that are found in this book, therefore, do not include alternative activities, extra challenge problems, or modified lessons that eliminate opportunities for students to learn and master every NGSS performance expectation that is associated with a Biology course. Instead, these 3D instructional materials are designed to provide differentiation for students who need it by embedding accommodations for students into each investigation. This approach to differentiation through embedded accommodation is informed by research in science education, special education, and gifted education as reviewed above. The broad guidelines that are outlined in the next section and the design of the instructional materials also ensure that the materials are:

1. compliant with any changes to state or federal laws and regulations governing the teaching of special populations and
2. flexible enough for compliance with individual education plans (IEPs) established for students.

Accommodations Already Embedded into the ADI Instructional Model

All the investigations that are included in this book include numerous accommodations for students who have special needs because they were created using the ADI instructional model. This instructional model was intentionally designed to differentiate the nature of instruction that happens over the course of a single lesson for all students rather than creating different lessons for different students.[79] During a single ADI investigation, for example, students obtain information in a variety of different ways (text, video, manipulatives, and discussion), use different modes to share what they know and how they know it (drawing, writing, talking with others, graphical, mathematical, and formal presentations), and experience different grouping formats (whole class, small group, and individual). This type of variation ensures that one mode of delivery, communication, or grouping format is not privileged more than any other. In addition, there are several accommodations built into each stage of the instructional model that are intended either to provide additional support when students need it or to provide opportunities for enrichment when students desire it (and/or when a teacher thinks enrichment is needed). Table 3 provides the specific supports and enrichments that are built into the ADI instructional model, and by extension, each investigation, by stage.

Additional Differentiation Based on Need

In addition to the numerous accommodations that are already embedded into the investigations included in this book, there are several ways teachers can differentiate a learning experience for students who need extra support or enrichment. These additional accommodations, however, should only be used when needed and should not be provided when they are no longer necessary. For example, teachers can read the texts in Stage 2 and have students follow along during the first investigation. As the teacher reads, they can stop at key points in the text and ask the students in the class to circle or highlight an idea that will be useful to revisit later. This type of accommodation helps struggling readers comprehend more of what is read. In the next investigation, the teacher can once again read the text aloud, but instead of telling students what is important in the text, the teacher can instruct the students in the class to work in small groups to identify the key ideas. These ideas can then be shared with the rest of the class. Like the first example, this accommodation is designed to support reading comprehension. However, unlike the first example, it provides more opportunity for the students in the class to do the cognitive labor. It is important for teachers to only use accommodations such as these as a temporary support while they continually push students to do more and more on their own with each subsequent lesson. This type of approach provides the support that students need while also providing a mechanism for continuous

---

[79] Callahan, Sampson, and Rivale, "Activating Bilingual English Language Learners' Strengths in Science: The Pedagogy of Argument Driven Inquiry (ADI)".

Table 3. Summary of the supports and enrichments available for students during each stage of the ADI Instructional model

Stage	Support	Enrichment
Task	Anchor lesson in a common experience regardless of background. Entry point has a "low floor" (so everyone can participate without first meeting specific prerequisites). Ability to share ideas through drawing.	Increase motivation to learn because there is a "real world" context (phenomenon). Entry point has a "high ceiling" (so students can create complex explanations or solutions to problems if desired).
Ideas	"Chunking of text" by ideas. Important vocabulary highlighted in text. Peer-to-peer discussion in small groups. Immersive reader allows students to change print size, translate, and use text-to-speech tools.	Opportunities to obtain and evaluate information provided in text. Choice in deciding what information is important or useful in a specific context.
Plan	Access to a graphic organizer. Peer-to-peer discussion in small groups. Feedback from teacher.	Opportunities for creativity, critical thinking, and non-routine problem-solving. More complex investigation plans. Peer review of plans written by other groups.
Do	Access to a graphic organizer. Peer-to-peer discussion in small groups. Hands-on activity. Feedback from teacher.	Opportunities for creativity, critical thinking, and non-routine problem-solving. Complex data collection and analysis methods.
Share	Access to a graphic organizer. Multiple ways to communicate information. Peer-to-peer discussion in small groups. Feedback from teacher. Opportunities for revision without grade-related penalty. Discussion scaffolds.	Opportunities for creativity, critical thinking, and debate. Peer review of arguments written by other groups. Complex presentation of evidence and justification.
Reflect	Review and reteach DCI and CCs from Ideas stage. Teacher feedback. Vocabulary after concept.	Opportunities to propose, support, challenge, and revise ideas about DCI, CC, and other related topics.
Report	Sentence stems to support writing. Peer-to-peer discussion in small groups. Feedback from others. Opportunities for revision without grade-related penalty.	Realistic writing task. Peer review of reports written by other students. Complex presentation of findings.

improvement. Table 4 provides some additional supports and enrichments that teachers can use during each stage of an investigation.

Table 4. Summary of the additional supports and enrichments that teachers can provide when needed for each stage of an ADI investigation

Stage	Support	Enrichment
Task	Preferential seating. Record shared ideas put forward by all groups on a wall for later reference.	Use a "Wonder Wall" for students to add new questions that will not be answered as part of the investigation that students can research or investigate when they have extra time.
Ideas	Add key terms to a word wall. Read the text aloud and then encouraging the students to annotate while following along. Read the text out loud and then discuss what to annotate as a whole group. Record shared ideas put forward by all groups on a wall for later reference.	Strategic and flexible grouping. Have the students read and annotate the text independently.
Plan	Provide strong and weak examples of plans and make the key differences explicit.	Have students create plans without graphic organizers.
Do	Strategic and flexible grouping. Additional time. Provide "just-in-time" lessons when students get stuck.	Have students share out their approach for analyzing data. Have students provide feedback to other groups.
Share	Provide strong and weak examples of arguments and make key differences explicit. Use "back-pocket questions" to stimulate peer-to-peer discussions. Have students give written instead of verbal feedback using "sticky notes."	Have students add a challenge to a counterclaim as part of their argument. Have students design and carry out a second investigation to validate their findings or to test alternative ideas that came up during discussion.
Reflect	Encourage students to share in small groups before sharing with the whole class. Use extended wait time after posing a question.	Have the students lead the whole-class discussion.
Report	Assign a scribe. Allow additional time to complete the report.	Require additional "gotta-haves" in the report.

Notes for "Future You"

Think about everything you have read in this chapter. Then write down some things (including the definitions of important terms) that you think will be helpful for you to remember when you use an ADI investigation in class. We all forget things, so do what you can now to help "future you".

Some things I want to make sure that I remember

Support for Emerging Multilingual Students When Using the ADI Instructional Model

Principles and Best Practices in the Context of Science Instruction

Approximately 5.3 million students in US public schools, or about 10.6% of all students in grades K–12, are considered emerging multilingual students.[80] While the term "emerging multilingual student" broadly captures any student learning an additional language, in the context of public K–12 education in the United States, it is often used to denote a student who is learning English. Given the number of students who are learning how to use the English language in US schools, it is important to think about how to best support these students in the context of science. Students who are learning English, like all students, come to school with rich experiences and ideas about how the world works, ideas shaped by their cultures and discussions with family and friends. These ideas and experiences are valuable and serve as a foundation for making sense of new ideas and experiences in science. Students who are learning English, however, will often struggle when asked to share these rich experiences and ideas inside the classroom using the English language. So how can we help these students develop greater proficiency in English? The simple answer is through its use. Research[81,82] suggests that for people to learn any new language, the following conditions must be met:

1. Continuous interactions with other people who have a higher level of proficiency in the new language so they can learn the "rules of the language" and receive feedback as they learn
2. A reason or a need to use the new language across the modalities of reading, writing, listening, and speaking.

Outside of schools in the United States, children often have continuous interactions with people who have a higher level of proficiency in "conversational" or "everyday" English. There is also a reason for them to use conversational or everyday English when reading, listening, and speaking. Children, as result, will often be able to develop high levels of proficiency in conversational or everyday English simply by living their lives. The English that is used and privileged in schools, however, is different. It is a specialized version, one of the many types of English that people develop to suit different situations or contexts. The version of English that is used and privileged in schools is often described as "academic" English. This form of English often takes longer for children to develop since they have less contact with people that use this form. Children also tend to have fewer opportunities to use academic English across the modalities of reading, writing, listening, and speaking outside of school.

The language of science is an even more particular version of that specialized academic English. Science focuses on certain types of relationships between ideas and patterns of thinking that lead to specific ways of reading, writing, and speaking. Students learn these patterns as they need them, and

---

[80] Condition of Education, "English Learners in Public Schools," *National Center for Education Statistics*, https://nces.ed.gov/programs/coe/indicator/cgf (accessed April 20, 2024).

[81] Brown, *Principles of Language Learning and Teaching*.

[82] Hamayan, "What is the role of culture in language learning?," 61-64.

they develop the language of science by "doing science." No student (including students who learned English as their first language), as a result, comes to school being highly proficient in "academic" English or the "language of science". To help address this issue, WIDA released a focus bulletin[83] that describes several design principles or "best practices" for creating a learning environment that enables all students, including emerging multilingual students, to learn academic English and the language of science at the same time. Among the most important of these principles or best practices include:

1. Science lessons should engage learners with relevant phenomena and encourage sensemaking activities that simultaneously combine previous experience with the learning of new content and language through firsthand 3D instruction.
2. Science lessons should elicit student ideas as a starting place for sensemaking and for co-constructing explanations of phenomena.
3. Science lessons should use the science and engineering practices to anchor content learning and language use and development in sense-making.
4. Teachers should do everything they can to ensure that students have agency and authority, so that their contributions are integral to sensemaking in science.

The ADI instructional model is well aligned with each of the design principles or "best practices" for helping all students learn academic English and the language of science at the same time. For example, the model places a heavy emphasis on children using the SEPs to make sense of phenomena. Students' contributions during an investigation therefore represent an ongoing effort to build, test, revise, and justify scientific ideas and models. This type of focus creates and keeps students' evolving meaning-making skills at the center of what it means to do science. This collective activity and meaning-making, both in the moment and across time, supports gradual shifts in English language use from "everyday" or "conversational" to "academic" and "science-specific". The ADI instructional model is also designed to position students with agency and authority. This type of positioning inside the classroom allows students to see that their ideas are important and useful. It also enables students to adopt new ways of speaking, listening, reading, and writing that reflect what is valuable or useful within that learning community. These opportunities in Biology are essential if we, as teachers, want to make science and language learning meaningful for children. Given this alignment with the design principles and practices, teachers can, therefore, use the ADI instructional model to support students in learning academic English and science at the same time, regardless of the student's current level of English language proficiency.

Additional Support for Students Learning Academic English and Science

Although the ADI instructional model is a good starting point for supporting emerging multilingual learners, it is important to recognize that each child starts at a different place when it comes to learning how to use academic English. It is therefore important to also think about what other supports that we, as teachers, can provide to individual students with different levels of academic English proficiency to make it easier for them to learn academic English and science at the same time. For example, students

[83] Rita MacDonald, David Crowther, and Jennifer Wilfrid, "Making Science Multilingual: Supporting Equity Through Design Principles," *WIDA*, June 2022, https://wida.wisc.edu/resources/making-science-multilingual-supporting-equity-through-design-principles.

whose proficiency in academic English is just beginning to develop often need ideas presented in multiple ways. Students whose "everyday" or "conversational" English is more developed, in contrast, usually need fewer supports and are more like their academic English-fluent peers when interacting with others during class. These students, however, still need to learn academic English vocabulary and new language patterns to express new and complex ideas in science. With this issue in mind, there are several general additional supports that teachers can provide emerging multilingual students to make it easier to learn during a science lesson that was designed using the ADI instructional model. These general supports include:

1. Share information in multiple ways—voice, print, pictures, hands-on materials—so that listening alone does not bear the full weight of meaning-making.
2. Provide opportunities to discover and discuss the relationship between meaning and language by asking questions such as: *How do you know that's what it's asking? Where does it say that? How can you be sure? How is that idea different from the other ones? What words tell you that? Which argument is more convincing, and why?*
3. Take time to discuss the relationship between linguistic choices and the disciplinary purposes it serves, such as being more precise, detailed, or objective.

In addition to these general, additional supports that can be used during any stage of an ADI investigation, we also provide specific recommendations for helping emerging multilingual students in the lesson plan for each investigation that are stage- and modality-specific. Be sure to read over these recommendations and use them as needed to make it easier for the emerging multilingual students in your classroom to not only participate in the investigation but also make meaningful contributions.

In conclusion, regardless of what supports you decide to provide emerging multilingual students to ensure they have a fair opportunity to learn, be sure to keep the following guidelines in mind if your goal is to help the students in your classroom learn academic language and science at the same time:

1. Emerging multilingual students' primary languages and cultures are assets that can leverage to enhance the learning of all students in class.
2. Emerging multilingual students learn best when learning experiences jointly build discipline-specific understanding and increased language proficiency.
3. Emerging multilingual students should be engaged in the same learning experiences as their peers. Reduced opportunities to learn content will hamper both the English language development and the disciplinary understanding of emerging multilingual students.
4. Learning experiences should focus on big questions grounded in authentic contexts. A narrow focus on vocabulary and memorization will hamper both the English language development and the disciplinary understanding of emerging multilingual students.
5. "Just-in-time" scaffolds will provide the additional support emerging multilingual students need to master disciplinary understanding and increase their proficiency in English during a lesson.

Notes for "Future You"

Think about everything you have read in this chapter. Then write down some things (including the definitions of important terms) that you think will be helpful for you to remember when you use an ADI investigation in class. We all forget things, so do what you can now to help "future you".

Some things I want to make sure that I remember

Ways to Create a Safer Learning Environment When Using the ADI Instructional Model

It is important to establish a safe learning environment for classroom or field investigations. A safe learning environment consists of properly attending to four aspects of safety:

1. Establishment of a physical environment promoting safe practices,
2. Proper use of personal safety equipment,
3. Creation of classroom rules for conduct during lab or field investigations, and
4. Proper disposal and storage of materials used during laboratory or field investigations.

The Physical Environment

In establishing the physical environment, the following steps should be taken.

1. At least one fire extinguisher should be present in the room. Proper signage indicating the location of the fire extinguisher should be displayed in both English and Spanish (as well as any other language that students speak as their primary language) and should be clearly visible from all workstations in the laboratory. Annual inspections of the fire extinguisher should be conducted by entities licensed by the state fire marshal.
2. All laboratories should have an eye wash station. Proper signage indicating the location of the eye wash station should be displayed in both English and Spanish (as well as any other language that students speak as their primary language) and should be clearly visible from all workstations in the laboratory. Eye wash stations should be inspected regularly as indicated by the manufacturer.
3. All laboratories should have a first aid kit. Proper signage indicating the location of the first aid kit should be displayed in both English and Spanish (as well as any other language that students speak as their primary language) and should be clearly visible from all workstations in the laboratory. The first aid kit should be replaced annually to ensure no components of the kit have passed their expiration date.
4. There should be a designated location for the storage of personal protective equipment, such as goggles, aprons, and non-latex gloves. Proper signage indicating the location of the storage of personal protective equipment should be displayed in both English and Spanish (as well as any other language that students speak as their primary language) and should be clearly visible from all workstations in the laboratory.
5. All workstations should be properly spaced for groups of four students and two adults to work concurrently. This will allow the teacher and any instructional aides to assist students in a way that does not pose a safety risk.

The Proper Use of Personal Protective Equipment

To ensure all students are safe during laboratory and field investigations, it is important to obtain and use personal protective equipment (PPE). At a minimum, the PPE students use must meet the ANSI/ISEA Z87.1D3 standard. The correct usage of this equipment is paramount for keeping students and others safe.

The following PPE should be stored and used when conducting investigations:

1. Goggles should be worn in any investigation where students are working with substances that pose a splash risk. This includes all liquids and powders. Goggles should be worn over students' eyes (and, if necessary, eyeglasses) and tightened to prevent gaps between the goggles and the skin around the eyes. Goggles should be cleaned after each investigation.
2. Aprons should be worn when working with any substances that pose a splash risk. This includes all liquids and powders.
3. Non-latex gloves should be worn during any investigation that includes liquids at or below room temperature.
4. For investigations where students work with substances above room temperature (e.g., hot water or hot glue), heat resistant gloves can be worn. Students should not wear gloves that are not rated as heat resistant.
5. All PPE should be put on prior to students entering the laboratory and working with materials.

The following PPE should be stored and used when conducting field investigations:

1. Students should wear long pants and long sleeves during field investigations to prevent any insect bites.
2. Insect repellent should also be used to prevent bites. Refer to the instructions on the insect repellent for proper application.
3. Sunscreen should be applied to any areas of skin not covered by clothing. Refer to the instructions on the sunscreen for proper application.
4. All students should bring a full water bottle during a field investigation.
5. All students should wear a name tag or another form of identification listing the student's name and a telephone number for the student's emergency contact.
6. The teacher should always bring a first aid kit and a whistle on a field investigation.
7. The teacher should bring with them a list of allergies students might have and any medications students might need should they have an adverse reaction to an allergen in the field.

The Creation of Classroom Rules and Procedures

Part of creating a safe learning environment is the creation of rules and procedures to govern laboratory or field investigations and design challenges. Rules and procedures are important because they provide structure and expectations for student behavior. A safety acknowledgment form is included in the last chapter of the book. We recommend reviewing this form with students and then sending it home for parents to review and sign within the first week of school. We also recommend reviewing this safety acknowledgment form with students before they begin working with materials in the Do stage of an investigation. Finally, we recommend posting classroom safety rules, such as the poster shown in Figure 18, where it is visible to students during an investigation. Be sure to review this set of safety rules prior to students beginning to work with materials in the Do stage of an investigation.

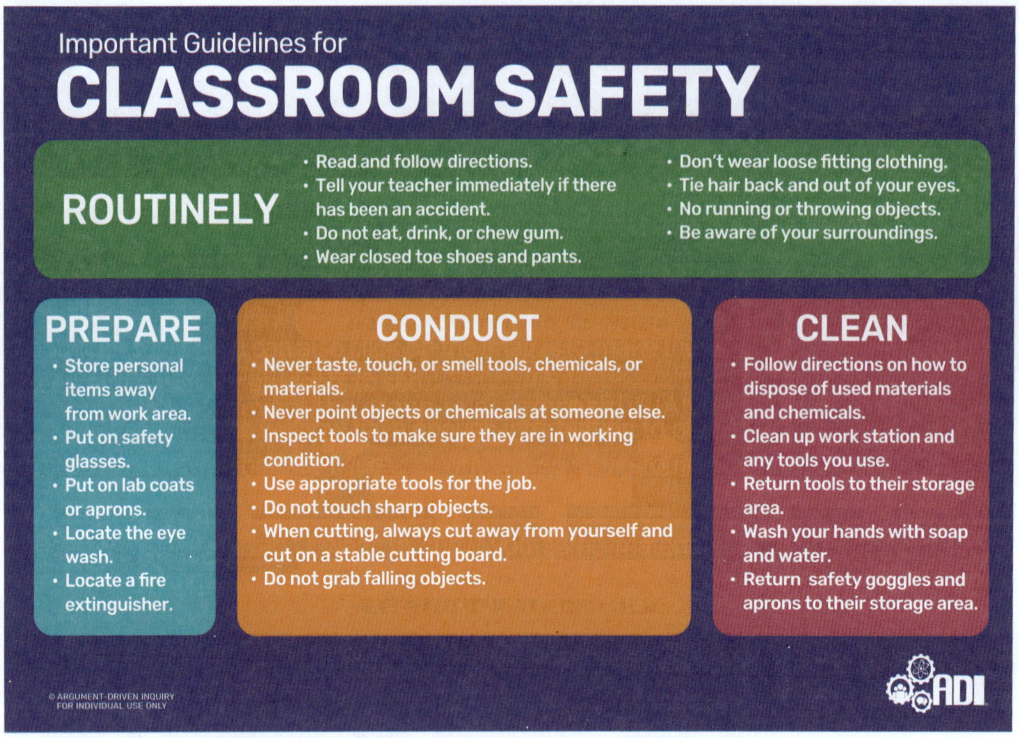

Figure 18. An example of a safety rules poster

Beyond the general rules for classroom laboratory safety, each investigation requires its own unique safety considerations. The teacher notes for each investigation includes a set of investigation-specific safety notes. For any investigation that requires the use of chemicals (e.g., iodine), a link to the materials safety data sheet (MSDS) for the specific chemical is included. The MSDS provides information on the use, storage, and disposal of chemicals. The MSDS also provides information on first aid measures should students be adversely exposed to a chemical. The teacher notes also include safety notes for the Do stage of the investigation. We recommend going over the investigation-specific safety notes with students before starting the Do stage. These notes should also be displayed on the classroom chalkboard/whiteboard or a screen that is visible to students from all workstations.

Cleaning Up After an Investigation

The final component of creating and establishing a safe laboratory environment is establishing rules and procedures for cleaning up materials. The most important rule for cleaning up after an investigation is that PPE must remain worn by students until all materials are stored properly, all consumables are disposed of properly, and any materials that needed to be cleaned are placed on a drying rack. The Materials and Preparation document accompanying each investigation has a section detailing the cleanup procedures for that investigation. For any investigation using chemicals, we provide general disposal information. However, the teacher should also review the MSDS prior to the investigation on how best to dispose of consumables. The teacher should also post any instructions for the disposal of consumables to the classroom chalkboard/whiteboard or a screen that is visible to students from every workstation.

In general, students should follow the following rules for cleaning up at the end of an investigation:

1. Properly dispose of any consumables.
2. Rinse with water any equipment that was in contact with consumable materials and put away. This includes equipment such beakers, test tubes, and stirring rods.
3. Clean up the workstation.
4. Wash hands with soap and water.
5. Return safety goggles and aprons to the storage location.

Notes for "Future You"

Think about everything you have read in this chapter. Then write down some things (including the definitions of important terms) that you think will be helpful for you to remember when you use an ADI investigation in class. We all forget things, so do what you can now to help "future you".

Some things I want to make sure that I remember

Assessment When Using the ADI Instructional Model

Purposes of Assessment

An assessment in education is simply a measurement of what someone (or a group of people) knows or can do at a given time. Educators can then use the measurement for several different purposes. If the measurement is used by an educator to determine if a student needs extra support, opportunities for enrichment, or is qualified to enter a specific program, then the assessment that was used to produce the measurement is described as *diagnostic* in nature. A measurement of what someone knows or can do at given time can also be used by educators to determine if a student has mastered the knowledge and skills that were a focus of a course, unit, or lesson. When a measurement is used in this way, the assessment that was used to generate the measurement is described as *summative* in nature. Diagnostic and summative assessments are often viewed as "assessments *of* learning" because they are used to determine the current level of mastery or performance and provide valuable information for educators who need to assign grades, make recommendations, or determine an appropriate placement. An assessment, however, can also be used by educators to help students learn. If an educator uses a measurement of what a student (or a group of students) knows or can do at a given time to guide future instructional decisions or to make changes to an existing lesson plan based on the needs of learners, then the assessment that was used to produce the measurement is described as *formative* in nature. Finally, a measurement can also be used by educators to provide feedback or advice to students about how to improve performance. An assessment that generates a measurement that is then used to determine what feedback to give to students is described as *educative* in nature. It is also important to note that educative assessments include a diagnostic component because they provide information on if and where students need extra support. However, educative assessments go beyond diagnostic assessments because they are used by a teacher to help a student do something better than before. Formative and educative assessments, as a result, are often viewed as "assessments *for* learning" because they are used to improve instruction and support ongoing changes in student thinking.

Assessment as Making Thinking Visible

One of the best ways to make an assessment useful, regardless of how the measurement that is generated by that assessment will be used in the future, is to help educators understand how students are thinking about content or a skill. Too often, assessments are designed to determine if students can or cannot complete a specific task, such as answering a question correctly, performing a calculation the right way, or providing an appropriate definition for a concept or term. When a student can complete the assigned task, educators assume that the student has an adequate understanding of the target content or skill and when a student is unable to complete the task, the student lacks the necessary understanding. There are several problems with making such an assumption. For example, students are often able to complete the tasks they are given as part of an assessment by simply guessing, eliminating answers that are obviously wrong, by mimicking something they were shown how to do previously, or by relying on content or skills that are not the target of the assessment. Students can also complete a task incorrectly but have a strong understanding of the target content or skills. For example, a student may complete a task incorrectly because that student read the directions incorrectly or made a simple mistake. The

ARGUMENT-DRIVEN INQUIRY.

student might also understand the target content or skill but decide that that target content or skill is not needed to complete the task. Unfortunately, most assessments that students complete provide us, as educators, with very little information about why a student completes a task a specific way or what that student was thinking about during the task. To address this issue, assessments should be seen as an effort to make student thinking about the content or skill visible.

When an assessment is designed to make student thinking about the content or skill visible, the task or tasks that a student is asked to complete during an assessment are intended to provide a window into not only what a student knows but also how they know and why they decided to complete a task in a specific way. With this type of information, educators can make a more valid and reliable inference about what students need or what they have learned. Educators, in other words, can figure out if a student can complete a task correctly and for the right reason. This type of information about student thinking also enables educators to make more informed decisions about what to do during the next lesson to better support a change in student thinking or how to help a student improve his or her performance over time. This type of information about student thinking is critical if educators want to use assessment to support learning rather than just using it assessment to document if students can perform at a certain level. We therefore think about assessment not only as a measurement of what students know or can do at a given point in time but also as an effort to make their thinking about content or skills visible so educators can use assessment for formative, educative, and summative purposes with confidence. In the following sections we will provide an overview of the various types of assessments that we use to make student thinking visible. These different types of assessments are also organized by their recommended purpose (formative, educative, or summative).

Formative Assessments

Embedded Performance Tasks

Students have numerous opportunities to make their thinking visible during an ADI investigation. For example, students create models to explain phenomena, investigation plans to share how they will use CCs and SEPs to carry out an investigation and make sense of data, and written arguments to share what they have figured out and how they can be sure that their ideas are valid or acceptable depending on the stage of the investigation. Figure 19 shows an example of an investigation plan (left) and a written argument (right). Notice how each one provides a window into student thinking at that point in time.

The models (Task stage), investigation plans (Plan stage), and arguments (Do stage) that students make are embedded performance tasks that can be used as formative assessments by teachers because they require students to use their current understanding of DCI, CCs, and SEPs. Teachers can therefore use these embedded assessments to quickly identify strengths and weaknesses in student thinking or performance and modify instruction as needed. As an added benefit, students do not even realize that their teacher is using such an embedded performance task as a formative assessment because they are "just doing science". These embedded assessments, as a result, can help reduce student test anxiety and will save valuable instructional class time that would normally be needed to administer a quiz or test.

Exit Tickets

Teachers can require students to fill out an exit ticket after completing each stage of an investigation. The ticket consists of one or two questions asking students to share what they learned, what they figured out,

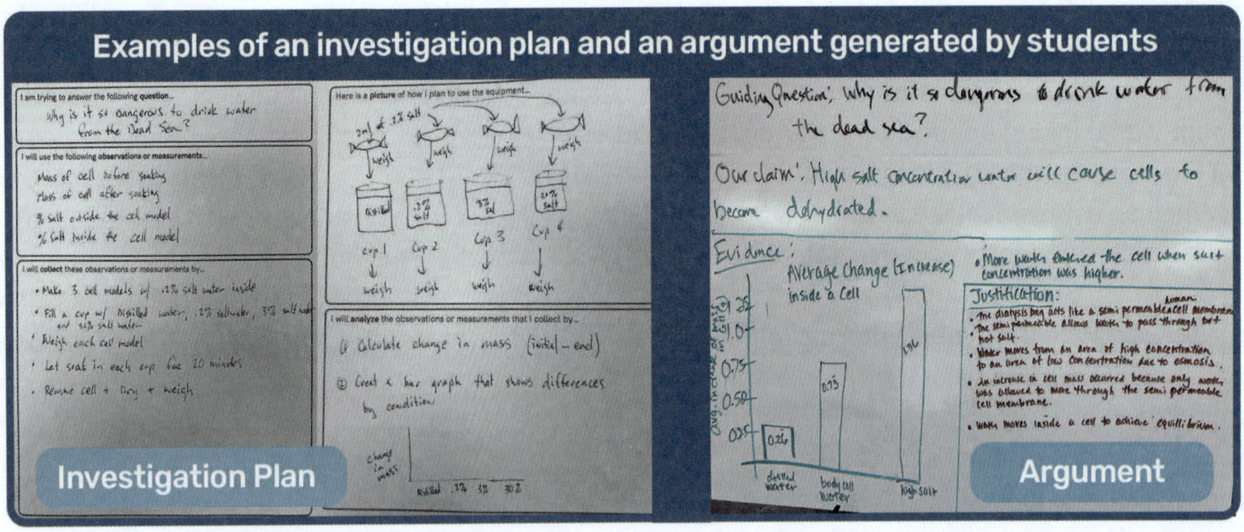

Figure 19. An example of an investigation plan (L) and an argument (R) generated by students

and what they are unsure about. Figure 20 is an example of a paper-based exit ticket. Exit ticket copy masters are included in the last chapter of the book (called **Other Useful Stuff**).

Exit Ticket Name: __________ Date: __________

My biggest takeaway from today...	I am now wondering...

Figure 20. An example of an exit ticket

The exit tickets, which only take a few minutes for students to complete, can be used as a formative assessment by teachers because they require students to share their thinking (and how it has or has not changed) during the investigation at key points in time. Teachers can therefore use these exit tickets to determine what students have picked up and any sources of confusion while there is still time to reteach. These exit tickets, as a result, help drive real-time adjustments and decisions about next steps.

Q&As without Scores

Q&As are short assessments that require students to make sense of authentic scenarios using DCI and CCs. These scenarios are often (but not always) framed as a phenomenon to explain. Students complete the assessment by answering a two-tiered item. The first tier requires students to provide an answer.

Often, students are given several potential answers to choose from, which can be images or words. The second tier requires students to explain their thinking or provide a reason for selecting an answer.

Teachers can assign these Q&As before or during an investigation to help make students' thinking visible. They can then use student responses to the two-tiered items to determine what students have selected and any sources of confusion while there is still time to reteach, just like an exit ticket. Q&As, as a result, can help drive real-time adjustments and decisions about next steps. The Q&As are found in the assessment library of the ADI Learning Hub (see chapter 9, **Resources for Teachers**).

Educative Assessments

Embedded Peer Review

Peer review is a key feature of an ADI investigation. At the very least, students participate in the peer review process twice during an ADI investigation: once during the Share stage as they share, critique, and revise their draft arguments, and then again during the Report stage when they share, critique, and revise their draft reports. Students can also peer review one another's investigation plans during the Plan stage of an investigation. An additional opportunity to participate in the peer review process during an ADI investigation is an excellent enrichment activity that can be used to push students further.

Peer review is an educative assessment because students make their thinking visible and are then given feedback about how to improve. They are encouraged to use this feedback to improve what they have created; how they are thinking about a DCI, CC, or an SEP; or how they make sense of the phenomenon at the heart of the investigation. They also have a chance to see stronger and weaker examples of products made by others for the same purpose and are given time to discuss what "counts" as high and low quality in this specific context. Over time, students will adopt new standards and norms that shape how they think about quality work and can be used to guide their performance on similar tasks during future investigations. This peer review process, as a result, provides a mechanism that leads to better understanding and performance over time.

Embedded Reflective Discussions

Students have an opportunity during the Reflect stage to share their ideas about a DCI or CC and how they use SEPs to plan and carry out an investigation. These reflective discussions give students an opportunity to apply what they know to explain a related phenomenon and to plan for how to improve the design of their investigations in the future. The reflective discussions are an educative assessment because students share their thinking and provide reasons for their ideas. If a student's (or group of students') ideas are not accurate from a scientific perspective or inconsistent with agreed-upon classroom norms or rules for investigations, their peers can provide further clarification, descriptions of alternative ways of thinking about the topic, and other "food for thought". Teachers can also fill any gaps identified in a student's ideas in a way that will promote and support ongoing changes in their thinking. These discussions, as a result, provide students with the feedback they need to learn.

Q&As with Feedback

The teacher can also use the feedback feature of the Q&As as an educative assessment. To accomplish this goal, teachers can require students to complete a Q&A in the ADI Learning Hub (see chapter 9, **Resources for Teachers**). The teacher can then review the answers given by the student and provide feedback. The teacher can set the assessment grading feature to "Feedback Only". The student can then access and read the feedback left by the teacher in the ADI Learning Hub. After viewing the feedback, the student can request to make a new submission.

Summative Assessments

Embedded Investigation Reports

The final draft of the investigation report is intended to be used by educators to determine if a student has mastered the knowledge and skills that were the focus of the investigation. These reports require each student to use DCIs, CCs, and SEPs to explain on their own and in writing how or why a phenomenon happens. They also provide an opportunity for students to demonstrate disciplinary writing skills, which are found in the Common Core Standards for ELA, and mastery of grade-level appropriate language development expectation, which are found in the WIDA English Language Development Standards (ELDS). These reports provide an authentic performance task for students to complete. These performance tasks are an excellent way to determine not only if students understand course-specific core concepts but also how well they can construct and present written arguments that justify an explanation of a phenomenon or a solution to problems using evidence acquired from the learning experience.

The PRG-TSR includes indicators that align with science DCIs, CCs, and SEPs, ELA literacy skills, and language development expectations. It is important to note that the indicators used to score the reports do not change depending on the report. This feature is intentional because it provides educators with a way to track growth in student performance over time by simply comparing the scores of a student (or class of students) on two or more investigation reports using the insight tool that is embedded into the ADI Learning Hub. Unlike the challenges that are associated with comparing test scores over time, which include having different numbers of items and a different focus each time, the scores on the reports are comparable because the same indicators are used to score the reports each time. This feature enables educators to see how students are improving and if specific students need extra support or enrichment.

Q&As with Scores

The Q&As can also be used as a summative assessment. They can be a rich source of information about students' understanding and retention of the DCIs and CCs that students are expected to master by the end of a unit (or grading period). Students can complete these assessments on paper or in the Learning Hub (see chapter 9, **Resources for Teachers**). The ADI Learning Hub includes an answer key for teachers to use when scoring paper-based tests. The ADI Learning Hub, however, will automatically score the answers (but not the reasons) when students complete one of these assessments in the ADI Learning Hub.

To use Q&A in the Learning Hub, the teacher can simply assign it to a class and select an open and close date and time. Students can post their answers using their account in the ADI Learning Hub. The teacher can then score the answers and provide feedback (if needed). The teacher can set the assessment to "Score Only" or "Score and Feedback". If the assessment is set to "Score Only", students will only be able to see the score they earned on the assessment (each answer of a two-tiered question is worth one point, and each reason is worth two points). Students, however, can view their submission and request to complete a "new submission". If the assessment is set to "Score and Feedback", students will only be able to see the score they earned on the assessment and any feedback left by their teacher.

Notes for "Future You"

Think about everything you have read in this chapter. Then write down some things (including the definitions of important terms) that you think will be helpful for you to remember when you use an ADI investigation in class. We all forget things, so do what you can now to help "future you".

Some things I want to make sure that I remember

Resources for Teachers

This book includes nine ADI investigations (see chapters 10–13). These investigations are not designed to replace an existing science curriculum but rather are designed to function as a tool that teachers can use to integrate more 3D instruction into their classroom. Each investigation is designed to function as a standalone lesson, giving you the flexibility to decide which ones to use and when to use them during the academic year. The investigations are organized by topic into different chapters in the book, but the topics can be taught in any order. All of the investigations are aligned with at least one NGSS performance expectation that is associated with Biology. In the remaining sections of this chapter, we will outline some of the resources that we have included to make it easier to start using the ADI instructional model. We will begin with the lesson plans for each investigation.

Lesson Plans

All eight investigations included in this book have an accompanying lesson plan. The lesson plans provide information on what the teacher and students will do during each stage of the investigation. Each lesson plan also includes:

- A general overview of the investigation, including the guiding question of the investigation and what students will figure out.
- A flowchart that shows what happens on each day of the investigation and how much time each activity should take to complete.
- The NGSS performance expectation and the WIDA ELDS for science that are addressed during the investigation.
- Cross-curricular connections to Common Core Standards for ELA.
- A description of the DCI, CC, and SEPs that students will use during the investigation.
- Information needed to prepare to deliver the investigation paper-based, paperless, and remote including lists of the hands-on materials that are needed for in-person delivery and where to find them.
- Instructions on how to prepare the materials prior to the Do stage of the investigation.
- Cleanup instructions, including information on waste disposal.
- Safety considerations for setting up and conducting the investigation.
- How much time each stage and activity within a stage should take.
- Detailed directions for what to say and do during each activity within a stage of an investigation.
- Ways to modify the activity for students with different needs.
- Ways to support emerging multilingual students at all levels of English language proficiency.

Student Handout Copy Masters

The instructional materials included in this book are reproducible copy masters designed to support students as they participate in an ADI investigation. The instructional materials needed for each investigation include an investigation handout and the peer review guide - teacher scoring rubric (PRG-TSR). Each student should be given a copy of the investigation handout at the beginning of each ADI investigation. This handout provides information about the phenomenon they will investigate and a guiding question for the students to answer. In addition, the handout provides space for students to design their investigation, record their observations and measurements, and create tables, graphs, or pictures to analyze the data they collect during the investigation. The handout also has space for the students to keep track of critiques, suggestions for improvement, and good ideas that arise during the argumentation session. To help support students as they learn how to write in this context, the handout includes sentence starters that they can use when writing the draft report. The PRG-TSR is designed to make the criteria that are used to judge the quality of an investigation report explicit. The PRG-TSR can be found in the book's last chapter (**Other Useful Stuff**). We recommend that you make one copy of the PRG-TSR for each student. Then, during the peer review stage, have the triads of reviewers fill out one PRG-TSR for each report that they review. The reviewers should rate the report on each criterion and then provide advice to the author about ways to improve it based on their rating. Once the review is complete, the author needs to revise his or her report and respond to the reviewers' rating and comments in the appropriate sections in the PRG-TSR. You should collect the PRG-TSR together with the first draft and the final report for a final evaluation. To score the report, simply fill out the "Teacher Score" column of the PRG-TSR and then total the scores. There is also space in at the bottom of the PRG-TSR for teacher feedback.

The ADI Learning Hub

The purchase of this book comes with access to the ADI Learning Hub. The ADI Learning Hub is a browser-based platform designed to make it easier for teachers to facilitate an ADI investigation. You can create a teacher account in the ADI Learning Hub by going to https://adilearninghub.com and clicking "create a free teacher account". With the free version of the ADI Learning Hub, teachers can facilitate an ADI investigation using the paper-based delivery mode. This means all students are in the same classroom and use handouts to complete the investigation. Teachers will then use the Presentation tool in the ADI Learning Hub to project directions, images, videos, and other useful information on a screen or large monitor during each stage of an investigation. This eliminates the need for teachers to create their own slide decks. Teachers can also download student handouts and assessments from the ADI Learning Hub.

You can upgrade your account if you wish to use the paperless or remote delivery modes. The paperless and remote delivery modes require students to have an account in the ADI Learning Hub. The *paperless* mode of delivery assumes that all students have their own computer or tablet (1:1), are in the same classroom, and will complete the investigation by submitting their work to the Learning Hub. This mode also enables students to use features such as the canvas, embedded simulations, and peer review tools built into the ADI Learning Hub. Students can still use hands-on materials for the Do stage with this mode. As with the paper-based mode, the teacher can use the Presentation tool to project directions, images, and other useful information on a screen or large monitor during each stage to help keep all students on track. Adding student accounts, however, also allows teachers to use the scoring and feedback tools built into the system. The *remote* delivery mode assumes that all students are in different locations, have access to a computer or tablet (1:1), and will complete the investigation by submitting their work in the Learning Hub. This mode enables students to work with other students in the class at a distance and gives them access to features such as collaboration boards and a canvas for drawing. To learn how to use the ADI Learning Hub for paperless or remote delivery, visit https://help.adilearninghub.com/knowledge.

The ADI Learning Hub also includes an Assessment Library. The Assessment Library includes hundreds of Q&As. As noted earlier, a Q&A requires students to make sense of an authentic scenario using DCI, CCs, and SEPs. These scenarios are often (but not always) framed as a phenomenon to explain. The Q&As consist of one or more parts. Each part of a Q&A focuses on a different scenario, or the same scenario as a previous part, but requires the use of a different combination of DCI, CCs, and SEPs. Students complete a part of a Q&A by answering a two-tiered item. The first tier requires students to provide an answer. Often, students are given several potential answers to choose from, which can be images or words. The second tier requires students to explain their thinking or provide a reason for selecting an answer. Teachers can assign a Q&A from the assessment library to be completed on "paper" or "paperless". The teacher must download a handout from the ADI Learning Hub for paper-based delivery. The teacher can then present the scenario using the Presentation tool and the students can complete the task on a downloadable handout. When the Q&A is set to paperless delivery, the students complete the entire task in the ADI Learning Hub. Students do not need an account in the Learning Hub to complete a Q&A in "paper" delivery mode, but they do need an account to complete it when it is assigned as "paperless". Teachers can also use the feedback features and automatic scoring and analysis tools built into the ADI Learning Hub when students complete the assessment in paperless mode. To learn how to use the Assessment Library in the ADI Learning Hub, visit https://help.adilearninghub.com/knowledge.

Investigation Kits

We have created kits for the investigations for each investigation that has a "hands-on" Do stage[84] to make it easier and less time-consuming for teachers to prepare before an investigation begins. These kits include materials for eight groups of students. These kits are available at https://shop.argumentdriveninquiry.com. It is important to note, however, that the purchase of one of our kits is not necessary if you already have all the necessary hands-on materials. The materials section of the lesson plan found before each investigation includes a full list of all materials needed for the Do stage.

Classroom Materials

We have created many different classroom materials that can make the facilitation of an in-person (paper-based or paperless) delivery of ADI investigation easier. Some of these materials are used by students during an investigation, such as whiteboards, easels, and feedback sticky notes. Other materials are used by the teacher to make the management of an active classroom simpler, such as group lanyards, or to help support ongoing changes in student thinking through scaffolding, such as posters, back pocket questions, and talk move table tents. These materials are available for purchase at https://shop.argumentdriveninquiry.com.

[84] Some investigations use a simulation for the Do stage. These simulations are embedded in the ADI Learning Hub.

Notes for "Future You"

Think about everything you have read in this chapter. Then write down some things that you think will be helpful for you to remember when you use an ADI investigation in class. We all forget things, so do what you can now to help "future you".

Some things I want to make sure that I remember

Investigations

10.1 Swimming in the Dead Sea

10.2 Tissue Repair

11.1 Elk Population Decline in Colorado

11.2 Bark Beetles and Climate Change in Washington

12.1 Chromosome Reduction in Reproductive Cells

12.2 Fruit Fly Body Color and Wing Shape

13.1 Antiobiotic Resistance in Bacteria

13.2 Cichlids of Lake Malawi

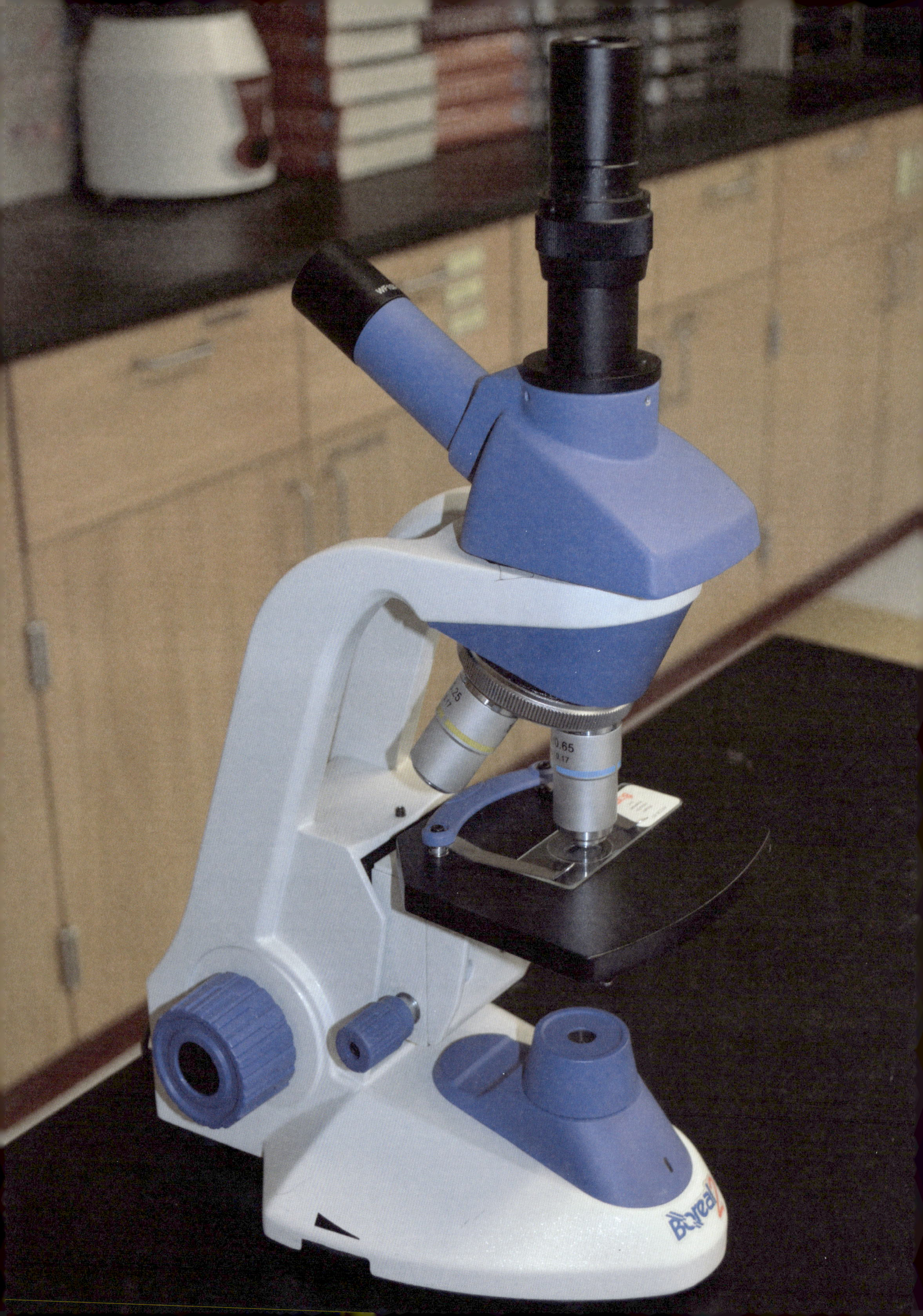

0.65

Swimming in the Dead Sea Lesson Plan

OVERVIEW

Description	Students figure out why drinking the saltwater from the Dead Sea is dangerous.
Guiding Question	Why is it so dangerous to drink water from the Dead Sea?
What Students Figure Out	The high concentration of salt in the Dead Sea creates a hypertonic solution with human blood cells causing the cells to lose a lot of water, which leads to dehydration.

STAGES

This investigation has seven stages. Each stage includes multiple activities. The flowchart below shows what happens during each stage and how much time is needed to complete each one.

Task (Day 1)
1 — **20 minutes**
Students are introduced to a phenomenon and a guiding question to answer. They also share their initial ideas.

Ideas (Day 1)
2 — **30 minutes**
Students read about DCI, a CC, and an SEP that they can use to help them figure out the phenomenon.

Plan (Day 2)
3 — **15 minutes**
Students formulate a plan collecting and analyzing the data they need to answer the guiding question.

Do (Day 2)
4 — **30 minutes**
Students carry out their plan, make sense of the data they collected, and answer the guiding question.

Share (Day 3)
5 — **45 minutes**
Students develop, share, and revise an evidence-based argument about what they figured out.

Reflect (Day 4)
6 — **20 minutes**
Students discuss a DCI and/or a CC and reflect on the SEPs they used during the investigation.

Report (Day 4)
7 — **25 minutes**
Students write investigation reports to share what they figured out during the investigation.

Report (Day 5)
7 — **45 minutes**
Students peer review the reports written by other students and offer feedback to improve.

Report (Day 6)
7 — **30 minutes**
Students revise their reports and then submit a final draft to their teacher for evaluation.

STANDARDS ALIGNMENT

NEXT GENERATION SCIENCE STANDARDS

Performance Expectation

Plan and conduct an investigation to provide evidence that feedback mechanisms maintain homeostasis (HS-LS1-3).

Core Ideas(s)	Crosscutting Concept(s)	Practices
Structure and Function (LS1.A)	Systems and System Models (CC4) Energy and Matter (CC5)	Asking Questions (SEP1) Developing and Using Models (SEP2) Planning and Carrying Out Investigations (SEP3) Analyzing and Interpreting Data (SEP4) Using Mathematics and CT (SEP5) Constructing Explanations (SEP6) Engaging in Argument from Evidence (SEP7) Obtaining, Evaluating, and Communicating Information (SEP8)

WIDA ENGLISH LANGUAGE DEVELOPMENT STANDARDS FOR SCIENCE

Explain	Argue
Interpret scientific explanations (ELD-SC.9-12.Explain.Interpretive) Construct scientific explanations (ELD-SC.9-12.Explain.Expressive)	Interpret scientific arguments (ELD-SC.9-12.Argue.Interpretive) Construct scientific arguments (ELD-SC.9-12.Argue.Expressive)

COMMON CORE ENGLISH LANGUAGE ARTS

Reading	Writing	Speaking and Listening
Key ideas and details (RST.9-10.1 and RST.9-10.2) Craft and structure (RST.9-10.4) Integration of knowledge and ideas (RST.9-10.7, RST.9-10.8, and RST.9-10.9) Range of reading (RST.9-10.10)	Text types and purposes (WHST.9-10.1 and WHST.9-10.2) Production and distribution of writing (WHST.9-10.4, WHST.9-10.5, and WHST.9-10.6)	Comprehension and collaboration (SL.9-10.1 and SL.9-10.3) Presentation of knowledge and ideas (SL.9-10.4 and SL.9-10.6)

THINGS TO KNOW BEFORE STARTING THE LESSON

The Core Ideas

Organisms are made up of one or more cells. Cells are surrounded by a membrane that creates a barrier between the cell and its environment. The membrane is semipermeable, meaning that some molecules and ions can pass through the membrane while others cannot. To maintain an even ratio of solute (molecules and ions) to water and keep the cell in equilibrium, cells add or remove water through the membrane via the process of osmosis.

Solutions can be described by their ratios. An isotonic solution is one that has the same concentration of solute and water inside and outside of the cell, meaning the cell does not gain or lose water. A hypotonic solution has a lower concentration of solute outside of the cell than inside the cell. Hypotonic cells gain water and expand. Lastly, a hypertonic solution is the opposite. It has a higher concentration of solute outside of the cell than inside. Hypertonic cells lose water and shrink.

The Crosscutting Concepts

A system is a group of related parts that make up a whole and carry out functions that they individually cannot. A cell is an example of a system. Because cells are very small and very complex, we must represent them using models. We can then use these models to test different ideas about how different parts of a cell interact and function with one another. For example, we can use a physical model of a cell to test how cells transfer matter with their surroundings.

The Science and Engineering Practices

A practice of science is to **ask and refine questions** that lead to descriptions and explanations of how the natural world works. Another practice of science is to **use and construct models** as helpful tools for representing ideas and explanations. These tools include diagrams, drawings, physical replicas, mathematical representations, analogies, and computer simulations. Scientists **plan and carry out investigations** in the field or a laboratory, working collaboratively and individually. Their investigations are systematic and require clarifying what counts as data and identifying variables or parameters. Scientific investigations produce **data that must be analyzed and interpreted**. Because data patterns and trends are not always obvious, scientists use a range of tools and practices to identify the significant features and patterns in the data. Scientists also use **mathematics and computational thinking** for a range of tasks such as constructing simulations; statistically analyzing data; and recognizing, expressing, and applying quantitative relationships. The products of doing science are **explanations**. Scientific explanations often specify variables and are used by scientists to describe and predict different natural phenomena. **Scientific argumentation**, or the process of sharing, supporting, critiquing, and revising claims based on evidence, is the process that scientists use to determine if an explanation is valid or acceptable. Finally, scientists must be able to **communicate clearly and persuasively** the ideas and methods they generate. Critiquing and communicating ideas individually and in groups is a critical professional activity.

PREPARATION

This lesson can be delivered paper-based, paperless, or remote. Use the paper-based delivery option for the investigation if students do not have 1:1 access to a computer and they are in the same classroom or physical space during the lesson. Use the paperless delivery option for the investigation if students have 1:1 access to a computer and they are in the same classroom or physical space during the lesson. Finally, if students are not in the same classroom or physical space during the lesson and have 1:1 access to a computer, use the remote delivery option.

Materials Needed

The materials needed by the teacher and each student in the class for each of the three modalities are listed in the table below.

Delivery Mode	Each Student Will Need	The Teacher Will Need
Paper-Based	• Copies of each student handout for the investigation (or a workbook) • A pencil	• A teacher account in the ADI Learning Hub • Access to a computer with high-speed internet and a browser (such as Chrome, Safari, Firefox, or Edge) • A projector or smart board (for displaying a computer screen)
Paperless	• A student account in the ADI Learning Hub • Access to a computer with high-speed internet and a browser (such as Chrome, Safari, Firefox, or Edge)	• A teacher account in the ADI Learning Hub • Access to a computer with high-speed internet and a browser (such as Chrome, Safari, Firefox, or Edge)
Remote	• A student account in the ADI Learning Hub • Access to a computer with high-speed internet and a browser (such as Chrome, Safari, Firefox, or Edge)	• A teacher account in the ADI Learning Hub • Access to a computer with high-speed internet and a browser (such as Chrome, Safari, Firefox, or Edge)

Additional Materials and Preparation Needed for In-Person Delivery

The materials needed for the in-person version of this investigation are listed in the table below. To purchase the **Swimming in the Dead Sea** kit from our store, please visit https://shop.argumentdriven-inquiry.com/. The **Swimming in the Dead Sea** kit comes with enough materials for eight groups.

Type of Material	Item	Quantity	
		For Entire Class	For Each Group
Consumables	Salt	26 oz	As needed
Equipment	Distilled water*	3 L	As needed
	Dialysis tubing	20 ft	2 ft
	Saltwater concentration stickers	3	As needed
	Pipette	8	1
	Funnel	8	1
	Measuring cups	32	8
	Scale	8	1
	Pitchers	3	As needed
	Spatula	1	As needed

* Not included in the kit

Preparation

Fill one pitcher with distilled water and label it either "Fresh water" or "Distilled water". Add 30 g of salt to the second pitcher and then add 1 L of distilled water to make a 3% saltwater solution. Label the second pitcher "Ocean water" or use the "Salt solution - medium concentration" sticker. In the third pitcher, mix 300 g of salt and 1 L of distilled water to make a 30% saltwater solution. Label the third pitcher "Dead Sea water" or use the sticker "Salt solution - high concentration". You can also make a low concentration saltwater solution by mixing 3 g of salt and 1 L of water to make a 0.3% saltwater solution. If you choose to make a 0.3% saltwater solution, use "Salt solution - low concentration" sticker to label the pitcher. The image below shows the materials that each group will need to design and carry out their investigation.

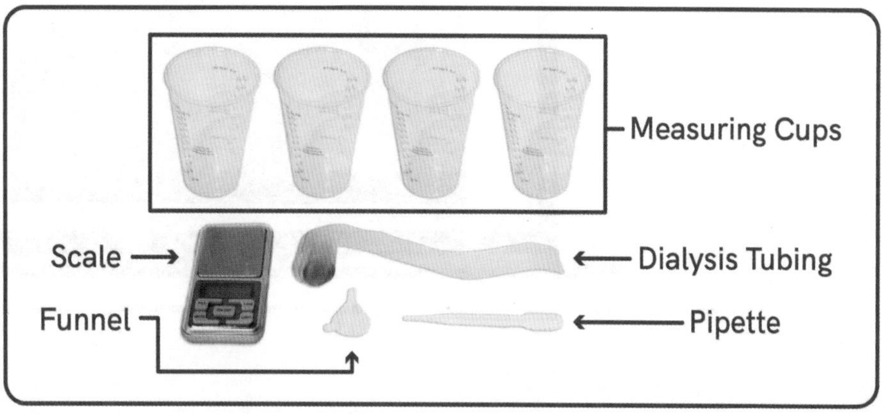

During the Investigation

Students can use the equipment to determine how saltwater concentration affects the rate of osmosis in physical models of cells. The rate of osmosis can be measured by tracking the change of mass of dialysis tubing after soaking it in different concentrations of saltwater. For best results, allow the bags to soak for at least 30 minutes and be sure that the outside of each bag is dry before weighing.

We also recommend that you use a set routine for distributing and collecting the materials during the investigation. The equipment for each group, for example, can be set up at each group's lab station before class begins, or one member from each group can collect them from a table or a cart when needed during class.

Safety Precautions

Remind students to follow all normal lab safety rules.

Laboratory Waste Disposal

The used dialysis tubing can be placed in a garbage can for disposal. The saltwater can be poured down the drain. Be sure to collect all the equipment at the end of the investigation and store it for later use.

FACILITATION NOTES

The facilitation notes for this investigation describe how to deliver this lesson in person. They are written specifically for the paper-based mode of delivery. These same facilitation notes, however, can also be used for a paperless delivery of the lesson by simply substituting the directions that mention paper with the entry into the ADI Learning Hub. If you are teaching this lesson using the remote mode of delivery, you only need to assign due dates and prerequisites for each stage. The students can then simply follow the directions built into the ADI Learning Hub as they move through the investigation.

ADI Go Digital

The digital version of this lesson is available in the **ADI Learning Hub**. Be sure to check out the built-in teacher presentation mode.

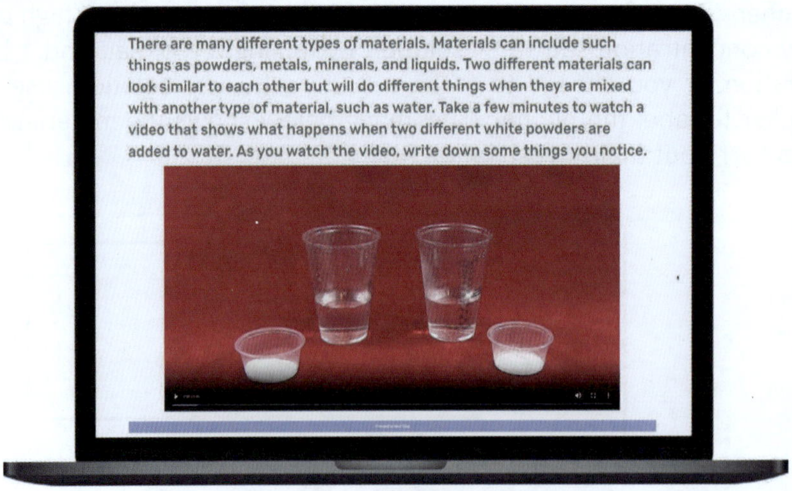

Swimming in the Dead Sea Lesson Plan

Stage 1 TASK

Total Time for the Stage	45 minutes
Grouping	Individual; Teams

Learning Goals

Students should make progress on their ability to do the following by the end of this stage of the lesson:

Science

1. Ask questions that arise from careful observation of phenomena to clarify and/or seek additional information (SEP1).
2. Develop and use a model to provide mechanistic accounts of phenomena (SEP2).
3. Evaluate the merits and limitations of different models of the same phenomenon (SEP2).
4. Ask questions that arise from examining models to clarify and/or seek additional information and relationships (SEP1).
5. Ask questions to clarify and refine a model (SEP1).

English Language Development

None

English Language Arts

1. Initiate and participate effectively in small-group collaborative discussions with diverse partners by building on others' ideas and expressing their own clearly (SL.9-10.1).
2. Present information clearly, concisely, and logically such that listeners can follow the line of reasoning, and the organization, substance, and style are appropriate to the purpose, audience, and task (SL.9-10.4).
3. Adapt speech to a variety of contexts and tasks, demonstrating command of formal English when indicated or appropriate (SL.9-10.6).

Activities within the Stage

Activity A	The Phenomenon
Time Needed for the Activity	5 minutes
Grouping	Teams

What to Do

1. Have students work in their assigned teams for this activity.
2. Begin the activity by giving out a copy of the Task handout to each student.
3. Show "The Phenomenon" page on a screen using the Presentation mode in the Learning Hub so everyone in the classroom can see it.
4. Read the introduction of the phenomenon out loud and then start the video. You can also have students read the introduction of the phenomenon and watch the video on their own devices.
5. Give the students a few moments to write down their observations on their handouts before moving on to the next activity.

What to Give Out

Task handout for
Swimming in the Dead Sea

What to Show

"The Phenomenon" page in the ADI Learning Hub

Teaching Tip

Instead of assigning fixed roles to students by name at the beginning of the investigation, give each member of a group a different number so you can easily assign and change roles as needed during the various investigation stages. For example, you might give directions such as "1s will share out for your group" or "3s, will you please come back here and pick up these materials for your group?" or "2s and 3s, you will present the argument for your group while 1s and 4s travel to the other groups to offer feedback."

Attending to Equity

Be sure to intentionally group emerging multilingual students at times with peers who know the same languages as they do and at other times with peers whose English language development is slightly more advanced. Thoughtful grouping that varies throughout an investigation allows emerging multilingual students to benefit from the different linguistic resources of their peers.

Activity B	Share What You Noticed
Time Needed for the Activity	5 minutes
Grouping	Teams

What to Do

1. Show the "Share What You Noticed" page on a screen using the Presentation mode in the Learning Hub so everyone in the classroom can see it.
2. Say, "Okay, please share what you noticed with the others at your table."
3. Give the students about two minutes to share at their tables.
4. Say, "Okay, let's take a few minutes to share with the class. Team X (call on a team at random), what did you talk about?"
5. Allow Team X to share something they noticed. If it is related to how high salt concentrations affect the human body, say, "That is really interesting. We are actually going to examine that in more detail during this investigation." If what is shared is not related to how high salt concentrations affect the human body, say, "That is really interesting. Team Y (call on another team at random), what did you notice?" and allow that team to share. Follow this pattern until a team mentions something related to why the swimmers could not drink any water from the Dead Sea. If no team mentions this, say, "You know what I found interesting? I saw that the swimmers had to wear special masks to prevent them from drinking any water from the Dead Sea. Did anyone notice that as well?"
6. Move on to the next activity.

What to Show

The "Share What You Noticed" page in the ADI Learning Hub

What to Look For

As soon as a group mentions something related to why the swimmers could not drink any water from the Dead Sea, move on to the next activity.

Attending to Equity

Be sure to provide students with opportunities to talk in small groups during this activity. Small-group structures offer all students, including emerging multilingual students, a chance to engage in sense-making with their peers and offers them space to use their verbal and nonverbal (e.g., writing, drawing pictures, and using gestures) resources to express their ideas. Small-group structures also give students a chance to learn from other students' uses of these resources.

Activity C	What You Need to Figure Out
Time Needed for the Activity	5 minutes
Grouping	Teams

What to Do

1. Show the "What You Need to Figure Out" page on a screen using the Presentation mode in the Learning Hub so everyone in the classroom can see it.
2. Read the task out loud and have students follow along on their handouts, on their devices, or on the screen.
3. Allow students to ask clarifying questions about the task or the guiding question.
4. Once everyone understands what they will need to figure out during the investigation, move on to the next activity.

What to Show

The "What You Need to Figure Out" page in the ADI Learning Hub

Teaching Tip

Have students highlight, circle, or place a star next to the guiding question. Marking the guiding question helps students find it later during the Plan stage.

Teacher Notes

Activity D	Share Your Ideas
Time Needed for the Activity	15 minutes
Grouping	Teams

What to Do

1. Show the "Share Your ideas" page on a screen using the Presentation mode in the Learning Hub so everyone in the classroom can see it.
2. Read the directions on the handout out loud and then give students five minutes to draw their initial representations of the science.
3. Say, "Okay, will everyone share their representations with the rest of your team? Please make sure everyone has time to talk. Also, be sure to ask one another questions about the scientific ideas that are included in each representation."
4. Give the students about 8-9 minutes to talk and share. Be sure to walk around and listen to the discussions.
5. Say, "Okay, I heard some interesting ideas as I walked around the room."
6. Move on to the next stage.

What to Show

The "Share Your Ideas" page in the ADI Learning Hub

What to Look For

Each student's initial representation should include a combination of symbols, diagrams, graphs, or words. It should also clearly show how a scientific idea can be used to help complete the task and answer the guiding question.

Attending to Equity

When students are tasked with expressing their ideas, be sure to encourage them to do so through multiple modes (speaking, writing, drawings, and gestures) and to use both content-specific and everyday registers. This process helps students fully express themselves, allowing peers to evaluate, question, and build on ideas. Also, make sure that every student in every team has a chance to share their initial representation.

Activity E	More Questions
Time Needed for the Activity	15 minutes
Grouping	Teams

What to Do

1. Show the "More Questions" page on a screen using the Presentation mode in the Learning Hub so everyone in the classroom can see it.
2. Say, "Okay, I heard a bunch of interesting questions as I was walking around the room. Will you work with your team to identify two or three questions you think are important to learn more about that came up as you all discussed the different representations you made? You can write these questions on your handout."
3. Give students about five minutes to talk and write down questions. Be sure to walk around and listen to the discussions.
4. Say, "Okay, let's take a few minutes to share. Team X (call on a team at random), will you share one question you have?"
5. Have each team share a question. Then say, "I want to share some useful scientific ideas with you. I bet this information will not only answer many of these questions but will also help you complete the task and answer the guiding question of the investigation."
6. Move on to the next stage.

What to Show

The "More Questions" page in the ADI Learning Hub

What to Look For

Each student should have 2-3 questions written on their handout.

Teaching Tip

When students are asked to create representations of the underlying scientific ideas, share these ideas with their peers, and then identify things they want to know more about, it sparks their curiosity and helps make the investigation meaningful. Also, identifying questions before reading (which students do during the Ideas stage) is an outstanding pre-reading activity that helps improve comprehension.

Swimming in the Dead Sea Lesson Plan

Stage 2 IDEAS

Total Time for the Stage	30 minutes
Grouping	Teams

Learning Goals

Students should make progress on their ability to do the following by the end of this stage of the lesson:

Science

1. Ask questions that arise from examining theory to clarify and/or seek additional information and relationships (SEP1).
2. Critically read scientific literature adapted for classroom use to determine the central ideas or conclusions and/or to obtain scientific and/or technical information to summarize complex concepts, processes, or information presented in a text by paraphrasing them in simpler but still accurate terms (SEP8).
3. Integrate sources of information presented in different media or formats (e.g., visually and quantitatively) as well as in words to address a scientific question (SEP8).
4. Gather, read, and evaluate scientific and/or technical information from multiple authoritative sources, assessing the evidence and usefulness of each source (SEP8).
5. Describe how feedback mechanisms maintain a living system's internal conditions within certain limits and mediate behaviors, allowing it to remain alive and function even as external conditions change (LS1.A).
6. Use physical, mathematical, and computer models to simulate systems and interactions between systems at different scales (CC4).
7. Describe changes of energy and matter in a system in terms of energy and matter flows into, out of, and within that system (CC5).

English Language Development

1. Paraphrase central ideas in complex evidence, concepts, processes, and information to help explain how or why a phenomenon occurs (ELD-SLC.9-12.Explain.Interpretive).

English Language Arts

1. Determine the central ideas or conclusions of a text; trace the text's explanation or depiction of a complex process, phenomenon, or concept; provide an accurate summary of the text (RST.9-10.2).
2. Determine the meaning of symbols, key terms, and other domain-specific words and phrases as they are used in a specific scientific or technical context (RST.9-10.4).
3. Read and comprehend scientific/technical texts (RST.9-10.10).
4. Initiate and participate effectively in small-group collaborative discussions with diverse partners by building on others' ideas and expressing their own clearly (SL.9-10.1).

Activities

Activity A	Discuss a Core Idea
Time Needed for the Activity	10 minutes
Grouping	Whole class/Teams

What to Do

1. Have students work in their assigned teams during this stage. Give out a copy of the Ideas handout to each student.
2. The first activity provides students with information about cells.
3. Show the "Read About a Core Idea You Can Use" page on a screen using the Presentation mode in the Learning Hub so everyone in the classroom can see it. You can also click on the images to make them full screen and easier for students to see.
4. Have students read the text on their own, either on their handouts or on the screen.
5. Give the students about three minutes to discuss the text with the others in their teams and develop a shared list of ideas that they think are important or useful.
6. Ask a team to share an idea they thought was useful. Have the remaining teams share a different idea until there are no more ideas to share.
7. Move on to the next activity.

What to Give Out

Ideas handout for
Swimming in the Dead Sea

What to Show

The "Read About a Core Idea You Can Use" page in the ADI Learning Hub

What to Look For

Students should write down some of the following ideas in the box labeled, "Some things I know from what I read" on their handouts:

- All cells are surrounded by a membrane.
- The cell membrane forms a semipermeable barrier between the cell and its environment, meaning that it allows some molecules and ions through the membrane but not others.
- Osmosis describes the movement of water molecules across the cell membrane.
- Cells add and remove water to ensure that the ratio, or concentration, of solutes to water remains in equilibrium.
- An isotonic solution has the same concentration of solute inside and outside the cell.
- A hypotonic solution has a lower concentration of solute outside the cell as opposed to inside the cell.
- Hypotonic cells absorb more water and expand to balance out the concentration of solute.
- A hypertonic solution has a higher concentration of solute inside the cell as opposed to outside the cell.
- Hypertonic cells release water and shrink to balance out the concentration of solute.
- Water from the Dead Sea has a salt concentration of 31.5% while water in a human blood cell has a salt concentration of 0.9%.

Teaching Tip

Students do not need to do a close read of this text or completely understand all the ideas in it before moving on to the next activity. They will be encouraged to revisit this text later in the investigation when creating their arguments and writing their reports. All students need to be able to do during this activity is to work with the other members of their group to identify one or two ideas that they think are important to keep in mind or are potentially helpful.

Activity B	Read About Another Core Idea
Time Needed for the Activity	10 minutes
Grouping	Teams

What to Do

1. The second activity provides students with information about systems and system models.
2. Show the "Read About Another Core Idea You Can Use" page on a screen using the Presentation mode in the Learning Hub so everyone in the classroom can see it. You can also click on the images to make them full screen and easier for students to see.
3. Have students read the text on their own, either on their handouts or on the screen.
4. Give the students about three minutes to discuss the text with the others in their teams and develop a shared list of ideas that they think are important or useful.
5. Ask a team to share an idea they thought was useful. Have the remaining teams share a different idea until there are no more ideas to share.
6. Move on to the next activity.

What to Show

The "Read About Another Core Idea You Can Use" page in the ADI Learning Hub

What to Look For

Students should write down some of the following ideas in the box labeled, "Some things I know from what I read" on their handouts:

- A system is a group of related parts that make up a whole and carry out functions that its individual parts cannot.
- A model is a representation of a system that allows scientists to test different ideas about the system functions.
- Models are limited representations because they can only represent certain parts or aspects of the system as opposed to the entire system.

Attending to Equity

There are many supports for helping students comprehend what they read already embedded into this activity (e.g., activating prior knowledge, providing a shared experience, making connections, synthesizing, and talking with peers). So, you might not need to provide much extra support. If you are concerned about some students not understanding what they are reading because of their scores on past reading comprehension tests, you can read it out loud as they follow along. As you read the text out loud, be sure to stop at each important idea and ask the students to put a star (or other annotation) next to it in the margin of their handouts. They can then discuss these ideas in their small groups.

Activity C	Read About a Final Core Idea
Time Needed for the Activity	10 minutes
Grouping	Teams

What to Do

1. The third activity provides students with information about tracking matter in systems.
2. Show the "Read about a Final Core Idea You Can Use" page on a screen using the Presentation mode in the Learning Hub so everyone in the classroom can see it. You can also click on the images to make them full screen and easier for students to see.
3. Have students read the text on their own, either on their handouts or on the screen.
4. Give the students about three minutes to discuss the text with the others in their teams and develop a shared list of ideas that they think are important or useful.
5. Ask a team to share an idea they thought was useful. Have the remaining groups share a different idea until there are no more ideas to share.
6. Move on to the next activity.

What to Show

The "Read About a Final Core Idea You Can Use" page in the ADI Learning Hub

What to Look For

Students should write down some of the following ideas in the box labeled, "Some things I know from what I read" on their handouts:

- Scientists often track the movement of matter into, out of, or within a system in order to understand how a change in the system's surroundings affects the system and its components.
- Scientists can track the movement of matter into or out of a system by measuring how the total weight of the system changes over time.

Attending to Equity

To support emerging multilingual students during this activity, be sure to create a word bank on the wall of high-frequency, high-need concrete vocabulary that is needed to communicate scientific ideas. The word bank highlights terms needed for basic communication in an academic context. It also helps with the spelling. You can also encourage your emerging multilingual students to:

1. Use strategic learning techniques such as concept mapping, drawing, memorizing, comparing, contrasting, and reviewing.
2. Decode (sound out) words using skills such as recognizing sound-letter relationships and identifying cognates, affixes, roots, and base words.
3. Use visual and contextual support and support from peers and teachers to enhance and confirm understanding of the texts.

Swimming in the Dead Sea Lesson Plan

Stage 3 PLAN

Total Time for the Stage	15 minutes
Grouping	Teams

Learning Goals

Students should make progress on their ability to do the following by the end of this stage of the lesson:

Science

1. Evaluate a question to determine if it is testable and relevant (SEP1).
2. Plan an investigation individually and collaboratively to produce data to serve as the basis for evidence as part of building and revising models (SEP3).
3. Consider possible variables or effects and evaluate the confounding investigation's design to ensure variables are controlled (SEP3).
4. Ask questions to determine relationships, including quantitative relationships, between independent and dependent variables (SEP1).
5. Decide on types, quantity, and accuracy of data needed and consider limitations on the precision of the data (e.g., number of trials, cost, risk and time), and refine the design accordingly (SEP3).
6. Select appropriate tools to collect, record, and analyze data (SEP3).

English Language Development

1. Define investigable questions on observations, information, and/or data about a phenomenon (ELD-SLC.9-12.Explain.Interpretive).
2. Develop reasoning to illustrate and/or predict the relationships between variables (ELD-SLC.9-12.Explain.Expressive).

English Language Arts

1. Initiate and participate effectively in small-group collaborative discussions with diverse partners by building on others' ideas and expressing their own clearly (SL.9-10.1).
2. Present information clearly, concisely, and logically such that listeners can follow the line of reasoning and the organization, substance, and style are appropriate to the purpose, audience, and task (SL.9-10.4).
3. Adapt speech to a variety of contexts and tasks, demonstrating command of formal English when indicated or appropriate (SL.9-10.6).

Activities

Activity A	Some Materials You Can Use
Time Needed for the Activity	5 minutes
Grouping	Teams

What to Do

1. Show the "Some Materials You Can Use" page on a screen using the Presentation mode in the Learning Hub so everyone in the classroom can see it. You can also click on the images to make them full screen and easier for students to see.
2. Say, "Okay, it sounds like we now have enough information to plan an investigation that will help us figure out why it is so dangerous to drink water from the Dead Sea?"
3. Say, "Before we get started, I want to show you some materials that we are going to use during our investigation. Will you all please come over to this table so I can show you what we have?"
4. Direct students to stand around the table where all the materials are set up.
5. Hold up the salt and distilled water.
6. Ask, "What do you think we could use these for?"
7. Allow students to share ideas. Then say, "We can add different amounts of salt to the water and create three solutions with different concentrations."
8. Hold up the saltwater concentration stickers.
9. Ask, "What do you think we could use these for?"
10. Allow students to share ideas. Then say, "We can use these to label the concentrations of our different solutions."
11. Hold up the measuring cups, funnels, pipettes, and spatula.
12. Ask, "What do you think we could use these for?"
13. Allow students to share ideas. Then say, "We can use these to measure out and then mix different amounts of salt and water."
14. Hold up the dialysis tubing.
15. Ask, "What do you think we could use this for?"
16. Allow students to share ideas. Then say, "We can use this to represent the cell membrane. We can fill it up with a saltwater concentration equal to that of a human red blood cell."
17. Hold up the scale.

What to Set Up

Place the following materials on a table so students can see them.

1. Salt
2. Distilled water
3. Dialysis tubing
4. Saltwater concentration stickers
5. Pipettes
6. Funnels
7. Measuring cups
8. Scale
9. Pitchers
10. Spatula

What to Show

The "Some Materials You Can Use" page in the ADI Learning Hub

(continued)

18. Ask, "What do you think we could use this for?"
19. Allow students to share ideas. Then say, "We can use this to track changes of mass for the system."
20. Say, "Okay, that is all the stuff we can use. Will you all head back to your teams so we can plan our investigation?"
21. Move on to the next activity.

Teaching Tip

This activity helps students see for themselves all the materials they will be able to use during the hands-on part of the investigation instead of just reading about them or looking at pictures. It is important to show each material to the students and then point out how they can use it during the investigation so they can participate in developing a plan for the investigation during the next activity.

Activity B	Make a Plan
Time Needed for the Activity	10 minutes
Grouping	Teams

What to Do

1. Have students work in their assigned teams during this activity. Give a copy of the Plan handout to each student.
2. Show the "Make a Plan" page on a screen using the Presentation mode in the Learning Hub so everyone in the classroom can see it.
3. Say, "Work as a team to create a plan for your investigation by filling out the chart on your handout. As part of your plan, be sure to include which tools and techniques you will use to solve problems."
4. Give the students at least 10 minutes to complete this task. As students work in small teams to create their investigation proposals, be sure to move from team to team to answer questions and offer advice.
5. When a team thinks that they have a good plan, read it over. If you agree, sign off at the bottom of their handouts to let them know that their plan has been approved. If you do not agree, tell them what they will need to add or change before you will approve it. Once they have made the changes, you can sign their handouts.
6. Once all the teams are done, you can move on to the next activity. You can also use an approved investigation plan as an "exit ticket" out of class.

What to Give Out

Plan handout for
Swimming in the Dead Sea

What to Show

The "Make a Plan" page in the ADI Learning Hub

What to Look For

The investigation plan should, at a minimum, include:

- The question they are trying to answer.
- How they will set up the available materials.
- What data they will collect using their setup.
- How they will analyze the data they collect.
- How they will use their data to answer the guiding question.

Teaching Tip

One way to do a quick check-in with a group during this activity is to use the "two questions, a nudge, and a goal" technique. This technique includes the following steps:

1. Sit down with a group of students.
2. Ask, "Would one of you tell me a little about what you all have been doing or thinking about?" (Question 1)
3. A student answers on behalf of the group.
4. Ask, "Would anyone else like to add anything to that?" (Question 2)
5. A different student provides additional information on behalf of the group.
6. Ask, "Have you thought about [a hint or something you want to highlight]?" (A nudge)
7. Say, "Okay, I'm going to check on the other groups. When I get back, I want to see [something that you want the group to accomplish]." (A goal)

When you have completed the check-in with one group, move on to the next group.

Attending to Equity

To support emerging multilingual students during this stage, be sure to create a word bank on the wall of high-frequency, high-need concrete vocabulary that is needed to communicate scientific processes and ideas. The word bank highlights terms needed for basic communication in an academic context. It also helps with the spelling. Also, you can encourage your emerging multilingual students to use modes (speaking, writing, drawings, and gestures) and to use both content-specific and everyday registers to discuss and create their plans. This process helps students fully express themselves, allowing peers to evaluate, question, and build on ideas.

Teacher Notes

Swimming in the Dead Sea Lesson Plan

Stage 4 DO

Total Time for the Stage	45 minutes
Grouping	Teams

Learning Goals

Students should make progress on their ability to do the following by the end of this stage of the lesson:

Science

1. Conduct an investigation individually and collaboratively to produce data to serve as the basis for evidence (SEP3).
2. Analyze data using tools and/or technologies to make valid and reliable scientific claims (SEP4).
3. Consider limitations of data analysis (e.g., measurement error and sample selection) when analyzing and interpreting data (SEP4).
4. Evaluate the impact of new data on a working explanation and/or model of a proposed process or system (SEP4).
5. Apply ratios, rates, percentages, and unit conversions in the context of complicated measurement problems involving quantities with derived or compound units (SEP5).
6. Make a quantitative and/or qualitative claim regarding the relationship between dependent and independent variables (SEP6).
7. Construct and revise an explanation based on valid and reliable evidence obtained from a variety of sources and the assumption that theories and laws that describe the natural world operate today as they did in the past and will continue to do so in the future (SEP6).
8. Apply scientific ideas, principles, and/or evidence to provide an explanation of phenomena taking into account possible unanticipated effects (SEP6).

English Language Development

1. Paraphrase central ideas in complex evidence, concepts, processes, and information to help explain how or why a phenomenon occurs (ELD-SLC.9-12.Explain.Interpretive).
2. Evaluate the extent to which reasoning, theory and/or models link evidence to claims and support conclusions (ELD-SLC.9-12.Explain.Interpretive).
3. Develop reasoning to illustrate the relationships between variables (ELD-SLC.9-12.Explain.Expressive).

English Language Arts

1. Initiate and participate effectively in small-group collaborative discussions with diverse partners by building on others' ideas and expressing their own clearly (SL.9-10.1).
2. Present information clearly, concisely, and logically such that listeners can follow the line of reasoning and the organization, substance, and style are appropriate to the purpose, audience, and task (SL.9-10.4).
3. Adapt speech to a variety of contexts and tasks, demonstrating command of formal English when indicated or appropriate (SL.9-10.6).

Activities

Activity A	Carry Out Your Plan
Time Needed for the Activity	25 minutes
Grouping	Teams

What to Do

1. Students should work in their assigned teams for this activity. Begin this activity by giving out a copy of the Do handout to each student.
2. Show the "Carry Out Your Plan" page on a screen using the Presentation mode in the Learning Hub so everyone in the classroom can see it.
3. Say, "Okay, it sounds like we are now ready to carry out our plans and collect the data or observations we need to answer the guiding question. You can get the materials and tools you need to use from this table." Point to the table where you have placed them.
4. Ask, "Does anyone have any questions about these materials or tools?"
5. Answer any questions that come up.
6. Say, "As you work, be sure to record your observations and measurements on your handout so you can keep track of everything you do. Does anyone have any final questions?"
7. Answer any questions that come up.
8. Say, "Okay, go ahead and get started. You have the next 25 minutes to work. I'll be available to help if any team needs me."
9. Give the students at least 25 minutes to carry out their plans and solve the problem. As students work in small teams, be sure to move from team to team to answer questions and offer advice.

What to Set Up

Place the following materials on a table so each group can use them.

1. Salt
2. Distilled water
3. Dialysis tubing
4. Saltwater concentration stickers
5. Pipettes
6. Funnels
7. Measuring cups
8. Scale
9. Pitchers
10. Spatula

What to Give Out

Do handout for **Swimming in the Dead Sea**

What to Show

The "Carry Out Your Plan" page in the ADI Learning Hub

Teaching Tip

One way to do a quick check-in with a group during this activity is to use the "two questions, a nudge, and a goal" technique. This technique includes the following steps:

1. Sit down with a group of students.
2. Ask, "Would one of you tell me a little about what you all have been doing or thinking about?" (Question 1)
3. A student answers on behalf of the group.
4. Ask, "Would anyone else like to add anything to that?" (Question 2)
5. A different student provides additional information on behalf of the group.
6. Ask, "Have you thought about [a hint or something you want to highlight]?" (A nudge)
7. Say, "Okay, I'm going to check on the other groups. When I get back, I want to see [something that you want the group to accomplish]." (A goal)

When you have completed the check-in with one group, move on to the next group.

Attending to Equity

Be sure to intentionally group emerging multilingual students at times with peers who know the same languages as they do and at other times with peers whose English language development is slightly more advanced. Thoughtful grouping that varies during different investigation (or activities within an investigation) allows emerging multilingual students to benefit from the different linguistic resources of their peers. We also recommend assessing emerging multilingual students based on how much they contribute to the group during this activity, rather than solely on what they write or draw on their handouts.

Teacher Notes

Activity B	Make Sense of Your Data
Time Needed for the Activity	20 minutes
Grouping	Teams

What to Do

1. Students should work in their assigned teams for this activity. Begin this activity by showing the "Make Sense of Your Data" page on a screen using the Presentation mode in the Learning Hub so everyone in the classroom can see it.
2. Say, "Okay, it sounds like we are now ready to make sense of the data we collected. One way to do this is to create a table, graph, or other visualization that shows a trend or change over time, a comparison between teams, or a relationship between things you tested. You can create a table, graph, or other visualization here on your handout."
3. Point to the box on the handout.
4. Say, "Please work together in your teams to figure out the best way to make sense of your data. Remember, you are looking for any trends, changes, differences, or relationships in the data that will help you answer the guiding question. As you work, be sure to use the space on your handout to perform any calculations that you need to do and create any representations that you need. Does anyone have any last questions?"
5. Answer any questions that come up.
6. Say, "Okay, go ahead and get started. You have the rest of the period to work. I'll be available to help if any team needs me."
7. Give the students at least 15 minutes to make sense of their data. As students work together in small teams, be sure to move from team to team to answer questions and offer advice.
8. Make sure students return the materials and tools to the table where they found them before dismissing class. Cleanup should take no longer than five minutes.

What to Show

The "Make Sense of Your Data" page in the ADI Learning Hub

What to Look For

One of the most useful ways to make sense of the data during this activity is to make a graph or table that tracks the mass of the dialysis tubing over time when it is soaked in different concentrations of saltwater.

Teaching Tip

Collecting and returning materials or tools to a central table works best if you assign the job to a specific number within each team. For example, you can say, "2s, please come up and grab the materials and tools that your team needs" at the start of the activity. Then you can say, "3s, please return the materials and tools your team used back to this table," at the end of the activity. Assigning jobs like this not only cuts down on how many students need to be up and moving around, but also allows you to hold specific students accountable for the materials and tools.

Attending to Equity

Encourage your emerging multilingual students to speak using learning strategies such as requesting assistance, employing nonverbal cues, and using synonyms and circumlocution (conveying ideas by defining or describing when exact English words are not known) during this activity. It is also helpful to point out and/or draw emerging multilingual learners' attention to elements of the English sound system in newly acquired vocabulary such as long and short vowels, silent letters, and consonant clusters as they come up naturally during this activity.

Teacher Notes

Swimming in the Dead Sea Lesson Plan

Stage 5 SHARE

Total Time Needed for the Stage	45 minutes
Grouping	Teams

Learning Goals

Students should make progress on their ability to do the following by the end of this stage of the lesson:

Science

1. Make and defend a claim based on evidence about the natural world that reflects scientific knowledge and student-generated evidence (SEP7).
2. Present an oral argument based on evidence (SEP7).
3. Evaluate the claims, evidence, and/or reasoning to determine the merits of arguments (SEP7).
4. Ask questions to clarify relationships, including quantitative relationships, between independent and dependent variables (SEP1).
5. Ask questions that challenge the interpretation of a dataset or the suitability of a method (SEP1).
6. Respectfully give critiques on scientific arguments by probing reasoning and evidence and challenging ideas and conclusions (SEP7).
7. Respectfully receive critiques on scientific arguments by responding thoughtfully to diverse perspectives and determining what additional information is required to resolve contradictions. (SEP7).
8. Communicate scientific information or ideas in multiple formats including orally, graphically, textually, and mathematically (SEP8).

English Language Development

1. Identify appropriate and sufficient evidence from data, models, and/or information from investigations of a phenomenon (ELD-SLC.9-12.Argue.Interpretive).
2. Compare reasoning and claims based on evidence from competing arguments (ELD-SLC.9-12. Argue.Interpretive).
3. Defend or refute a claim based on data and evidence (ELD-SLC.9-12.Argue.Expressive).
4. Establish and maintain an appropriate tone and stance (ELD-SLC.9-12.Argue.Expressive).
5. Signal logical relationships among reasoning, evidence, data, and/or models when making and defending a claim, counterclaim, and/or rebuttal (ELD-SLC.9-12.Argue.Expressive).

English Language Arts

1. Initiate and participate effectively in small-group collaborative discussions with diverse partners by building on others' ideas and expressing their own clearly (SL.9-10.1).
2. Integrate multiple sources of information presented in diverse media or formats (e.g., visually, quantitatively, and orally) evaluating the credibility and accuracy of each source (SL.9-10.2).
3. Evaluate a speaker's reasoning and use of evidence and rhetoric, identifying any fallacious reasoning or exaggerated or distorted evidence (SL.9-10.3).

(continued)

4. Present information clearly, concisely, and logically such that listeners can follow the line of reasoning and the organization, substance, and style are appropriate to the purpose, audience, and task (SL.9-10.4).
5. Make strategic use of media (e.g., textual, graphical, and visual) in presentations to enhance understanding of findings, reasoning, and evidence and to add interest (SL.9.10.5).
6. Adapt speech to a variety of contexts and tasks, demonstrating command of formal English when indicated or appropriate (SL.9-10.6).

Activities

Activity A	Create a Draft Argument
Time Needed for the Activity	15 minutes
Grouping	Teams

What to Do

1. Students should work in their assigned teams for this activity. Begin this activity by giving out a copy of the Share handout to each student.
2. Show the "Create a Draft Argument" page on a screen using the Presentation mode in the Learning Hub so everyone in the classroom can see it.
3. Say, "Okay, it sounds like we have all solved the task. I would now like you to create a draft argument to share what you figured out with the other teams. Your argument should include everything in this image here." Point to the image of a science-based argument projected on the screen. Be sure to go over what should be included in each component of the argument and where students can find the information they need.
4. Say, "I would like you to make your draft argument on a whiteboard. You can use the materials on this table to make your argument." Point to the table where you have placed the materials they need.
5. Ask, "Does anyone have any questions?"
6. Answer any questions that come up.
7. Say, "I'm going to give you exactly 15 minutes to get this done. Your draft argument doesn't need to be perfect. We just need something down on the whiteboards so we can share our ideas."
8. Give each team of students 15 minutes to create a draft argument. As students work in small teams, be sure to move from team to team to answer questions and offer advice.

What to Set Up

Place the following items on a table so students can use them during this stage:

- 24″ x 36″ whiteboard
- 3-4 dry-erase pens in different colors
- Eraser or cloth towel

What to Give Out

Share handout for **Swimming in the Dead Sea**

What to Show

The "Create a Draft Argument" page in the ADI Learning Hub

What to Look For

Students should include core ideas that they read about during the Ideas stage in their justification.

Teaching Tip

Be sure to plant some different ideas with each group about how to make a strong argument as you move from group to group to check in, answer questions, and offer advice. This strategy helps to ensure that there will be some really strong examples of the evidence and justification components of the argument for students to see and discuss during the next activity. For example, you could help two groups have a really strong evidence component by telling one group, "I think a graph would be really helpful here" and telling another group, "I think your evidence would be stronger if you explained, in words, what this graph shows." You could help a third group have a really strong justification by saying, "Let's brainstorm some ideas that will help you defend your choice of evidence," or "Let's go back to our Ideas handout and see if there are any concepts we can include."

It is important that you make sure that all the arguments are different at this point in the investigation by planting different ideas with different groups because the students will have nothing to critique or discuss during the next activity if they all have the same argument. The goal during this activity is to set the stage for the next activity by ensuring that the various components of a strong argument can be found on different whiteboards within the class. These strong components will then be taken up by all the groups because they reflect good ideas. Be sure to give students a voice and choice during this activity, so they can make mistakes and learn from them.

Attending to Equity

It is often helpful to assign and change roles at least once during this activity. For example, you might assign the 1s to be "in charge of writing" at the beginning of the activity. You can then easily switch roles about five minutes into the activity by saying, "1s, please hand the pens to the 2s. 2s, you are now in charge of writing for the group." By changing roles, you can ensure that all students have a chance to participate and contribute their ideas to the creation of a draft argument.

ADI Go Digital

The **ADI Learning Hub** has tons of features that allow students to share their thinking with each other when they are doing the investigation paperless or remote.

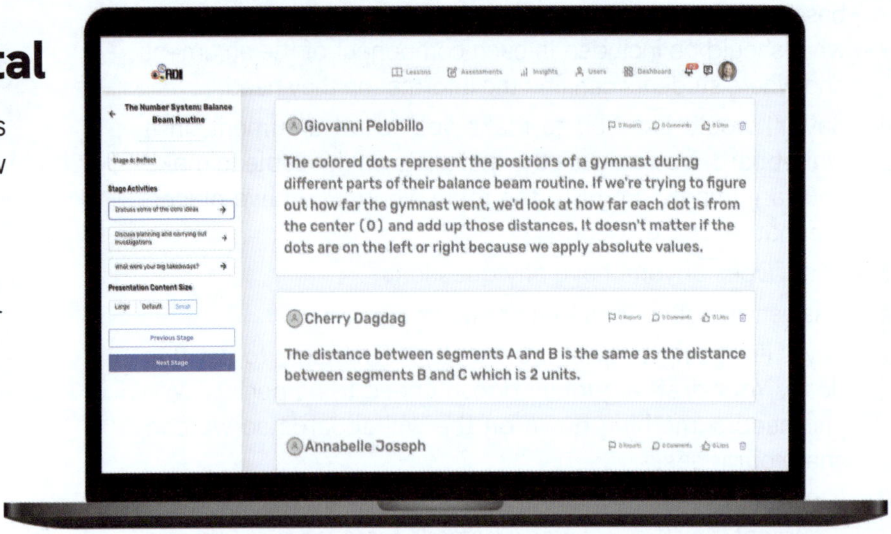

Activity B	Present Your Draft Argument
Time Needed for the Activity	15 minutes
Grouping	Teams

What to Do

1. Students should work in their assigned teams for this activity.
2. Show the "Present Your Draft Argument" page on a screen using the Presentation mode in the Learning Hub so everyone in the classroom can see it.
3. Say, "We are now going to share our arguments. Please set up your whiteboards so everyone can see them." Allow students time to set up their whiteboards. They should be set up on stands or chairs, so they are facing toward the outside of the room.
4. Say, "Okay, before we get started, I want to explain what we are going to do next. I'm going to ask some of you to present your arguments to your classmates. If you are presenting your argument, your job is to share your team's claim, evidence, and justification of the evidence. The rest of you will be travelers. If you are a traveler, your job is to listen to the presenters, ask them questions if you do not understand something, and then offer them some suggestions about ways to make their argument better. Do you have any questions about these instructions?"
5. Answer any questions that come up.
6. Say, "As you talk to one another, make sure you write down any good ideas or suggestions that are important in the space on your handout, so you don't forget about them." Point to the box labeled, "Some possible ways to improve our argument" on the handout.
7. Say, "After we have a chance to learn from one another, I'm going to give you some time to revise your arguments and make them better. Do you have any questions about what we are doing?"
8. Answer any questions that come up.
9. Say, "Okay. Let's get started. Travelers, move one team to the left. Don't move to the next team until I tell you to move. Presenters, go ahead and share your argument with the travelers when they get there."
10. Give each team of presenters and travelers about 3-4 minutes to talk.
11. Tell the students, "Okay. Let's move on to the next argument. Travelers, move one team to the left."
12. Again, give each team of presenters and travelers about 3-4 minutes to talk.

(continued)

What to Show

The "Present Your Draft Argument" page in the ADI Learning Hub

What to Look For

The critiques and feedback that students give one another during this activity should:

- Be shared in a way that is kind (respectful),
- Provide information about what, exactly, needs to be changed or improved (specific), and
- Offer a suggestion about how to make something better rather than just saying something is wrong (helpful).

In addition, students should be focusing their critiques on the science included in the arguments, not the appearance of the arguments.

13. Repeat this process until each team has had their argument read and critiqued three times. Then tell the students to return to their workstations.
14. Be sure to move between workstations during this activity to keep everyone involved and model what it means to communicate and critique scientific ideas and reasoning.

Attending to Equity

Be sure to encourage students to provide feedback that is respectful, specific, and helpful during this activity. It is important to encourage student-to-student talk because it helps students develop speaking and listening skills and enables them to make sense of what and how they know. It also provides a context for emerging multilingual students to use both productive (speaking and writing) and receptive (listening and reading) language and learn science through meaningful interactions and scientific sense-making.

Teaching Tip

There are two types of formats for sharing and critiquing arguments. The first format is called a *gallery walk.* In this format, each group sets up their argument at their workstation so others can see it. The entire group then moves to a different workstation. While at a different workstation, the group is expected to read the argument at that workstation (which was created by a different group) and offer critiques and suggestions for improvement. The students should move to a different workstation after a few minutes to read and critique an argument written by a different group. Students should also be encouraged to keep a record of good ideas or potential ways to improve their own arguments as they move from workstation to workstation to offer critiques and feedback. After repeating this process in the gallery walk 3-4 times, every group will have had an opportunity to read and critique three or four different arguments and receive feedback from three or four different groups.

The second format is called a *modified gallery walk.* In the modified gallery walk format, one or two members of the group stay at their workstation to share their group's ideas (these students are called presenters) while the other members of their group go to different groups, one group at a time, to listen to and critique the arguments developed by their classmates (these students are called travelers). The travelers should visit at least three different workstations during the argumentation session. We also recommend that the presenters keep a record of the critiques made by their classmates and any suggestions for improvement. The travelers should also be encouraged to keep a record of good ideas or potential ways to improve their own arguments as they travel from group to group.

Attending to Equity

Another good way to have students, especially emerging multilingual learners, give one another feedback is to have them use four different color sticky notes to make comments. Give each group a single pink, blue, yellow, and green sticky note. Then direct the students to use the pink sticky note to suggest "something to add" to the argument, the blue sticky note to suggest "something to change" in the argument, the yellow sticky note to highlight "something they like" in the argument, and the green sticky note to ask a question about "something they don't understand". This strategy is useful because everyone in the classroom will be actively engaged during this activity, and everyone will have a chance to see different arguments. This format, however, does not give students a chance to support, critique, and challenge the ideas, solutions, scientific reasoning of other groups through talk because two different groups are never at the same workstation at the same time.

Activity C	Revise Your Argument
Time Needed for the Activity	15 minutes
Grouping	Teams

What to Do

1. Students should work in their assigned teams for this activity.
2. Show the "Revise Your Argument" page on a screen using the Presentation mode in the Learning Hub so everyone in the classroom can see it.
3. Say, "I'm now going to give you all about 10 minutes to revise your argument. Please return to your original teams."
4. Wait for the students to move back into their original teams.
5. Say, "Okay, take a few minutes to talk in your teams and determine what you want to change to make your argument better. Once you have decided what to change, go ahead and make the changes to your whiteboard."
6. Ask, "Do you have any questions about what you need to do?"
7. Answer any questions that come up.
8. Tell the students, "Okay. Let's get started."
9. Give the students 10 minutes to work in their teams on their arguments. As they work, move from team to team to check in, ask probing questions, and offer a suggestion if a team gets stuck.
10. When students are finished with their final arguments, have them take a picture of it or have them put the boards in a safe spot so they can refer to the picture or the boards during the Report stage.

What to Show

The "Revise Your Argument" page in the ADI Learning Hub

What to Look For

A strong argument includes the following information:

- A claim that states that the high concentration of salt in the Dead Sea creates a hypertonic solution with human blood cells, causing a loss of water that leads to dehydration.
- A graph or table that tracks the mass of the dialysis tubing over time when it is soaked in different concentrations of saltwater.
- An explanation of the graph or table.
- An explanation of the chosen evidence that uses core ideas from the investigation.

Teaching Tip

It is important to give students time to meet with their original groups so they can discuss what they learned by interacting with individuals from the other groups. At this point in the process, students often realize that their initial problem-solving model was flawed in some way. It is therefore important to encourage students to "redo the science" as needed. At the end of this activity, each group should have a final argument that is much better than their original one.

Swimming in the Dead Sea Lesson Plan

Stage 6 REFLECT

Total Time for the Stage	25 minutes
Grouping	Whole class, Teams

Learning Goals

Students should make progress on their ability to do the following by the end of this stage of the lesson:

Science

1. Ask questions that arise from careful observation of phenomena, or unexpected results, to clarify and/or seek additional information as needed (SEP1).
2. Describe how feedback mechanisms maintain a living system's internal conditions within certain limits and mediate behaviors, allowing it to remain alive and function even as external conditions change (LS1.A).
3. Use physical, mathematical, and computer models to simulate systems and interactions between systems at different scales (CC4).
4. Describe changes of energy and matter in a system in terms of energy and matter flows into, out of, and within that system (CC5).

English Language Development

1. Paraphrase central ideas in complex evidence, concepts, processes, and information to help explain how or why a phenomenon occurs (ELD-SLC.9-12.Explain.Interpretive).
2. Use scientific explanations about a phenomenon in a different context or situation (ELD-SLC.9-12.Explain.Expressive).

English Language Arts

1. Participate effectively in small-group collaborative discussions with diverse partners on texts by building on others' ideas and expressing their own clearly and persuasively (SL.9-10.1).

Activities

Activity A	Discuss a Core Idea
Time Needed for the Activity	10 minutes
Grouping	Whole class/Teams

What to Do

1. Students should be in small teams for this activity. However, they do not need to be in the same teams as they were during previous stages.
2. Begin this activity by giving out a copy of the Reflect handout to each student.
3. Show the "Discuss a Core Idea" page on a screen using the Presentation mode in the Learning Hub so everyone in the classroom can see it.
4. Say, "Okay. Now that we have all revised our arguments and made them better, let's take some time to discuss some of the core ideas you used during this investigation and to justify the evidence in your arguments so we are all on the same page. Look at the picture on the screen."
5. Pause for about 30 seconds.
6. Say, "Take a few minutes to talk in your teams and see if you can use what you know about cells, systems and system models to explain what is happening in the picture."
7. Give the students at least three minutes to talk in their small teams.
8. Say, "Okay. Let's share with the whole class. Team A (or whichever team you want to share first), what did you talk about?"
9. Allow the first team to share.
10. Say, "Interesting. Would anyone like to add to, challenge, or refine the ideas shared by Team A?"
11. Allow students to respond.

What to Give Out

Reflect handout for **Swimming in the Dead Sea**

What to Show

The "Discuss a Core Idea" page in the ADI Learning Hub

Explanation of the Image

The image that the students discuss during this activity shows three different samples of human blood cells.

What to Look For During the Discussion

Students who understand the content will likely mention some, if not all, of the following ideas when providing an explanation for the image:

- All cells are surrounded by a membrane.
- The cell membrane forms a semipermeable barrier between the cell and its environment, meaning that it allows some molecules and ions through the membrane but not others.
- Osmosis describes the movement of water molecules across the cell membrane.
- Cells add and remove water to ensure that the ratio, or concentration, of solutes to water remains in equilibrium.
- An isotonic solution has the same concentration of solute inside and outside of the cell.
- A hypotonic solution has a lower concentration of solute outside of the cell as opposed to inside the cell.
- Hypotonic cells absorb more water and expand to balance out the concentration of solute.
- A hypertonic solution has a higher concentration of solute inside the cell as opposed to outside the cell.
- Hypertonic cells release water and shrink to balance out the concentration of solute.

(*continued*)

12. Say, "Okay. Team B (or whichever team you want to share second), what did you talk about?"
13. Allow the second team to share.
14. Say, "Interesting. Would anyone like to add to, challenge, or refine the ideas shared by Team B?"
15. Allow students to respond.
16. Keep up this pattern of questioning until all teams have shared. Be sure to keep encouraging students to talk about the content until they reach a correct and consensual explanation for what is shown in the image. If they are unable to reach a consensus or reach consensus using an inaccurate idea, be sure to provide an accurate explanation for the image before moving on to the next activity.

Attending to Equity

Be sure to intentionally group emerging multilingual students at times with peers who know the same languages as they do and at other times with peers whose English language development is slightly more advanced. Also, be sure to ask students how they have seen the core idea in their everyday lives; this will make the topic of discussion more engaging and relatable.

Teaching Tip

This activity is a discussion, not a presentation about what the students "should have seen" or "should have learned" during the investigation. The more students talk during this activity, the more meaningful the experience will be for them and the more you can learn about their thinking. Be sure to use talk goals and moves to help facilitate the discussion. Below are some examples of talk goals that you can use to help facilitate the discussion and some questions that you can ask the students to help accomplish each goal. These talk goals and questions are only suggestions.

Goal 1: Encourage students to share, expand on, or clarify an idea.
- What do you see going on here?
- What do you think is happening?
- How might we explain that?
- Can you tell me more about that?
- What do you mean by that?
- Can you give me an example of that?
- Can you say that in another way?

Goal 2: Encourage students to support an idea.
- Why do you think that?
- What is your evidence for that?
- How did you arrive at that conclusion?
- What makes you say that?
- How do you know that?

Goal 3: Encourage students to think with others.
- Would anyone like to add to that idea?
- What should we keep or change about that idea?
- Can anyone else give me some examples of that?
- Can anyone take that idea and push it a little further?
- Does anyone want to respond to that idea?
- Do you agree or disagree with that idea, and why?
- What do we think about that idea?
- Does that fit with what we have figured out so far?
- Is there another way to do it?

Goal 4: Encourage students to think about what they know and don't know yet.
- I'm confused now. What have we figured out so far?
- What do we need to figure out now?
- So, what do we agree on right now, and what are we still trying to figure out?
- What are we sure and not so sure about?
- What questions do you have?

Activity B	Discuss the Practices
Time Needed for the Activity	10 minutes
Grouping	Whole class/Teams

What to Do

1. Show the first page of "Discuss the Practices" on a screen using the Presentation mode in the Learning Hub so everyone in the classroom can see it.
2. Say, "Okay. Now that we have talked about some of the core ideas that we used during our investigation, let's take a few minutes to talk about the science practices that we used during the investigation. An example of a scientific practice used was analyzing and interpreting data. On the screen there is a question that I want you all to think about: What other things did you do during your investigation that you feel were good examples of a scientific practice or attempting to figure out something in a scientific way?"
3. Pause for about 30 seconds.
4. Say, "Take a few minutes to talk about this question in your teams."
5. Give the students at least three minutes to talk in their small teams.
6. Say, "Okay. Let's share with the whole class. Team A (or whichever team you want to share first), what did you talk about?"
7. Allow the first team to share.
8. Say, "Okay. Team B (or whichever team you want to share second), what did you talk about?"
9. Allow the second team to share.
10. Continue asking teams until every team has a chance to share.
11. Show the second page of "Discuss the Practices" on a screen using the Presentation mode in the Learning Hub so everyone in the classroom can see it.
12. Say, "Here is another question that I want you all to think about: What things did you do during your investigation that were not good examples of a scientific practice or not very scientific?"
13. Pause for about 30 seconds.

What to Show First

The first page of "Discuss the Practices" in the ADI Learning Hub

What to Show Second

The second page of "Discuss the Practices" in the ADI Learning Hub

What to Show Third

The third page of "Discuss the Practices" in the ADI Learning Hub

(continued)

14. Say, "Take a few minutes to talk about this question in your team."
15. Give the students at least three minutes to talk in their small teams.
16. Say, "Okay. Let's share with the whole class. Team A (or whichever team you want to share first), what did you talk about?"
17. Allow the first team to share.
18. Say, "Okay. Team B (or whichever team you want to share second), what did you talk about?"
19. Allow the second team to share.
20. Continue asking teams until every team has a chance to share.
21. Show the third page of "Discuss the Practices" on a screen using the Presentation mode in the Learning Hub so everyone in the classroom can see it.
22. Say, "Here is a final question that I want you all to think about: What should we all do during our next investigation to make sure we are using scientific practices and being scientific in the way we try to figure something out?"
23. Pause for about 30 seconds.
24. Say, "Take a few minutes to talk about this question in your team."
25. Give the students at least three minutes to talk in their small teams.
26. Say, "Okay. Let's share with the whole class. Team A (or whichever team you want to share first), what did you talk about?"
27. Allow the first team to share.
28. Say, "Okay. Team B (or whichever team you want to share second), what did you talk about?"
29. Allow the second team to share.
30. Continue asking teams until every team has a chance to share.
31. Once everyone has shared, offer any suggestions or advice about what they can do to improve their next investigation and then move on to the next stage.

Teaching Tip

A great way to wrap up this activity is for the students to contribute to a list of class norms or "gotta-haves" to keep in mind when planning and carrying out an investigation in science. To create this list, each group can suggest for a potential norm or "gotta-have" to add to the list at the end of this activity, then the class votes on the one that they think is most important. The most important one gets added to the list. The list then serves as reminder for students during the Plan and Do stages of subsequent investigations. Over time, the list will grow and each addition to the list will be more nuanced than the last as students develop expertise with scientific processes.

Attending to Equity

Be sure to give all students, especially emerging multilingual students, enough time to talk in small groups before asking anyone to share with the entire class during this activity. Small-group structures offer all students, including emerging multilingual students, a chance to engage in sense-making with their peers as well as space to use their verbal and nonverbal (e.g., writing, drawing pictures, and using gestures) resources to express their ideas. Small-group structures also give students a chance to learn from other students' uses of these resources. When students are tasked with expressing their ideas, be sure that they know it is okay for them to do so in a mode or combination of modes (speaking, writing, drawings, and gestures).

Swimming in the Dead Sea Lesson Plan

Stage 7 REPORT

Total Time Needed for the Stage	100 minutes (over three days)
Grouping	Individual, Teams

Learning Goals

Students should make progress on their ability to do the following by the end of this stage of the lesson:

Science

1. Plan and conduct an investigation to provide evidence that feedback mechanisms maintain homeostasis (HS-LS1-3).
2. Make and defend a claim based on evidence about the natural world that reflects scientific knowledge and student-generated evidence (SEP7).
3. Construct a written argument based on evidence (SEP7).
4. Evaluate the claims, evidence, and/or reasoning to determine the merits of arguments (SEP7).
5. Respectfully give critiques on scientific arguments by and challenging ideas and conclusions (SEP7).
6. Communicate scientific information or ideas in multiple formats including graphically and textually (SEP8).

English Language Development

1. Identify appropriate and sufficient evidence from data, models, and/or information from investigations of a phenomenon (ELD-SLC.9-12.Argue.Interpretive).
2. Compare reasoning and claims based on evidence from competing arguments (ELD-SLC.9-12.Argue.Interpretive).
3. Defend or refute a claim based on data and evidence (ELD-SLC.9-12.Argue.Expressive).
4. Establish and maintain an appropriate tone and stance (ELD-SLC.9-12.Argue.Expressive).
5. Signal logical relationships among reasoning, evidence, data, and/or models when making and defending a claim (ELD-SLC.9-12.Argue.Expressive).

English Language Arts

1. Write routinely over extended time frames (time for reflection and revision) and shorter time frames for a range of discipline-specific tasks, purposes, and audiences (WHST.9-10.10).
2. Write arguments focused on discipline-specific content (WHST.9-10.1).
3. Write informative/explanatory texts, including the narration of scientific procedures/experiments or technical processes (WHST.9-10.2).
4. Produce clear and coherent writing in which the development, organization, and style are appropriate to task, purpose, and audience (WHST.9-10.4).
5. Strengthen writing as needed by revising, editing, rewriting, and focusing on addressing what is most significant for a specific purpose and audience (WHST.9-10.5).

(*continued*)

6. Determine the central ideas or conclusions of a text; trace the text's explanation or depiction of a complex process, phenomenon, or concept; provide an accurate summary of the text (RST.9-10.2).
7. Determine the meaning of symbols, key terms, and other domain-specific words and phrases as they are used in a specific scientific or technical context (RST.9-10.4).
8. Analyze the structure of the relationships among concepts in a text, including relationships among key terms (RST.9-10.5).
9. Assess the extent to which the reasoning and evidence in a text support the author's claim (RST.9-10.8).
10. Compare and contrast findings presented in a text to those from other sources (including their own experiments), noting when the findings support or contradict other accounts (RST. 9-10.9).

Activities

Activity A	Draft Report
Time Needed for the Activity	20 minutes
Grouping	Individual

What to Do

1. Have students work on their own during this activity. Give a copy of the Report handout to each student.
2. Show the "Create a Draft Report" page on a screen using the Presentation mode in the Learning Hub so everyone in the classroom can see it.
3. Say, "You are now going to write an investigation report to share what you figured out. The first part of the report is called the 'Introduction'. In this section, you should explain to the reader what you were investigating, why you were investigating it, and what question you were trying to answer. All this information can be found on your Task handout."
4. Hold up the Task handout.
5. Ask the students, "Do you have any questions about what you need to do?"
6. Answer any questions that come up.
7. Say, "Here are some sentence starters to help you begin writing."
8. Hold up the Report handout and point to the sentence starters.
9. Tell the students, "Okay, let's write."
10. Give the students five minutes to write the "Introduction" section of the report. As they work, move from student to student to check in, ask probing questions, and offer a suggestion if a student gets stuck.

What to Give Out

Report handout for **Swimming in the Dead Sea**

What to Show

The "Create a Draft Report" page in the ADI Learning Hub

(continued)

11. Say, "Okay, let's move on to the second part of the report. The second part of the report is called the 'Method'. In this section, you should explain to the reader what you did during the investigation to complete the task and answer the guiding question. All this information can be found in the Plan handout. Remember that you all planned and carried out different investigations, so do not assume that the reader will know what you did."
12. Ask the students, "Do you have any questions about what you need to do?"
13. Answer any questions that come up.
14. Say, "Here are some sentence starters to help you begin writing."
15. Hold up the Report handout and point to the sentence starters.
16. Tell the students, "Okay, let's write."
17. Give the students five minutes to write the "Method" section of the report. As they work, move from student to student to check in, ask probing questions, and offer a suggestion if a student gets stuck.
18. Say, "Okay, let's move on to the last part of the report. The third part of the report is called the 'Argument'. In this section, you should explain to the reader what you figured out and how you know you are correct. All this information can be found on your Do handout and on your team's whiteboard."
19. Hold up the Do handout and point to the boards (or remind them that they have a picture of their board).
20. Ask the students, "Do you have any questions about what you need to do?"
21. Answer any questions that come up.
22. Say, "Here are some sentence starters to help you begin writing."
23. Hold up the Report handout and point to the sentence starters.
24. Tell the students, "Okay, let's write."
25. Give the students 10 minutes to write the "Argument" section of the report. As they work, move from student to student to check in, ask probing questions, and offer a suggestion if a student gets stuck.
26. Collect all the draft reports before moving on to the next activity. Tell students to mark their draft reports with a symbol that they can use to identify it later. They should NOT write their names on their draft reports.

Teaching Tip

Writing is an important component of this model. When a student writes the report on their own, it not only helps the student learn to write in the context of science (an important literacy skill) but also allows that student to write to learn (develop a better understanding of scientific concepts by writing about them). There are many writing supports (e.g., sentence starters) already embedded into this activity, so you might not need to provide much extra support for your students.

This activity can be done completely in class or started in class and then finished as homework. Do not assign the entire activity as homework unless you know that every student in your class has dedicated time, space, and resources to write outside of class and everyone understands and is comfortable with the report writing process.

If you are concerned about students feeling overwhelmed, you can develop a class "gotta-have-it" list of information that needs to be included in each section of the report. Students can then refer to this list as they write each section.

Don't worry if students do not produce a perfect report at this point. All the reports will be reviewed and revised before they are turned in to you. The reports should therefore be viewed as a starting point. Be sure to encourage students to "just get something down on paper" so "we can work on it together."

Attending to Equity

Emerging multilingual students' access to instructional tasks that require complex thinking, such as writing a report, is enhanced when linguistic complexity and instructional support match their current level of language proficiency. Therefore, it is important to scaffold the writing of the reports for the emerging multilingual students based on their current English language proficiency level. The use of sentence starters to model written language production skills can help emerging multilingual students develop English language skills because the development of academic language and academic content knowledge are interrelated processes. However, it is important that such scaffolds be used purposefully and removed when no longer needed.

ADI Go Digital

Use the **ADI Learning Hub** to go paperless. When you create a class and add students to it, students can complete all the stages right in the system.

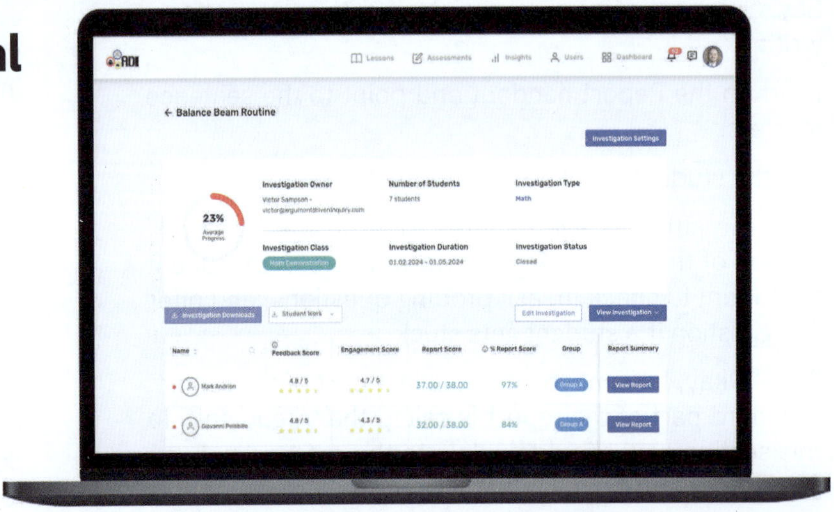

Activity B	Peer Review the Reports
Time Needed for the Activity	45 minutes
Grouping	Teams

What to Do

1. Students should work in teams of three for this activity. These can be completely different teams.
2. Show the "Peer Review Some Draft Reports" page on a screen using the Presentation mode in the Learning Hub so everyone in the classroom can see it.
3. Say, "We are now going to review our reports to find ways to make them better."
4. If possible, use a document camera to project the peer review guide (PRG) on a screen.
5. Say, "We are going to use this peer review guide to give one another feedback." Point to the image.
6. Say, "I'm going to ask you to work in teams of three to do this. I will give each team a draft report to read. You three will then read the report together. Once you are done reading the report, I want you to answer each of the questions on the peer review guide." Point to the review questions on the image of the PRG.
7. Tell the students, "You can check No, Almost, or Yes after each question." Point to the checkboxes on the image of the PRG.
8. Say, "This will be your rating for this part of the report. Make sure you agree on the rating you give the author. If you mark No or Almost, then you need to tell the author what he or she needs to do to get a Yes." Point to the space for the reviewer feedback on the image of the PRG.
9. Say, "It is really important for you to give the authors feedback that is helpful. That means you need to tell them exactly what they need to do to make their reports better."
10. Ask the students, "Do you have any questions about what you need to do?"
11. Answer any questions that come up.
12. Tell the students, "Please get into your teams of three."
13. Allow the students time to move into their triads.
14. Say, "Okay, I'm now going to give you one report to read." Give out one draft report to each triad. Make sure that the report you give a triad was not written by one of the students in that triad. Give each triad one PRG to fill out.

(continued)

What to Give Out

Peer Review Guide (one per group per review, so for this investigation you need the same number of peer review guides as the number of draft reports that will be reviewed)

What to Show

The "Peer Review Some Draft Reports" page in the ADI Learning Hub

What to Look For

Make sure students are reaching consensus about what rating to give each criterion. Also, be sure that students are giving one another feedback that is specific, detailed, and helpful when they think an aspect of a report needs to be improved.

15. Tell the students, "Okay, I'm going to give you 15 minutes to read the report I gave you and to fill out the peer review guide. Go ahead and get started."
16. Give the students 15 minutes to work. As they work, move around from team to team to check in and see how things are going, answer questions, and offer advice.
17. After 15 minutes pass, tell the students, "Okay, time is up. Please give me the report and the peer review guide that you filled out."
18. Collect the draft reports and the PRGs, or collect the workbooks if students are using them.
19. Say, "Okay, I am now going to give you a different report to read and a new peer review guide to fill out." Give out one more report to each triad. Make sure that the report you give a triad was not written by one of the students in that triad. Give each triad one new PRG to fill out as a team.
20. Say, "Okay, I'm going to give you 15 minutes to read this new report and to fill out the peer review guide. Go ahead and get started."
21. Give the students 15 minutes to work. As they work, move around from team to team to check in and see how things are going, answer questions, and offer advice.
22. After 15 minutes pass, tell the students, "Okay, time is up. Please give me the report and the peer review guide that you filled out."
23. Repeat this process for the remaining draft reports. Each triad should have one more report to review (for a total of three reviews).
24. Collect the last draft reports and PRGs. Set all the draft reports on a table in separate stacks. Each stack should have a draft report on top (with the identification symbol facing up so students can find their reports) and the PRG on the bottom.

Teaching Tip

The peer review process is one of the best ways for students to learn how to write. When students read and review different reports, they have an opportunity to (1) see examples of texts written by others for the same purpose, (2) discuss what counts as quality and why some reports are stronger than others, (3) discuss ways to strengthen a report, and (4) pick up things that they can do to improve their reports. This helps the students learn new ways to organize and present information, which will help them write better on subsequent reports.

Attending to Equity

One helpful scaffold is to give students sentence stems for what to write in their feedback to the author. Ideal feedback is respectful, specific about what needs to be changed, and helpful in that it suggests how the author should go about making that change. You can review the writing of the emerging multilingual students in your class if they have limited English language proficiency, but make sure that all emerging multilingual students participate in the peer review process. You will see tremendous growth in their writing skills as they review their peers' writing and discuss what makes some reports stronger than others.

Activity C	Revise the Reports
Time Needed for the Activity	25 minutes
Grouping	Individuals

What to Do

1. Students should work on their own for this activity.
2. Show the "Reflect on the Feedback" page on a screen using the Presentation mode in the Learning Hub so everyone in the classroom can see it.
3. Say, "Please come pick up your reports from over here on this table." Point to the table.
4. Give the students a few minutes to pick up their reports and return to their seats.
5. Say, "Please take a few minutes to read over the peer review guide. You should use it to figure out what you need to change in your report and how you will change it."
6. Allow the students to read the PRG.
7. Show the "Create a Final Report" page on a screen using the Presentation mode in the Learning Hub so everyone in the classroom can see it.
8. Say, "Okay, let's revise our reports. I would like you to rewrite your report. You can use your draft report as a starting point but use the feedback on the peer review guide to make it better."
9. Ask the students, "Do you have any questions about what you need to do?"
10. Answer any questions that come up.
11. Tell the students, "Okay, let's write." Allow about 20 minutes for the students to revise their reports.
12. After about 20 minutes, give the following directions, "Okay, time's up. I will now come around and collect your first draft, final report, and peer review guide."
13. Collect all the first drafts, PRGs, and final reports.
14. Use the "Teacher Score" columns in the PRG to grade the final reports. Be sure to give the students feedback about their writing in the "Teacher Comments" section.

What to Show First

The "Reflect on the Feedback" page in the ADI Learning Hub

What to Show Second

The "Create a Final Report" page in the ADI Learning Hub

Attending to Equity

Students don't always use the feedback they receive from others to improve the quality of their reports. Make sure you move from student to student and encourage them to make changes based on the feedback. If students are not sure how to revise their reports, model how to use feedback. For example, you might say, "Okay, based on the peer review guide, I need to do more to describe what I did to collect data. I bet if I add information about ________, readers will have a better idea of what I did."

Teaching Tip

The peer review guide (PRG) is designed to be used with any ADI investigation, thus allowing you to use the same scoring rubric throughout the entire year. This is beneficial for several reasons. First, the criteria for what counts as a high-quality report do not change from investigation to investigation. Students, therefore, quickly learn what is expected from them when they write a report, and you do not have to spend valuable class time explaining the various components of the PRG each time students write a report. Second, the PRG makes it clear which components of a report need to be improved next time because the grade is not based on a holistic evaluation of the report. The PRG format allows students to see which aspects of their writing are strong enough and which aspects need improvement. Finally, and perhaps most importantly, the PRG provides you with a standardized measure of student performance that can be compared over multiple reports across semesters, thus allowing you to track improvement over time.

Teacher Notes

Swimming in the Dead Sea

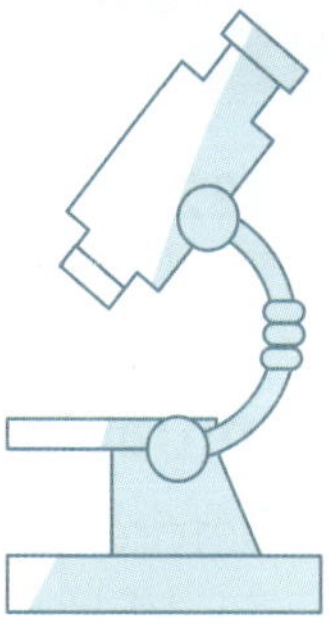

TASK

The Phenomenon

The Dead Sea is a large lake located at the lowest point on Earth. Take a few minutes to watch a video about a team of international swimmers that completed a historic 17-km swim across the Dead Sea from Jordan to Israel in 2016. As you watch the video, write down some things you notice in the space below.

Things I **noticed...**

The Task

The video you watched highlighted the dangers of drinking water from the Dead Sea. It is dangerous to drink water from the Dead Sea because the concentration of salt in the water is about 31.5%, and drinking water with so much salt in it can lead to dehydration. In fact, it is so dangerous that the swimmers in the video had to wear special masks to prevent them from accidentally drinking any of the water during their swim. People, however, often spend long periods of time swimming in the ocean, which has a salt concentration of about 3.5%, without needing to wear special equipment to keep them from accidentally drinking the water. Your goal in this investigation is to figure out why it is more dangerous to drink water from the Dead Sea than it is to drink water from any other sea or oceanq. The guiding question of this investigation is:

Why is it so dangerous to drink water from the Dead Sea?

Initial Ideas

Before you start this investigation, take a few minutes to think about how drinking salt water could lead to dehydration. Then draw a model in the space below that shows how you think drinking water with a high salt concentration could lead to dehydration. Be sure to include labels to help explain your thinking.

Some **ideas** that I have...

Share your model with the rest of your group. Once everyone has had a chance to share their ideas, list some questions that your group has about this phenomenon.

Some **questions** we have now...

Swimming in the Dead Sea

IDEAS

Some Ideas You Can Use: Cells

All cells are surrounded by a membrane. This **cell membrane** forms a barrier between the cell and its environment. The cell membrane is referred to as **semipermeable** because some molecules or ions can cross it while others cannot. In cells, water cannot simply move across the cell membrane. However, special openings in the membrane allow for water molecules, but not other particles, to easily move through the membrane so cells can take in or remove water as needed. **Osmosis** is the movement of water molecules across a semipermeable cell membrane.

Generally speaking, cells will add or remove water in a way that ensures that the concentrations of water and solute (the particles that are dissolved) inside and outside the cell are balanced or in **equilibrium**. An **isotonic** solution is a solution that has the same concentration of molecules or ions and water inside and outside of a cell. A cell in an isotonic solution will not gain or lose water. A **hypotonic** solution, in contrast, has a lower concentration of solute (and a higher concentration of water molecules) outside the cell than inside the cell. A cell in a hypotonic solution will gain water and expand. Finally, a **hypertonic** solution is a solution that has a higher concentration of solute (and lower concentration of water) outside the cell than inside the cell. When a cell is in a hypertonic solution, it will lose water and shrink. The image below shows the movement of the water when a cell is in in isotonic, hypotonic, and hypertonic solutions.

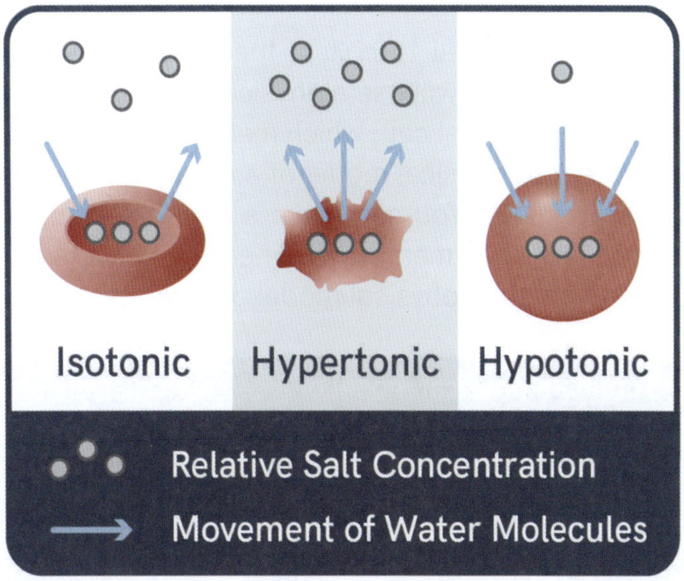

Water from the Dead Sea is 31.5% salt and ocean water is 3.5% salt. The concentration of salt inside a human blood cell is 0.9%. A cell placed in water of the Dead Sea or ocean water will be in a hypertonic solution because both types of water have a greater salt concentration than what is found inside the cell. Cells in Dead Sea water or ocean water, as a result, will lose water.

ARGUMENT-DRIVEN INQUIRY

Some **useful ideas** from what I read...

Some Ideas You Can Use: Systems and System Models

A cell is an example of a system. A **system** is a group of related parts that make up a whole and can carry out functions that its individual parts cannot. A system is often described in terms of its parts and their interactions. Scientists often use models to help understand a system. A **model** is a representation of a system. Models are often used to test different ideas about how the system functions. Models, however, are limited in that they only represent certain parts or aspects of the system under study.

You will use a physical model of a cell to collect the data you need to answer the guiding question during this investigation. You will use a physical model for two reasons: (1) a physical model of a cell is much larger than a real cell, which makes the process of data collection much easier; and (2) you can create your physical model of a cell in any way you see fit, which makes it easier to carry out an experiment. You can make a physical model of a cell using **dialysis tubing**, which behaves much like a cell membrane. The image below shows an example of how you can use dialysis tubing to create a physical model of a cell.

How to create a physical model of a cell with dialysis tubing

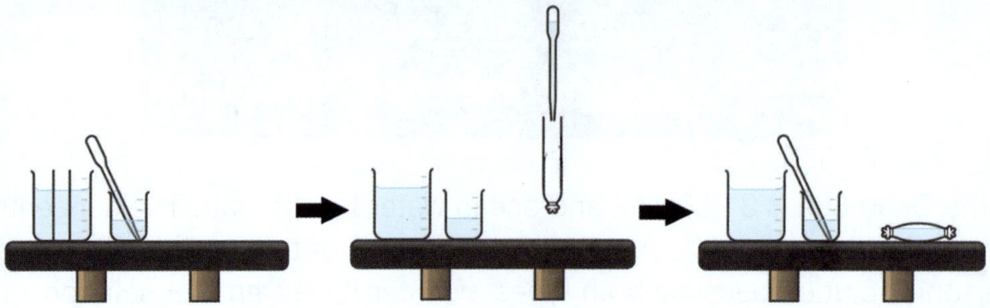

Some useful ideas from what I read...

Some Ideas You Can Use: Tracking Matter in Systems

When attempting to understand how that system works, scientists often need to track how much and the types of matter that move into, out of, or within a system. They also track the movement of matter into, out of, or within a system when they want to understand how a change in a system's surroundings will affect the system or the interactions within it.

Scientists can track the movement of matter into or out of a system by measuring how the total weight of that system changes over time. Scientists can use a change in the total weight of a system to track the movement of matter into or out of the system because all matter has mass. For example, a 1-mL sample of water weighs 1 g, and a 2.8-mL (or cm^3) sample of salt, such as the one shown in the picture on the left, weighs 6 g.

A cell model can be thought of as a system. You can therefore figure out what is happening in a cell model by tracking how its weight changes over time. For example, if molecules of water that started outside the cell model move into the cell model, then the total weight of the cell model will increase. If molecules of water that started inside the cell model move out of it, then the total weight of the cell model will decrease. The total weight of the cell model will remain the same if (a) matter only moves within the cell model but does not move into or out of it or (b) the same amount of the same type of matter moves into the cell model as what moves out of it.

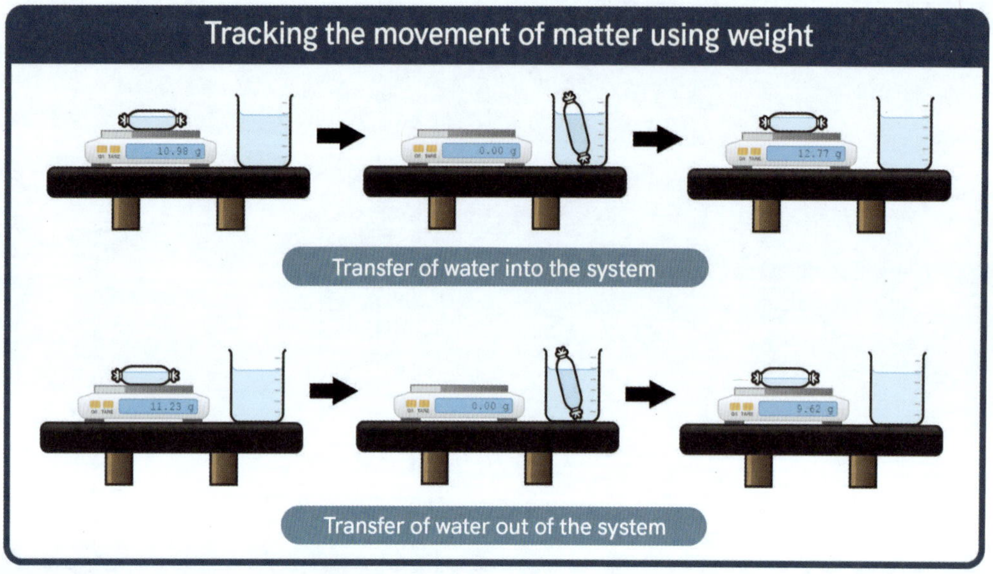

Some useful ideas from what I read...

Swimming in the Dead Sea

PLAN

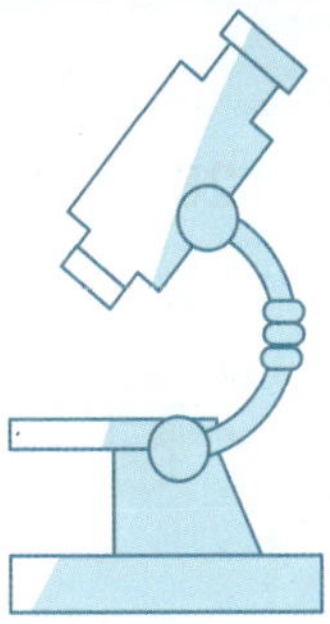

Plan Your Investigation

Prepare a plan for your investigation by filling out the chart below.

I am trying to answer the following **question**...

I will use the following **observations or measurements**...

I will **collect** these observations or measurements by...

Here is a **picture** of how I plan to use the equipment...

I will **analyze** the observations or measurements that I collect by...

I approve of this investigation plan.

Teacher's signature

Date

Swimming in the Dead Sea

DO

Collect Data

Keep a record of what you observe or measure during your investigation.

My **observations or measurements...**

Analyze Your Data

Use the space below to create any tables or graphs that will help you make sense of the data you collected so you can answer the guiding question.

Swimming in the Dead Sea

SHARE

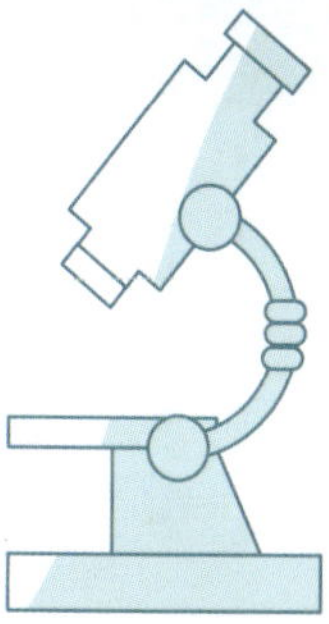

Draft Argument

Develop an argument on a whiteboard. It should include:

1. A *claim* that answers the guiding question,
2. *Evidence* in support of the claim that provides an analysis of the data and a description of what the analysis means,
3. A *justification of the evidence* that explains why your group thinks the evidence is important.

Our Guiding Question:

Claim:

Evidence:

Justification of the Evidence:

Argumentation Session

Share your argument with your classmates. In the space below, be sure to keep track of any ideas that you can use to revise your argument and make it better.

Some possible ways to **improve** our argument...

Swimming in the Dead Sea

REFLECT

Reflective Discussion

In the space below, you can keep track of any ideas from the discussion that you think are important or will be useful in the future.

Some important **ideas...**

Swimming in the Dead Sea

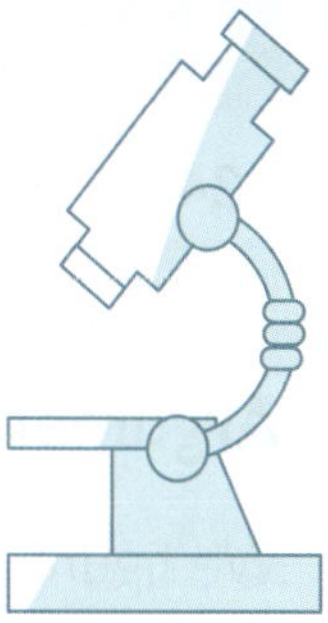

REPORT

Draft Report

Prepare a report to share what you figured out during your investigation.

Introduction

We have been studying ______________________________ in class. Before we started this investigation, we explored ______________________________

____

____

We noticed ______________________________

____

____

Our goal for this investigation was to figure out ______________________________

____

____

The guiding question was ______________________________

____

Method

To answer this question, ______________________________

____

____

____

____

____

____

____

____

____

Argument

We figured out ______________________________

____

____

____

The graph below includes information about ______________________________

____

____

____

This analysis suggests ______________________________

____

____

____

This evidence is based on several important scientific ideas. The first one is ______________________________

____

____

____

____

____

____

Tissue Repair Lesson Plan

OVERVIEW

Description	Students figure out how the human body makes new cells from existing cells.
Guiding Question	How do our bodies make new cells from our existing cells so our wounds can heal?
What Students Figure Out	New cells are created in a process called mitosis where old cells replicate their genetic material and then divide into a new and an old cell.

STAGES

This investigation has seven stages. Each stage includes multiple activities. The flowchart below shows what happens during each stage and how much time is needed to complete each one.

Task (Day 1)
1 — **20 minutes**
Students are introduced to a phenomenon and a guiding question to answer. They also share their initial ideas.

Ideas (Day 1)
2 — **30 minutes**
Students read about DCI, a CC, and an SEP that they can use to help them figure out the phenomenon.

Plan (Day 2)
3 — **15 minutes**
Students formulate a plan collecting and analyzing the data they need to answer the guiding question.

Do (Day 2)
4 — **30 minutes**
Students carry out their plan, make sense of the data they collected, and answer the guiding question.

Share (Day 3)
5 — **45 minutes**
Students develop, share, and revise an evidence-based argument about what they figured out.

Reflect (Day 4)
6 — **20 minutes**
Students discuss a DCI and/or a CC and reflect on the SEPs they used during the investigation.

Report (Day 4)
7 — **25 minutes**
Students write investigation reports to share what they figured out during the investigation.

Report (Day 5)
7 — **45 minutes**
Students peer review the reports written by other students and offer feedback to improve.

Report (Day 6)
7 — **30 minutes**
Students revise their reports and then submit a final draft to their teacher for evaluation.

STANDARDS ALIGNMENT

NEXT GENERATION SCIENCE STANDARDS

Performance Expectation

Use the periodic table as a model to predict the relative properties of elements based on the patterns of electrons in the outermost energy level of atoms (HS-PS1-1).

Core Ideas(s)	Crosscutting Concept(s)	Practices
Growth and Development of Organisms (LS1.B)	Systems and System Models (CC4)	Asking Questions (SEP1) Developing and Using Models (SEP2) Planning and Carrying Out Investigations (SEP3) Analyzing and Interpreting Data (SEP4) Using Mathematics and CT (SEP5) Constructing Explanations (SEP6) Engaging in Argument from Evidence (SEP7) Obtaining, Evaluating, and Communicating Information (SEP8)

WIDA ENGLISH LANGUAGE DEVELOPMENT STANDARDS FOR SCIENCE

Explain	Argue
Interpret scientific explanations (ELD-SC.9-12.Explain.Interpretive) Construct scientific explanations (ELD-SC.9-12.Explain.Expressive)	Interpret scientific arguments (ELD-SC.9-12.Argue.Interpretive) Construct scientific arguments (ELD-SC.9-12.Argue.Expressive)

COMMON CORE ENGLISH LANGUAGE ARTS

Reading	Writing	Speaking and Listening
Key ideas and details (RST.9-10.1 and RST.9-10.2) Craft and structure (RST.9-10.4) Integration of knowledge and ideas (RST.9-10.7, RST.9-10.8, and RST.9-10.9) Range of reading (RST.9-10.10)	Text types and purposes (WHST.9-10.1 and WHST.9-10.2) Production and distribution of writing (WHST.9-10.4, WHST.9-10.5, and WHST.9-10.6)	Comprehension and collaboration (SL.9-10.1 and SL.9-10.3) Presentation of knowledge and ideas (SL.9-10.4 and SL.9-10.6)

THINGS TO KNOW BEFORE STARTING THE LESSON

The Core Ideas

All living things are made up of one or more cells. These cells are in turn made up of organelles, which are structures that carry out specific functions within the cell. All organelles are suspended in a type of matter called the cytoplasm. One organelle is the cell membrane. The cell membrane separates the cell's interior from the cell's surroundings while also regulating what moves in and out of the cell. Another common type of organelle is the nucleus. The nucleus holds the genetic information of each cell. Cells use their genetic information to function properly and to make new cells. A cell's genetic information is stored in molecules called deoxyribonucleic acid (DNA). The DNA is tightly coiled and packaged into chromosomes. Each chromosome consists of a single molecule of DNA. Cells in the human body consist of 46 chromosomes, or 23 pairs, that are organized by size and appearance.

The Crosscutting Concepts

A system is a group of parts that interact with one another to carry out some function that the individual parts cannot achieve alone. A cell is a small but highly complex system. To better understand cells and their parts, we can construct a model. A model is a physical, visual, or computer-based representation of a system that we can test and use to make predictions about the system.

The Science and Engineering Practices

A practice of science is to **ask and refine questions** that lead to descriptions and explanations of how the natural world works. Another practice of science is to **use and construct models** as helpful tools for representing ideas and explanations. These tools include diagrams, drawings, physical replicas, mathematical representations, analogies, and computer simulations. Scientists **plan and carry out investigations** in the field or a laboratory, working collaboratively and individually. Their investigations are systematic and require clarifying what counts as data and identifying variables or parameters. Scientific investigations produce **data that must be analyzed and interpreted**. Because data patterns and trends are not always obvious, scientists use a range of tools and practices to identify the significant features and patterns in the data. Scientists also use **mathematics and computational thinking** for a range of tasks such as constructing simulations; statistically analyzing data; and recognizing, expressing, and applying quantitative relationships. The products of doing science are **explanations**. Scientific explanations often specify variables and are used by scientists to describe and predict different natural phenomena. **Scientific argumentation**, or the process of sharing, supporting, critiquing, and revising claims based on evidence is the process that scientists use to determine if an explanation is valid or acceptable. Finally, scientists must be able to **communicate clearly and persuasively** the ideas and methods they generate. Critiquing and communicating ideas individually and in groups is a critical professional activity.

PREPARATION

This lesson can be delivered paper-based, paperless, or remote. Use the paper-based delivery option for the investigation if students do not have 1:1 access to a computer and they are in the same classroom or physical space during the lesson. Use the paperless delivery option for the investigation if students have 1:1 access to a computer and they are in the same classroom or physical space during the lesson. Finally, if students are not in the same classroom or physical space during the lesson and have 1:1 access to a computer, use the remote delivery option.

Materials Needed

The materials needed by the teacher and each student in the class for each of the three modalities are listed in the table below.

Delivery Mode	Each Student Will Need	The Teacher Will Need
Paper-Based	• Copies of each student handout for the investigation (or a workbook) • A pencil	• A teacher account in the ADI Learning Hub • Access to a computer with high-speed internet and a browser (such as Chrome, Safari, Firefox, or Edge) • A projector or smart board (for displaying a computer screen)
Paperless	• A student account in the ADI Learning Hub • Access to a computer with high-speed internet and a browser (such as Chrome, Safari, Firefox, or Edge)	• A teacher account in the ADI Learning Hub • Access to a computer with high-speed internet and a browser (such as Chrome, Safari, Firefox, or Edge)
Remote	• A student account in the ADI Learning Hub • Access to a computer with high-speed internet and a browser (such as Chrome, Safari, Firefox, or Edge)	• A teacher account in the ADI Learning Hub • Access to a computer with high-speed internet and a browser (such as Chrome, Safari, Firefox, or Edge)

Additional Materials and Preparation Needed for In-Person Delivery

The materials needed for the in-person version of this investigation are listed in the table below. To purchase the **Tissue Repair** kit from our store, please visit https://shop.argumentdriveninquiry.com/. The **Tissue Repair** kit comes with enough materials for eight groups.

Type of Material	Item	Quantity	
		For Entire Class	For Each Group
Equipment	Card 1: Event A	8	1
	Card 2: Event B	8	1
	Card 3: Event C	8	1
	Card 4: Event D	8	1
	Card 5: Event E	8	1
	Card 6: Appearance of Chromosomes	8	1
	Red pop beads	192	24
	Yellow pop beads	384	48
	Hook-and-loop cable ties, 4″	32	4
	Plastic cups	8	1

Preparation

Create a set of "existing cells making new cells" cards for each group. Each set of cards should include one copy of cards 1-5 to make a total of 5 cards in each set. Leave card 6 out of the set. This card is an "extra clue" you can give students if they do not figure out that the chromosomes must replicate as part of cell division to provide new cells with the same genetic information. Place 24 red pop beads, 48 yellow pop beads, and eight 4-inch hook-and-loop cable ties in each cup. Label each cup with each's groups name or number.

During the Investigation

Students can use the cards to determine how the appearance of a cell changes as it divides. Examples of the cards are shown in the images below.

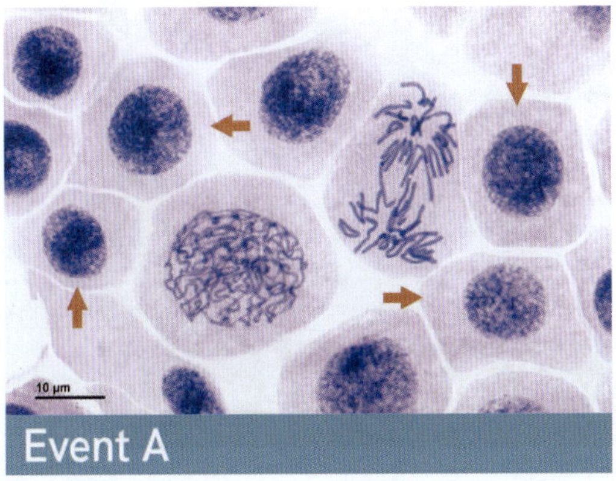

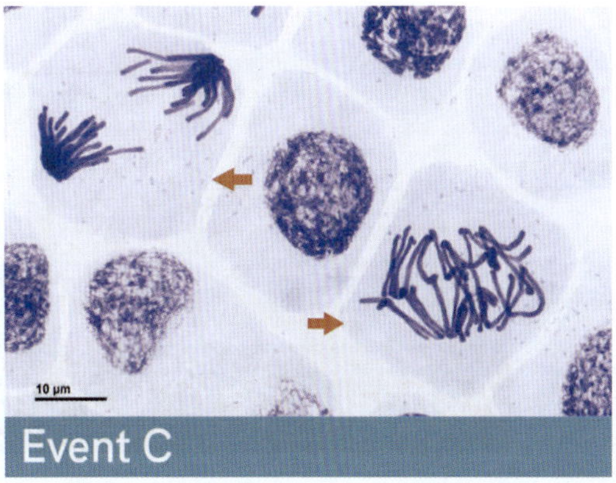

Students can then use the pop beads and hook-and-loop cable ties to create a physical model of chromosomes. They can then use the physical model to figure out what the chromosomes are doing during the various events of cell division. The image below is an example of a physical model. The names of each stage of mitosis are included in this image; however, students will likely use the same event letters as on the cards since they have not been formally introduced to each stage of mitosis at this time.

Example of a physical model of the process of mitosis

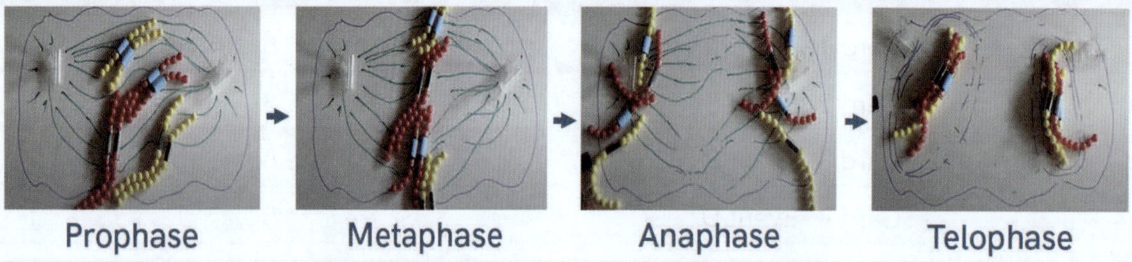

We recommend that you use a set routine for distributing and collecting the materials during the investigation. The equipment for each group, for example, can be set up at each group's lab station before class begins, or one member from each group can collect them from a table or a cart when needed during class.

Additional Tips

Students often do not realize the chromosomes are attached to their replicas, which are often described as the sister chromatids, during the prophase and metaphase stages of mitosis. If a group of students gets stuck because they are unable to explain how cells do not lose genetic information when they divide, you can give them a copy of card 6. This card clearly shows each chromosome and its replica. An image of this card is provided below.

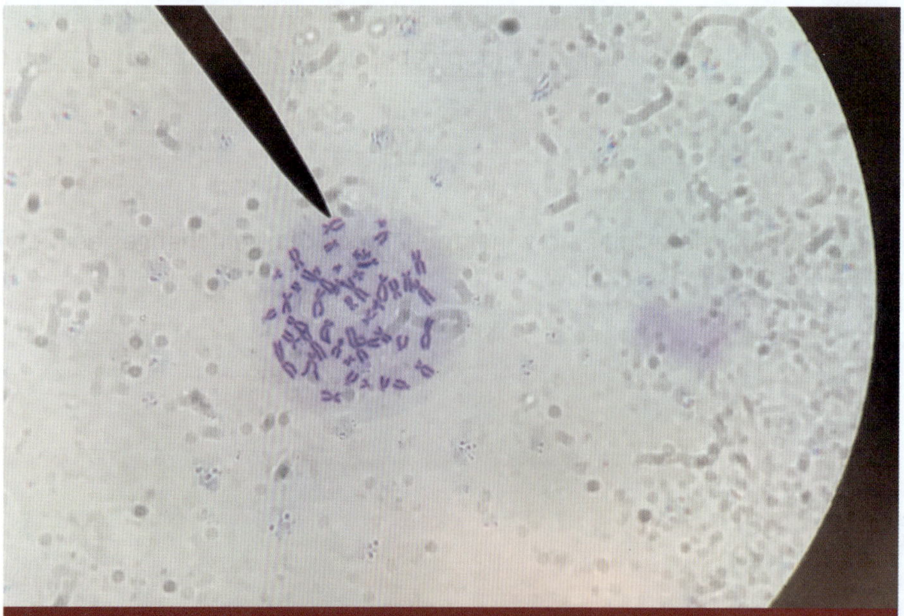

Appearance of Chromosomes prior to Event D

Safety Precautions

Remind students to follow all normal lab safety rules.

Laboratory Waste Disposal

There is no waste produced during this activity. Collect the equipment at the end of the investigation and store for later use.

FACILITATION NOTES

The facilitation notes for this investigation describe how to deliver this lesson in person. They are written specifically for the paper-based mode of delivery. These same facilitation notes, however, can also be used for a paperless delivery of the lesson by simply substituting the directions that mention paper with the entry into the ADI Learning Hub. If you are teaching this lesson using the remote mode of delivery, you only need to assign due dates and prerequisites for each stage. The students can then simply follow the directions built into the ADI Learning Hub as they move through the investigation.

ADI Go Digital

The digital version of this lesson is available in the **ADI Learning Hub**. Be sure to check out the built-in teacher presentation mode.

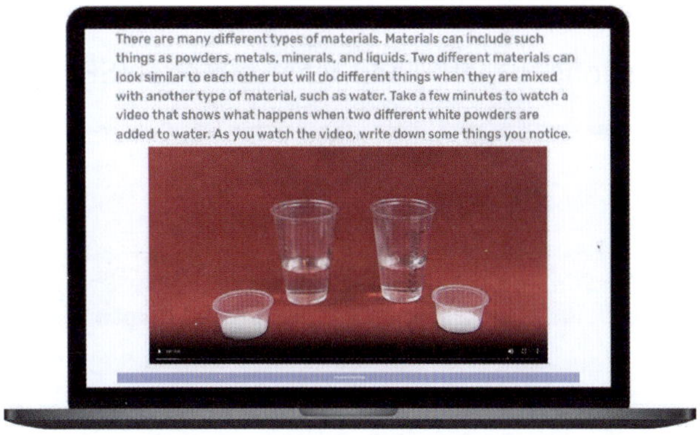

Teacher Notes

Tissue Repair Lesson Plan

Stage 1 TASK

Total Time for the Stage	45 minutes
Grouping	Individual; Teams

Learning Goals

Students should make progress on their ability to do the following by the end of this stage of the lesson:

Science

1. Ask questions that arise from careful observation of phenomena to clarify and/or seek additional information (SEP1).
2. Develop and use a model to provide mechanistic accounts of phenomena (SEP2).
3. Evaluate the merits and limitations of different models of the same phenomenon (SEP2).
4. Ask questions that arise from examining models to clarify and/or seek additional information and relationships (SEP 1).
5. Ask questions to clarify and refine a model (SEP1).

English Language Development

None

English Language Arts

1. Initiate and participate effectively in small-group collaborative discussions with diverse partners by building on others' ideas and expressing their own clearly (SL.9-10.1).
2. Present information clearly, concisely, and logically such that listeners can follow the line of reasoning and the organization, substance, and style are appropriate to the purpose, audience, and task (SL.9-10.4).
3. Adapt speech to a variety of contexts and tasks, demonstrating command of formal English when indicated or appropriate (SL.9-10.6).

Activities within the Stage

Activity A	The Phenomenon
Time Needed for the Activity	5 minutes
Grouping	Teams

What to Do

1. Have students work in their assigned teams for this activity.
2. Begin the activity by giving out a copy of the Task handout to each student.
3. Show "The Phenomenon" page on a screen using the Presentation mode in the Learning Hub so everyone in the classroom can see it.
4. Read the introduction of the phenomenon out loud and then start the video. You can also have students read the introduction of the phenomenon and watch the video on their own devices.
5. Give the students a few moments to write down their observations on their handouts before moving on to the next activity.

What to Give Out

Task handout for **Tissue Repair**

What to Show

"The Phenomenon" page in the ADI Learning Hub

Teaching Tip

Instead of assigning fixed roles to students by name at the beginning of the investigation, give each member of a group a different number so you can easily assign and change roles as needed during the various investigation stages. For example, you might give directions such as "1s will share out for your group" or "3s, will you please come back here and pick up these materials for your group?" or "2s and 3s, you will present the argument for your group while 1s and 4s travel to the other groups to offer feedback."

Attending to Equity

Be sure to intentionally group emerging multilingual students at times with peers who know the same languages as they do and at other times with peers whose English language development is slightly more advanced. Thoughtful grouping that varies throughout an investigation allows emerging multilingual students to benefit from the different linguistic resources of their peers.

Activity B	Share What You Noticed
Time Needed for the Activity	5 minutes
Grouping	Teams

What to Do

1. Show the "Share What You Noticed" page on a screen using the Presentation mode in the Learning Hub so everyone in the classroom can see it.
2. Say, "Okay, please share what you noticed with the others at your table."
3. Give the students about two minutes to share at their tables.
4. Say, "Okay, let's take a few minutes to share with the class. Team X (call on a team at random), what did you talk about?"
5. Allow Team X to share something they noticed. If it is related to cellular reproduction, say, "That is really interesting. We are actually going to examine that in more detail during this investigation." If what is shared is not related to cellular reproduction, say, "That is really interesting. Team Y (call on another team at random), what did you notice?" and allow that team to share. Follow this pattern until a team mentions something related to how the human body creates new cells to heal a cut. If no team mentions this, say, "You know what I found interesting? I saw that human body created new skin cells to heal the injury. Did anyone notice that as well?"
6. Move on to the next activity.

What to Show

The "Share What You Noticed" page in the ADI Learning Hub

What to Look For

As soon as a group mentions something related to how the human body creates new cells to heal a cut, move on to the next activity.

Attending to Equity

Be sure to provide students with opportunities to talk in small groups during this activity. Small-group structures offer all students, including emerging multilingual students, a chance to engage in sense-making with their peers and offers them space to use their verbal and nonverbal (e.g., writing, drawing pictures, and using gestures) resources to express their ideas. Small-group structures also give students a chance to learn from other students' uses of these resources.

Activity C	What You Need to Figure Out
Time Needed for the Activity	5 minutes
Grouping	Teams

What to Do

1. Show the "What You Need to Figure Out" page on a screen using the Presentation mode in the Learning Hub so everyone in the classroom can see it.
2. Read the task out loud and have students follow along on their handouts, on their devices, or on the screen.
3. Allow students to ask clarifying questions about the task or the guiding question.
4. Once everyone understands what they will need to figure out during the investigation, move on to the next activity.

What to Show

The "What You Need to Figure Out" page in the ADI Learning Hub

Teaching Tip

Have students highlight, circle, or place a star next to the guiding question. Marking the guiding question helps students find it later during the Plan stage.

Teacher Notes

Activity D	Share Your Ideas
Time Needed for the Activity	15 minutes
Grouping	Teams

What to Do

1. Show the "Share Your ideas" page on a screen using the Presentation mode in the Learning Hub so everyone in the classroom can see it.
2. Read the directions on the handout out loud and then give students five minutes to draw their initial representations of the science.
3. Say, "Okay, will everyone share their representations with the rest of your team? Please make sure everyone has time to talk. Also, be sure to ask one another questions about the scientific ideas that are included in each representation."
4. Give the students about 8-9 minutes to talk and share. Be sure to walk around and listen to the discussions.
5. Say, "Okay, I heard some interesting ideas as I walked around the room."
6. Move on to the next stage.

What to Show

The "Share Your Ideas"page in the ADI Learning Hub

What to Look For

Each student's initial representation should include a combination of symbols, diagrams, graphs, or words. It should also clearly show how a scientific idea can be used to help complete the task and answer the guiding question.

Attending to Equity

When students are tasked with expressing their ideas, be sure to encourage them to do so through multiple modes (speaking, writing, drawings, and gestures) and to use both content-specific and everyday registers. This process helps students fully express themselves, allowing peers to evaluate, question, and build on ideas. Also, make sure that every student in every team has a chance to share their initial representation.

Activity E	More Questions
Time Needed for the Activity	15 minutes
Grouping	Teams

What to Do

1. Show the "More Questions" page on a screen using the Presentation mode in the Learning Hub so everyone in the classroom can see it.
2. Say, "Okay, I heard a bunch of interesting questions as I was walking around the room. Will you work with your team to identify two or three questions that you think are important to learn more about that came up as you all discussed the different representations you made? You can write these questions on your handout."
3. Give students about five minutes to talk and write down questions. Be sure to walk around and listen to the discussions.
4. Say, "Okay, let's take a few minutes to share. Team X (call on a team at random), will you share one question you have?"
5. Have each team share a question. Then say, "I want to share some useful scientific ideas with you. I bet this information will not only answer many of these questions but will also help you complete the task and answer the guiding question of the investigation."
6. Move on to the next stage.

What to Show

The "More Questions" page in the ADI Learning Hub

What to Look For

Each student should have 2-3 questions written on their handout.

Teaching Tip

When students are asked to create representations of the underlying scientific ideas, share these ideas with their peers, and then identify things they want to know more about, it sparks their curiosity and helps make the investigation meaningful. Also, identifying questions before reading (which students do during the Ideas stage) is an outstanding pre-reading activity that helps improve comprehension.

Tissue Repair Lesson Plan

Stage 2 IDEAS

Total Time for the Stage	30 minutes
Grouping	Teams

Learning Goals

Students should make progress on their ability to do the following by the end of this stage of the lesson:

Science

1. Ask questions that arise from examining theory to clarify and/or seek additional information and relationships (SEP1).
2. Critically read scientific literature adapted for classroom use to determine the central ideas or conclusions and/or to obtain scientific and/or technical information to summarize complex concepts, processes, or information presented in a text by paraphrasing them in simpler but still accurate terms (SEP8).
3. Integrate sources of information presented in different media or formats (e.g., visually and quantitatively) as well as in words to address a scientific question (SEP8).
4. Gather, read, and evaluate scientific and/or technical information from multiple authoritative sources, assessing the evidence and usefulness of each source (SEP8).
5. Describe how individual cells in multicellular organisms grow and then divide in a process called mitosis (LS1.B).
6. Use models to simulate systems and interactions within and between systems at different scales (CC4).

English Language Development

1. Paraphrase central ideas in complex evidence, concepts, processes, and information to help explain how or why a phenomenon occurs (ELD-SLC.9-12.Explain.Interpretive).

English Language Arts

1. Determine the central ideas or conclusions of a text; trace the text's explanation or depiction of a complex process, phenomenon, or concept; provide an accurate summary of the text (RST.9-10.2).
2. Determine the meaning of symbols, key terms, and other domain-specific words and phrases as they are used in a specific scientific or technical context (RST.9-10.4).
3. Read and comprehend scientific/technical texts (RST.9-10.10).
4. Initiate and participate effectively in small-group collaborative discussions with diverse partners by building on others' ideas and expressing their own clearly (SL.9-10.1).

Activities

Activity A	Read About a Core Idea
Time Needed for the Activity	10 minutes
Grouping	Teams

What to Do

1. Have students work in their assigned teams during this stage. Give out a copy of the Ideas handout to each student.
2. The first activity provides students with information about cell structures.
3. Show the "Read About a Core Idea You Can Use" page on a screen using the Presentation mode in the Learning Hub so everyone in the classroom can see it. You can also click on the images to make them full screen and easier for students to see.
4. Have students read the text on their own, either on their handouts or on the screen.
5. Give the students about three minutes to discuss the text with the others in their teams and develop a shared list of ideas that they think are important or useful.
6. Ask a team to share an idea they thought was useful. Have the remaining teams share a different idea until there are no more ideas to share.
7. Move on to the next activity.

What to Give Out

Ideas handout for **Tissue Repair**

What to Show

The "Read About a Core Idea You Can Use" page in the ADI Learning Hub

What to Look For

Students should write down some of the following ideas in the box labeled, "Some things I know from what I read" on their handouts:

- All living things on Earth are made up of one or more cells that contain different organelles.
- An organelle is any structure within a cell that carries out a specific function.
- Organelles are suspended in a material called cytoplasm.
- The cell membrane is an organelle that separates the cell's interior from its surroundings and regulates what can and cannot pass into and out of the cell.
- The nucleus is an organelle that contains a cell's genetic information, which tells the cell how to function and reproduce.
- A cell's genetic information is stored in molecules called deoxyribonucleic acid (DNA).
- Each molecule of DNA is coiled tightly and packaged into a chromosome.
- The human body consists of 46 chromosomes, or 23 pairs.

Teaching Tip

Students do not need to do a close read of this text or completely understand all the ideas in it before moving on to the next activity. They will be encouraged to revisit this text later in the investigation when creating their arguments and writing their reports. All students need to be able to do during this activity is to work with the other members of their group to identify one or two ideas that they think are important to keep in mind or are potentially helpful.

Activity B	Read About a Core Idea
Time Needed for the Activity	10 minutes
Grouping	Teams

What to Do

1. The second activity provides students with information about cells creating new cells.
2. Show the "Read About Another Core Idea You Can Use" page on a screen using the Presentation mode in the Learning Hub so everyone in the classroom can see it. You can also click on the images to make them full screen and easier for students to see.
3. Have students read the text on their own, either on their handouts or on the screen.
4. Give the students about three minutes to discuss the text with the others in their teams and develop a shared list of ideas that they think are important or useful.
5. Ask a team to share an idea they thought was useful. Have the remaining teams share a different idea until there are no more ideas to share.
6. Move on to the next activity.

What to Show

The "Read About Another Core Idea You Can Use" page in the ADI Learning Hub

What to Look For

Students should write down some of the following ideas in the box labeled, "Some things I know from what I read" on their handouts:

- During some stages of reproduction, the chromosomes cannot be seen in the cells, but they can be seen in other stages.

Attending to Equity

There are many supports for helping students comprehend what they read already embedded into this activity (e.g., activating prior knowledge, providing a shared experience, making connections, synthesizing, and talking with peers). So, you might not need to provide much extra support. If you are concerned about some students not understanding what they are reading because of their scores on past reading comprehension tests, you can read it out loud as they follow along. As you read the text out loud, be sure to stop at each important idea and ask the students to put a star (or other annotation) next to it in the margin of their handouts. They can then discuss these ideas in their small groups.

Activity C	Read About a Final Core Idea
Time Needed for the Activity	10 minutes
Grouping	Teams

What to Do

1. The third activity provides students with information about using models to represent a process.
2. Show the "Read about a Final Core Idea You Can Use" page on a screen using the Presentation mode in the Learning Hub so everyone in the classroom can see it. You can also click on the images to make them full screen and easier for students to see.
3. Have students read the text on their own, either on their handouts or on the screen.
4. Give the students about three minutes to discuss the text with the others in their teams and develop a shared list of ideas that they think are important or useful.
5. Ask a team to share an idea they thought was useful. Have the remaining teams share a different idea until there are no more ideas to share.
6. Move on to the next activity.

What to Show

The "Read About a Final Core Idea You Can Use" page in the ADI Learning Hub

What to Look For

Students should write down some of the following ideas in the box labeled, "Some things I know from what I read" on their handouts:

- Models are used to represent an object, event, or process.
- Models can be physical, visual, or a computer-based simulation.
- A visual model includes images, symbols, words, and numbers to help explain a process.
- A visual model provides a way to make predictions about what will happen in the future.

Attending to Equity

To support emerging multilingual students during this activity, be sure to create a word bank on the wall of high-frequency, high-need concrete vocabulary that is needed to communicate scientific ideas. The word bank highlights terms needed for basic communication in an academic context. It also helps with the spelling. You can also encourage your emerging multilingual students to:

1. Use strategic learning techniques such as concept mapping, drawing, memorizing, comparing, contrasting, and reviewing.
2. Decode (sound out) words using skills such as recognizing sound-letter relationships and identifying cognates, affixes, roots, and base words.
3. Use visual and contextual support and support from peers and teachers to enhance and confirm understanding of the texts.

Tissue Repair Lesson Plan

Stage 3 PLAN

Total Time for the Stage	15 minutes
Grouping	Teams

Learning Goals

Students should make progress on their ability to do the following by the end of this stage of the lesson:

Science

1. Evaluate a question to determine if it is testable and relevant (SEP1).
2. Plan an investigation individually and collaboratively to produce data to serve as the basis for evidence as part of building and revising models (SEP3).
3. Consider possible variables or effects and evaluate the confounding investigation's design to ensure variables are controlled (SEP3).
4. Ask questions to determine relationships, including quantitative relationships, between independent and dependent variables (SEP1).
5. Decide on types, quantity, and accuracy of data needed and consider limitations on the precision of the data (e.g., number of trials, cost, risk, and time), and refine the design accordingly (SEP3).
6. Select appropriate tools to collect, record, and analyze data (SEP3).

English Language Development

1. Define investigable questions on observations, information, and/or data about a phenomenon (ELD-SLC.9-12.Explain.Interpretive).
2. Develop reasoning to illustrate and/or predict the relationships between variables (ELD-SLC.9-12.Explain.Expressive).

English Language Arts

1. Initiate and participate effectively in small-group collaborative discussions with diverse partners by building on others' ideas and expressing their own clearly (SL.9-10.1).
2. Present information clearly, concisely, and logically such that listeners can follow the line of reasoning and the organization, substance, and style are appropriate to the purpose, audience, and task (SL.9-10.4).
3. Adapt speech to a variety of contexts and tasks, demonstrating command of formal English when indicated or appropriate (SL.9-10.6).

Activities

Activity A	Some Materials You Can Use
Time Needed for the Activity	5 minutes
Grouping	Teams

What to Do

1. Show the "Some Materials You Can Use" page on a screen using the Presentation mode in the Learning Hub so everyone in the classroom can see it. You can also click on the images to make them full screen and easier for students to see.
2. Say, "Okay, it sounds like we now have enough information to plan an investigation that will help us figure out the properties of the two unknown elements based on the properties of the 30 known elements."
3. Say, "Before we get started, I want to show you some materials that we are going to use during our investigation. Will you all please come over to this table so I can show you what we have?"
4. Direct students to stand around the table where all the materials are set up.
5. Hold up the five event cards.
6. Ask, "What do you think we could use these for?"
7. Allow students to share ideas. Then say, "We can use these cards to see what cells look like at different stages of reproduction."
8. Hold up the red pop beads, yellow pop beads, and hook-and-loop cable ties.
9. Ask, "What do you think we could use these for?"
10. Allow students to share ideas. Then say, "We can use these to make a model of a chromosome."
11. Hold up the plastic cups.
12. Ask, "What do you think we could use these for?"
13. Allow students to share ideas. Then say, "We can use these to hold the extra beads and ties."
14. Say, "Okay, that is all the stuff we can use. Will you all head back to your teams so we can plan our investigation?"
15. Move on to the next activity.

What to Set Up

Place the following materials on a table so students can see them:

1. Event A-E cards
2. Red pop beads
3. Yellow pop beads
4. Hook-and-loop cable ties
5. Plastic cups

What to Show

The "Some Materials You Can Use" page in the ADI Learning Hub

Teaching Tip

This activity helps students see for themselves all the materials they will be able to use during the hands-on part of the investigation instead of just reading about them or looking at pictures. It is important to show each material to the students and then point out how they can use it during the investigation so they can participate in developing a plan during the next activity.

Activity B	Make a Plan
Time Needed for the Activity	10 minutes
Grouping	Teams

What to Do

1. Have students work in their assigned teams during this activity. Give a copy of the Plan handout to each student.

2. Show the "Make a Plan" page on a screen using the Presentation mode in the Learning Hub so everyone in the classroom can see it.

3. Say, "Work as a team to create a plan for your investigation by filling out the chart on your handout. As part of your plan, be sure to include which tools and techniques you will use to solve problems."

4. Give the students at least 10 minutes to complete this task. As students work in small teams to create their investigation proposals, be sure to move from team to team to answer questions and offer advice.

5. When a team thinks that they have a good plan, read it over. If you agree, sign off at the bottom of their handouts to let them know that their plan has been approved. If you do not agree, tell them what they will need to add or change before you will approve it. Once they have made the changes, you can sign their handouts.

6. Once all the teams are done, you can move on to the next activity. You can also use an approved investigation plan as an "exit ticket" out of class.

What to Give Out

Plan handout for **Tissue Repair**

What to Show

The "Make a Plan" page in the ADI Learning Hub

What to Look For

The investigation plan should, at a minimum, include:

- The question they are trying to answer.
- How they will set up their materials.
- What data they will collect using their setup.
- How they will analyze the data they collect.
- How they will use their data to answer the guiding question.

Teaching Tip

One way to do a quick check-in with a group during this activity is to use the "two questions, a nudge, and a goal" technique. This technique includes the following steps:

1. Sit down with a group of students.
2. Ask, "Would one of you tell me a little about what you all have been doing or thinking about?" (Question 1)
3. A student answers on behalf of the group.
4. Ask, "Would anyone else like to add anything to that?" (Question 2)
5. A different student provides additional information on behalf of the group.
6. Ask, "Have you thought about [a hint or something you want to highlight]?" (A nudge)
7. Say, "Okay, I'm going to check on the other groups. When I get back, I want to see [something that you want the group to accomplish]." (A goal)

When you have completed the check-in with one group, move on to the next group.

Attending to Equity

To support emerging multilingual students during this stage, be sure to create a word bank on the wall of high-frequency, high-need concrete vocabulary that is needed to communicate scientific processes and ideas. The word bank highlights terms needed for basic communication in an academic context. It also helps with the spelling. Also, you can encourage your emerging multilingual students to use modes (speaking, writing, drawings, and gestures) and to use both content-specific and everyday registers to discuss and create their plans. This process helps students fully express themselves, allowing peers to evaluate, question, and build on ideas.

Teacher Notes

Tissue Repair Lesson Plan

Stage 4 DO

Total Time for the Stage	45 minutes
Grouping	Teams

Learning Goals

Students should make progress on their ability to do the following by the end of this stage of the lesson:

Science

1. Conduct an investigation individually and collaboratively to produce data to serve as the basis for evidence (SEP3).
2. Analyze data using tools and/or technologies to make valid and reliable scientific claims (SEP4).
3. Consider limitations of data analysis (e.g., measurement error and sample selection) when analyzing and interpreting data (SEP4).
4. Evaluate the impact of new data on a working explanation and/or model of a proposed process or system (SEP4).
5. Make a quantitative and/or qualitative claim regarding the relationship between dependent and independent variables (SEP6).
6. Construct and revise an explanation based on valid and reliable evidence obtained from a variety of sources and the assumption that theories and laws that describe the natural world operate today as they did in the past and will continue to do so in the future (SEP6).
7. Apply scientific ideas, principles, and/or evidence to provide an explanation of phenomena taking into account possible unanticipated effects (SEP6).

English Language Development

1. Paraphrase central ideas in complex evidence, concepts, processes, and information to help explain how or why a phenomenon occurs (ELD-SLC.9-12.Explain.Interpretive).
2. Evaluate the extent to which reasoning, theory and/or models link evidence to claims and support conclusions (ELD-SLC.9-12.Explain.Interpretive).
3. Develop reasoning to illustrate the relationships between variables (ELD-SLC.9-12.Explain.Expressive).

English Language Arts

1. Initiate and participate effectively in small-group collaborative discussions with diverse partners by building on others' ideas and expressing their own clearly (SL.9-10.1).
2. Present information clearly, concisely, and logically such that listeners can follow the line of reasoning and the organization, substance, and style are appropriate to the purpose, audience, and task (SL.9-10.4).
3. Adapt speech to a variety of contexts and tasks, demonstrating command of formal English when indicated or appropriate (SL.9-10.6).

Activities

Activity A	Carry Out Your Plan
Time Needed for the Activity	25 minutes
Grouping	Teams

What to Do

1. Students should work in their assigned teams for this activity. Begin this activity by giving out a copy of the Do handout to each student.
2. Show the "Carry Out Your Plan" page on a screen using the Presentation mode in the Learning Hub so everyone in the classroom can see it.
3. Say, "Okay, it sounds like we are now ready to carry out our plans and collect the data or observations we need to answer the guiding question. You can get the materials and tools you need to use from this table." Point to the table where you have placed them.
4. Ask, "Does anyone have any questions about these materials or tools?"
5. Answer any questions that come up.
6. Say, "As you work, be sure to record your observations and measurements on your handout so you can keep track of everything you do. Does anyone have any final questions?"
7. Answer any questions that come up.
8. Say, "Okay, go ahead and get started. You have the next 25 minutes to work. I'll be available to help if any team needs me."
9. Give the students at least 25 minutes to carry out their plans and solve the problem. As students work in small teams, be sure to move from team to team to answer questions and offer advice.

What to Set Up

Place the following materials on a table so each group can use them:

1. Event A-E cards
2. Red pop beads
3. Yellow pop beads
4. Hook-and-loop cable ties
5. Plastic cups

Keep the sixth card, Appearance of Chromosomes, on hand in case students need a hint.

What to Give Out

Do handout for **Tissue Repair**

What to Show

The "Carry Out Your Plan" page in the ADI Learning Hub

Teaching Tip

One way to do a quick check-in with a group during this activity is to use the "two questions, a nudge, and a goal" technique. This technique includes the following steps:

1. Sit down with a group of students.
2. Ask, "Would one of you tell me a little about what you all have been doing or thinking about?" (Question 1)
3. A student answers on behalf of the group.
4. Ask, "Would anyone else like to add anything to that?" (Question 2)
5. A different student provides additional information on behalf of the group.
6. Ask, "Have you thought about [a hint or something you want to highlight]?" (A nudge)
7. Say, "Okay, I'm going to check on the other groups. When I get back, I want to see [something that you want the group to accomplish]." (A goal)

When you have completed the check-in with one group, move on to the next group.

Attending to Equity

Be sure to intentionally group emerging multilingual students at times with peers who know the same languages as they do and at other times with peers whose English language development is slightly more advanced. Thoughtful grouping that varies during different investigation (or activities within an investigation) allows emerging multilingual students to benefit from the different linguistic resources of their peers. We also recommend assessing emerging multilingual students based on how much they contribute to the group during this activity, rather than solely on what they write or draw on their handouts.

Teacher Notes

Activity B	Make Sense of Your Data
Time Needed for the Activity	20 minutes
Grouping	Teams

What to Do

1. Students should work in their assigned teams for this activity. Begin this activity by showing the "Make Sense of Your Data" page on a screen using the Presentation mode in the Learning Hub so everyone in the classroom can see it.
2. Say, "Okay, it sounds like we are now ready to make sense of the data we collected. One way to do this is to create a table, graph, or other visualization that shows a trend or change over time, a comparison between teams, or a relationship between things you tested. You can create a table, graph, or other visualization here on your handout."
3. Point to the box on the handout.
4. Say, "Please work together in your teams to figure out the best way to make sense of your data. Remember, you are looking for any trends, changes, differences, or relationships in the data that will help you answer the guiding question. As you work, be sure to use the space on your handout to perform any calculations that you need to do and create any representations that you need. Does anyone have any last questions?"
5. Answer any questions that come up.
6. Say, "Okay, go ahead and get started. You have the rest of the period to work. I'll be available to help if any team needs me."
7. Give the students at least 15 minutes to make sense of their data. As students work together in small teams, be sure to move from team to team to answer questions and offer advice.
8. Make sure students return the materials and tools to the table where they found them before dismissing class. Cleanup should take no longer than five minutes.

What to Show

The "Make Sense of Your Data" page in the ADI Learning Hub

What to Look For

One of the most useful ways to make sense of the data during this activity is to make a diagram or model that shows the different phases of cellular reproduction and the location of the chromosomes.

Teaching Tip

Collecting and returning materials or tools to a central table works best if you assign the job to a specific number within each team. For example, you can say, "2s, please come up and grab the materials and tools that your team needs" at the start of the activity. Then you can say, "3s, please return the materials and tools your team used back to this table," at the end of the activity. Assigning jobs like this not only cuts down on how many students need to be up and moving around, but also allows you to hold specific students accountable for the materials and tools.

Attending to Equity

Encourage your emerging multilingual students to speak using learning strategies such as requesting assistance, employing nonverbal cues, and using synonyms and circumlocution (conveying ideas by defining or describing when exact English words are not known) during this activity. It is also helpful to point out and/or draw emerging multilingual learners' attention to elements of the English sound system in newly acquired vocabulary such as long and short vowels, silent letters, and consonant clusters as they come up naturally during this activity.

Teacher Notes

Tissue Repair Lesson Plan

Stage 5 SHARE

Total Time Needed for the Stage	45 minutes
Grouping	Teams

Learning Goals

Students should make progress on their ability to do the following by the end of this stage of the lesson:

Science

1. Make and defend a claim based on evidence about the natural world that reflects scientific knowledge and student-generated evidence (SEP7).
2. Present an oral argument based on evidence (SEP7).
3. Evaluate the claims, evidence, and/or reasoning to determine the merits of arguments (SEP7).
4. Ask questions to clarify relationships, including quantitative relationships, between independent and dependent variables (SEP1).
5. Ask questions that challenge the interpretation of a data set or the suitability of a method (SEP1).
6. Respectfully give critiques on scientific arguments by probing reasoning and evidence and challenging ideas and conclusions (SEP7).
7. Respectfully receive critiques on scientific arguments by responding thoughtfully to diverse perspectives and determining what additional information is required to resolve contradictions. (SEP7).
8. Communicate scientific information or ideas in multiple formats including orally, graphically, textually, and mathematically (SEP8).

English Language Development

1. Identify appropriate and sufficient evidence from data, models, and/or information from investigations of a phenomenon (ELD-SLC.9-12.Argue.Interpretive).
2. Compare reasoning and claims based on evidence from competing arguments (ELD-SLC.9-12.Argue.Interpretive).
3. Defend or refute a claim based on data and evidence (ELD-SLC.9-12.Argue.Expressive).
4. Establish and maintain an appropriate tone and stance (ELD-SLC.9-12.Argue.Expressive).
5. Signal logical relationships among reasoning, evidence, data, and/or models when making and defending a claim, counterclaim, and/or rebuttal (ELD-SLC.9-12.Argue.Expressive).

English Language Arts

1. Initiate and participate effectively in small-group collaborative discussions with diverse partners by building on others' ideas and expressing their own clearly (SL.9-10.1).
2. Integrate multiple sources of information presented in diverse media or formats (e.g., visually, quantitatively, and orally) evaluating the credibility and accuracy of each source (SL.9-10.2).
3. Evaluate a speaker's reasoning and use of evidence and rhetoric, identifying any fallacious reasoning or exaggerated or distorted evidence (SL.9-10.3).

(continued)

4. Present information clearly, concisely, and logically such that listeners can follow the line of reasoning and the organization, substance, and style are appropriate to the purpose, audience, and task (SL.9-10.4).
5. Make strategic use of media (e.g., textual, graphical, and visual) in presentations to enhance understanding of findings, reasoning, and evidence and to add interest (SL.9.10.5).
6. Adapt speech to a variety of contexts and tasks, demonstrating command of formal English when indicated or appropriate (SL.9-10.6).

Activities

Activity A	Create a Draft Argument
Time Needed for the Activity	15 minutes
Grouping	Teams

What to Do

1. Students should work in their assigned teams for this activity. Begin this activity by giving out a copy of the Share handout to each student.
2. Show the "Create a Draft Argument" page on a screen using the Presentation mode in the Learning Hub so everyone in the classroom can see it.
3. Say, "Okay, it sounds like we have all solved the task. I would now like you to create a draft argument to share what you figured out with the other teams. Your argument should include everything in this image here." Point to the image of a science-based argument projected on the screen. Be sure to go over what should be included in each component of the argument and where students can find the information they need.
4. Say, "I would like you to make your draft argument on a whiteboard. You can use the materials on this table to make your argument." Point to the table where you have placed the materials they need.
5. Ask, "Does anyone have any questions?"
6. Answer any questions that come up.
7. Say, "I'm going to give you exactly 15 minutes to get this done. Your draft argument doesn't need to be perfect. We just need something down on the whiteboards so we can share our ideas."
8. Give each team of students 15 minutes to create a draft argument. As students work in small teams, be sure to move from team to team to answer questions and offer advice.

What to Set Up

Place the following items on a table so students can use them during this stage:

- 24″ x 36″ whiteboard
- 3-4 dry-erase pens in different colors
- Eraser or cloth towel

What to Give Out

Share handout for **Tissue Repair**

What to Show

The "Create a Draft Argument" page in the ADI Learning Hub

What to Look for

Students should include core ideas that they read about during the Ideas stage in their justification.

Teaching Tip

Be sure to plant some different ideas with each group about how to make a strong argument as you move from group to group to check in, answer questions, and offer advice. This strategy helps to ensure that there will be some really strong examples of the evidence and justification components of the argument for students to see and discuss during the next activity. For example, you could help two groups have a really strong evidence component by telling one group, "I think a graph would be really helpful here" and telling another group, "I think your evidence would be stronger if you explained, in words, what this graph shows." You could help a third group have a really strong justification by saying, "Let's brainstorm some ideas that will help you defend your choice of evidence," or "Let's go back to our Ideas handout and see if there are any concepts we can include."

It is important you make sure that all the arguments are different at this point in the investigation by planting different ideas with different groups because the students will have nothing to critique or discuss during the next activity if they all have the same argument. The goal during this activity is to set the stage for the next activity by ensuring that the various components of a strong argument can be found on different whiteboards within the class. These strong components will then be taken up by all the groups because they reflect good ideas. Be sure to give students a voice and choice during this activity, so they can make mistakes and learn from them.

Attending to Equity

It is often helpful to assign and change roles at least once during this activity. For example, you might assign the 1s to be "in charge of writing" at the beginning of the activity. You can then easily switch roles about five minutes into the activity by saying, "1s, please hand the pens to the 2s. 2s, you are now in charge of writing for the group." By changing roles, you can ensure that all students have a chance to participate and contribute their ideas to the creation of a draft argument.

ADI Go Digital

The **ADI Learning Hub** has tons of features that allow students to share their thinking with each other when they are doing the investigation paperless or remote.

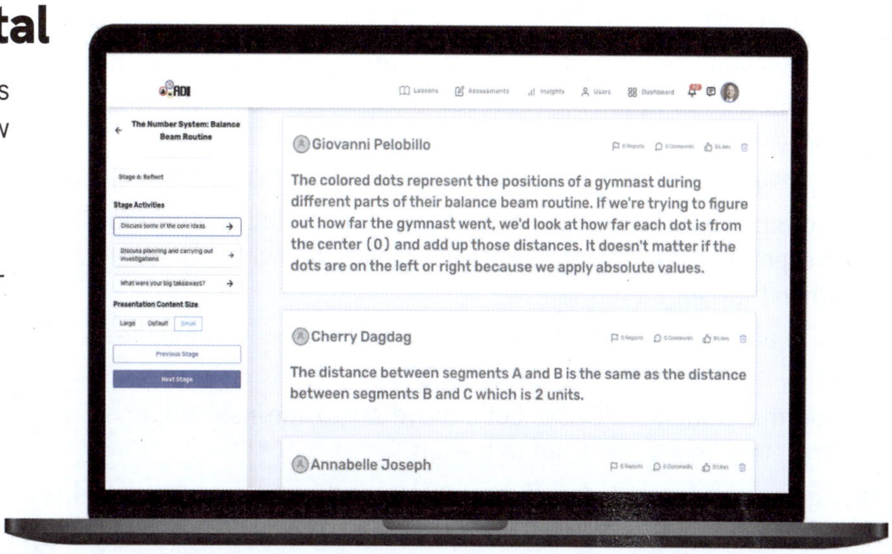

Activity B	Present Your Draft Argument
Time Needed for the Activity	15 minutes
Grouping	Teams

What to Do

1. Students should work in their assigned teams for this activity.
2. Show the "Present Your Draft Argument" page on a screen using the Presentation mode in the Learning Hub so everyone in the classroom can see it.
3. Say, "We are now going to share our arguments. Please set up your whiteboards so everyone can see them." Allow students time to set up their whiteboards. They should be set up on stands or chairs, so they are facing toward the outside of the room.
4. Say, "Okay, before we get started, I want to explain what we are going to do next. I'm going to ask some of you to present your arguments to your classmates. If you are presenting your argument, your job is to share your team's claim, evidence, and justification of the evidence. The rest of you will be travelers. If you are a traveler, your job is to listen to the presenters, ask them questions if you do not understand something, and then offer them some suggestions about ways to make their argument better. Do you have any questions about these instructions?"
5. Answer any questions that come up.
6. Say, "As you talk to one another, make sure you write down any good ideas or suggestions that are important in the space on your handout, so you don't forget about them." Point to the box labeled, "Some possible ways to improve our argument" on the handout.
7. Say, "After we have a chance to learn from one another, I'm going to give you some time to revise your arguments and make them better. Do you have any questions about what we are doing?"
8. Answer any questions that come up.
9. Say, "Okay. Let's get started. Travelers, move one team to the left. Don't move to the next team until I tell you to move. Presenters, go ahead and share your argument with the travelers when they get there."
10. Give each team of presenters and travelers about 3-4 minutes to talk.
11. Tell the students, "Okay. Let's move on to the next argument. Travelers, move one team to the left."
12. Again, give each team of presenters and travelers about 3-4 minutes to talk.

What to Show

The "Present Your Draft Argument" page in the ADI Learning Hub

What to Look For

The critiques and feedback that students give one another during this activity should:

- Be shared in a way that is kind (respectful),
- Provide information about what, exactly, needs to be changed or improved (specific), and
- Offer a suggestion about how to make something better rather than just saying something is wrong (helpful).

In addition, students should be focusing their critiques on the science included in the arguments, not the appearance of the arguments.

(continued)

13. Repeat this process until each team has had their argument read and critiqued three times. Then tell the students to return to their workstations.
14. Be sure to move between workstations during this activity to keep everyone involved and model what it means to communicate and critique scientific ideas and reasoning.

Attending to Equity

Be sure to encourage students to provide feedback that is respectful, specific, and helpful during this activity. It is important to encourage student-to-student talk because it helps students develop speaking and listening skills and enables them to make sense of what and how they know. It also provides a context for emerging multilingual students to use both productive (speaking and writing) and receptive (listening and reading) language and learn science through meaningful interactions and scientific sense-making.

Teaching Tip

There are two types of formats for sharing and critiquing arguments. The first format is called a *gallery walk.* In this format, each group sets up their argument at their workstation so others can see it. The entire group then moves to a different workstation. While at a different workstation, the group is expected to read the argument at that workstation (which was created by a different group) and offer critiques and suggestions for improvement. The students should move to a different workstation after a few minutes to read and critique an argument written by a different group. Students should also be encouraged to keep a record of good ideas or potential ways to improve their own arguments as they move from workstation to workstation to offer critiques and feedback. After repeating this process in the gallery walk 3-4 times, every group will have had an opportunity to read and critique three or four different arguments and receive feedback from three or four different groups.

The second format is called a *modified gallery walk.* In the modified gallery walk format, one or two members of the group stay at their workstation to share their group's ideas (these students are called presenters) while the other members of their group go to different groups, one group at a time, to listen to and critique the arguments developed by their classmates (these students are called travelers). The travelers should visit at least three different workstations during the argumentation session. We also recommend that the presenters keep a record of the critiques made by their classmates and any suggestions for improvement. The travelers should also be encouraged to keep a record of good ideas or potential ways to improve their own arguments as they travel from group to group.

Attending to Equity

Another good way to have students, especially emerging multilingual learners, give one another feedback is to have them use four different color sticky notes to make comments. Give each group a single pink, blue, yellow, and green sticky note. Then direct the students to use the pink sticky note to suggest "something to add" to the argument, the blue sticky note to suggest "something to change" in the argument, the yellow sticky note to highlight "something they like" in the argument, and the green sticky note to ask a question about "something they don't understand". This strategy is useful because everyone in the classroom will be actively engaged during this activity, and everyone will have a chance to see different arguments. This format, however, does not give students a chance to support, critique, and challenge the ideas, solutions, scientific reasoning of other groups through talk because two different groups are never at the same workstation at the same time.

Activity C	Revise Your Argument
Time Needed for the Activity	15 minutes
Grouping	Teams

What to Do

1. Students should work in their assigned teams for this activity.
2. Show the "Revise Your Argument" page on a screen using the Presentation mode in the Learning Hub so everyone in the classroom can see it.
3. Say, "I'm now going to give you all about 10 minutes to revise your argument. Please return to your original teams."
4. Wait for the students to move back into their original teams.
5. Say, "Okay, take a few minutes to talk in your teams and determine what you want to change to make your argument better. Once you have decided what to change, go ahead and make the changes to your whiteboard."
6. Ask, "Do you have any questions about what you need to do?"
7. Answer any questions that come up.
8. Tell the students, "Okay. Let's get started."
9. Give the students 10 minutes to work in their teams on their arguments. As they work, move from team to team to check in, ask probing questions, and offer a suggestion if a team gets stuck.
10. When students are finished with their final arguments, have them take a picture of it or have them put the boards in a safe spot so they can refer to the picture or the boards during the Report stage.

What to Show

The "Revise Your Argument" page in the ADI Learning Hub

What to Look For

A strong argument includes the following information:

- A claim that states something similar to, "New cells are created in a process called mitosis where old cells replicate their genetic material and then divide into a new and an old cell."
- A diagram or model that shows the different phases of cellular reproduction and the location of the chromosomes.
- An explanation of the diagram or model.
- An explanation of the chosen evidence that uses core ideas from the investigation.

Teaching Tip

It is important to give students time to meet with their original groups so they can discuss what they learned by interacting with individuals from the other groups. At this point in the process, students often realize that their initial problem-solving model was flawed in some way. It is therefore important to encourage students to "redo the science" as needed. At the end of this activity, each group should have a final argument that is much better than their original one.

Tissue Repair Lesson Plan

Stage 6 REFLECT

Total Time for the Stage	25 minutes
Grouping	Whole class, Teams

Learning Goals

Students should make progress on their ability to do the following by the end of this stage of the lesson:

Science

1. Ask questions that arise from careful observation of phenomena, or unexpected results, to clarify and/or seek additional information as needed (SEP1).
2. Describe how individual cells in multicellular organisms grow and then divide in a process called mitosis (LS1.B).
3. Use models to simulate systems and interactions within and between systems at different scales (CC4).

English Language Development

1. Paraphrase central ideas in complex evidence, concepts, processes, and information to help explain how or why a phenomenon occurs (ELD-SLC.9-12.Explain.Interpretive).
2. Use scientific explanations about a phenomenon in a different context or situation (ELD-SLC.9-12.Explain.Expressive).

English Language Arts

1. Participate effectively in small-group collaborative discussions with diverse partners on texts by building on others' ideas and expressing their own clearly and persuasively (SL.9-10.1)

Activities

Activity A	Discuss a Core Idea
Time Needed for the Activity	10 minutes
Grouping	Whole class/Teams

What to Do

1. Students should be in small teams for this activity. However, they do not need to be in the same teams as they were during previous stages.
2. Begin this activity by giving out a copy of the Reflect handout to each student.
3. Show the "Discuss a Core Idea" page on a screen using the Presentation mode in the Learning Hub so everyone in the classroom can see it.
4. Say, "Okay. Now that we have all revised our arguments and made them better, let's take some time to discuss some of the core ideas you used during this investigation and to justify the evidence in your arguments so we are all on the same page. Look at the picture on the screen."
5. Pause for about 30 seconds.
6. Say, "Take a few minutes to talk in your teams and see if you can use what you know about cell structures and cells creating new cells to explain what is happening in the picture."
7. Give the students at least three minutes to talk in their small teams.
8. Say, "Okay. Let's share with the whole class. Team A (or whichever team you want to share first), what did you talk about?"
9. Allow the first team to share.
10. Say, "Interesting. Would anyone like to add to, challenge, or refine the ideas shared by Team A?"

What to Give Out

Reflect handout for **Tissue Repair**

What to Show

The "Discuss a Core Idea" page in the ADI Learning Hub

Explanation of the Image

The image that the students discuss during this activity shows a sample of cells in different stages of mitosis.

What to Look For During the Discussion

Students who understand the content will likely mention some, if not all, of the following ideas when providing an explanation for the image:

- All living things on Earth are made up of one or more cells that contain different organelles.
- An organelle is any structure within a cell that carries out a specific function.
- Organelles are suspended in a material called cytoplasm.
- The cell membrane is an organelle that separates the cell's interior from its surroundings and regulates what can and cannot pass into and out of the cell.
- The nucleus is an organelle that contains a cell's genetic information, which tells the cell how to function and reproduce.
- A cell's genetic information is stored in molecules called deoxyribonucleic acid (DNA).
- Each molecule of DNA is coiled tightly and packaged into a chromosome.
- The human body consists of 46 chromosomes, or 23 pairs.

(*continued*)

11. Allow students to respond.
12. Say, "Okay. Team B (or whichever team you want to share second), what did you talk about?"
13. Allow the second team to share.
14. Say, "Interesting. Would anyone like to add to, challenge, or refine the ideas shared by Team B?"
15. Allow students to respond.
16. Keep up this pattern of questioning until all teams have shared. Be sure to keep encouraging students to talk about the content until they reach a correct and consensual explanation for what is shown in the image. If they are unable to reach a consensus or reach consensus using an inaccurate idea, be sure to provide an accurate explanation for the image before moving on to the next activity.

Attending to Equity

Be sure to intentionally group emerging multilingual students at times with peers who know the same languages as they do and at other times with peers whose English language development is slightly more advanced. Also, be sure to ask students where they have seen the core idea in their everyday lives; this will make the topic of discussion more engaging and relatable.

Teaching Tip

This activity is a discussion, not a presentation about what the students "should have seen" or "should have learned" during the investigation. The more students talk during this activity, the more meaningful the experience will be for them and the more you can learn about their thinking. Be sure to use talk goals and moves to help facilitate the discussion. Below are some examples of talk goals that you can use to help facilitate the discussion and some questions that you can ask the students to help accomplish each goal. These talk goals and questions are only suggestions.

Goal 1: Encourage students to share, expand on, or clarify an idea.

- What do you see going on here?
- What do you think is happening?
- How might we explain that?
- Can you tell me more about that?
- What do you mean by that?
- Can you give me an example of that?
- Can you say that in another way?

Goal 2: Encourage students to support an idea.

- Why do you think that?
- What is your evidence for that?
- How did you arrive at that conclusion?
- What makes you say that?
- How do you know that?

Goal 3: Encourage students to think with others.

- Would anyone like to add to that idea?
- What should we keep or change about that idea?
- Can anyone else give me some examples of that?
- Can anyone take that idea and push it a little further?
- Does anyone want to respond to that idea?
- Do you agree or disagree with that idea, and why?
- What do we think about that idea?
- Does that fit with what we have figured out so far?
- Is there another way to do it?

Goal 4: Encourage students to think about what they know and don't know yet.

- I'm confused now. What have we figured out so far?
- What do we need to figure out now?
- So, what do we agree on right now, and what are we still trying to figure out?
- What are we sure and not so sure about?
- What questions do you have?

Activity B	Discuss the Practices
Time Needed for the Activity	10 minutes
Grouping	Whole class/Teams

What to Do

1. Show the first page of "Discuss the Practices" on a screen using the Presentation mode in the Learning Hub so everyone in the classroom can see it.
2. Say, "Okay. Now that we have talked about some of the core ideas that we used during our investigation, let's take a few minutes to talk about the science practices that we used during the investigation. An example of a scientific practice used was analyzing and interpreting data. On the screen there is a question that I want you all to think about: What other things did you do during your investigation that you feel were good examples of a scientific practice or attempting to figure out something in a scientific way?"
3. Pause for about 30 seconds.
4. Say, "Take a few minutes to talk about this question in your teams."
5. Give the students at least three minutes to talk in their small teams.
6. Say, "Okay. Let's share with the whole class. Team A (or whichever team you want to share first), what did you talk about?"
7. Allow the first team to share.
8. Say, "Okay. Team B (or whichever team you want to share second), what did you talk about?"
9. Allow the second team to share.
10. Continue asking teams until every team has a chance to share.
11. Show the second page of "Discuss the Practices" on a screen using the Presentation mode in the Learning Hub so everyone in the classroom can see it.
12. Say, "Here is another question that I want you all to think about: What things did you do during your investigation that were not good examples of a scientific practice or not very scientific?"
13. Pause for about 30 seconds.

What to Show First

The first page of "Discuss the Practices" in the ADI Learning Hub

What to Show Second

The second page of "Discuss the Practices" in the ADI Learning Hub

What to Show Third

The third page of "Discuss the Practices" in the ADI Learning Hub

(continued)

14. Say, "Take a few minutes to talk about this question in your team."
15. Give the students at least three minutes to talk in their small teams.
16. Say, "Okay. Let's share with the whole class. Team A (or whichever team you want to share first), what did you talk about?"
17. Allow the first team to share.
18. Say, "Okay. Team B (or whichever team you want to share second), what did you talk about?"
19. Allow the second team to share.
20. Continue asking teams until every team has a chance to share.
21. Show the third page of "Discuss the Practices" on a screen using the Presentation mode in the Learning Hub so everyone in the classroom can see it.
22. Say, "Here is a final question that I want you all to think about: What should we all do during our next investigation to make sure we are using scientific practices and being scientific in the way we try to figure something out?"
23. Pause for about 30 seconds.
24. Say, "Take a few minutes to talk about this question in your team."
25. Give the students at least three minutes to talk in their small teams.
26. Say, "Okay. Let's share with the whole class. Team A (or whichever team you want to share first), what did you talk about?"
27. Allow the first team to share.
28. Say, "Okay. Team B (or whichever team you want to share second), what did you talk about?"
29. Allow the second team to share.
30. Continue asking teams until every team has a chance to share.
31. Once everyone has shared, offer any suggestions or advice about what they can do to improve their next investigation and then move on to the next stage.

Teaching Tip

A great way to wrap up this activity is for the students to contribute to a list of class norms or "gotta-haves" to keep in mind when planning and carrying out an investigation in science. To create this list, each group can suggest for a potential norm or "gotta-have" to add to the list at the end of this activity, then the class votes on the one that they think is most important. The most important one gets added to the list. The list then serves as reminder for students during the Plan and Do stages of subsequent investigations. Over time, the list will grow and each addition to the list will be more nuanced than the last as students develop expertise with scientific processes.

Attending to Equity

Be sure to give all students, especially emerging multilingual students, enough time to talk in small groups before asking anyone to share with the entire class during this activity. Small-group structures offer all students, including emerging multilingual students, a chance to engage in sense-making with their peers as well as space to use their verbal and nonverbal (e.g., writing, drawing pictures, and using gestures) resources to express their ideas. Small-group structures also give students a chance to learn from other students' uses of these resources. When students are tasked with expressing their ideas, be sure that they know is okay for them to do so in a mode or combination of modes (speaking, writing, drawings, and gestures).

Tissue Repair Lesson Plan

Stage 7 REPORT

Total Time Needed for the Stage	100 minutes (over three days)
Grouping	Individual, Teams

Learning Goals

Students should make progress on their ability to do the following by the end of this stage of the lesson:

Science

1. Use a model to illustrate the role of cellular division (mitosis) and differentiation in producing and maintaining complex organisms (HS-LS1-4).
2. Make and defend a claim based on evidence about the natural world that reflects scientific knowledge and student-generated evidence (SEP7).
3. Construct a written argument based on evidence (SEP7).
4. Evaluate the claims, evidence, and/or reasoning to determine the merits of arguments (SEP7).
5. Respectfully give critiques on scientific arguments by and challenging ideas and conclusions (SEP7).
6. Communicate scientific information or ideas in multiple formats including graphically and textually (SEP8).

English Language Development

1. Identify appropriate and sufficient evidence from data, models, and/ or information from investigations of a phenomenon (ELD-SLC.9-12.Argue.Interpretive).
2. Compare reasoning and claims based on evidence from competing arguments (ELD-SLC.9-12.Argue.Interpretive).
3. Defend or refute a claim based on data and evidence (ELD-SLC.9-12.Argue.Expressive).
4. Establish and maintain an appropriate tone and stance (ELD-SLC.9-12.Argue.Expressive).
5. Signal logical relationships among reasoning, evidence, data, and/or models when making and defending a claim (ELD-SLC.9-12.Argue.Expressive).

English Language Arts

1. Write routinely over extended time frames (time for reflection and revision) and shorter time frames for a range of discipline-specific tasks, purposes, and audiences (WHST.9-10.10).
2. Write arguments focused on discipline-specific content (WHST.9-10.1).
3. Write informative/explanatory texts, including the narration of scientific procedures/ experiments or technical process (WHST.9-10.2).
4. Produce clear and coherent writing in which the development, organization, and style are appropriate to task, purpose, and audience (WHST.9-10.4).
5. Strengthen writing as needed by revising, editing, rewriting, and focusing on addressing what is most significant for a specific purpose and audience (WHST.9-10.5).

(continued)

6. Determine the central ideas or conclusions of a text; trace the text's explanation or depiction of a complex process, phenomenon, or concept; provide an accurate summary of the text (RST.9-10.2).
7. Determine the meaning of symbols, key terms, and other domain-specific words and phrases as they are used in a specific scientific or technical context (RST.9-10.4).
8. Analyze the structure of the relationships among concepts in a text, including relationships among key terms (RST.9-10.5).
9. Assess the extent to which the reasoning and evidence in a text support the author's claim (RST.9-10.8).
10. Compare and contrast findings presented in a text to those from other sources (including their own experiments), noting when the findings support or contradict other accounts (RST.9-10.9).

Activities

Activity A	Draft Report
Time Needed for the Activity	20 minutes
Grouping	Individual

What to Do

1. Have students work on their own during this activity. Give a copy of the Report handout to each student.
2. Show the "Create a Draft Report" page on a screen using the Presentation mode in the Learning Hub so everyone in the classroom can see it.
3. Say, "You are now going to write an investigation report to share what you figured out. The first part of the report is called the 'Introduction'. In this section, you should explain to the reader what you were investigating, why you were investigating it, and what question you were trying to answer. All this information can be found on your Task handout."
4. Hold up the Task handout.
5. Ask the students, "Do you have any questions about what you need to do?"
6. Answer any questions that come up.
7. Say, "Here are some sentence starters to help you begin writing."
8. Hold up the Report handout and point to the sentence starters.
9. Tell the students, "Okay, let's write."
10. Give the students five minutes to write the "Introduction" section of the report. As they work, move from student to student to check in, ask probing questions, and offer a suggestion if a student gets stuck.

What to Give Out

Report handout for **Tissue Repair**

What to Show

The "Create a Draft Report" page in the ADI Learning Hub

(continued)

11. Say, "Okay, let's move on to the second part of the report. The second part of the report is called the 'Method'. In this section, you should explain to the reader what you did during the investigation to complete the task and answer the guiding question. All this information can be found in the Plan handout. Remember that you all planned and carried out different investigations, so do not assume that the reader will know what you did."
12. Ask the students, "Do you have any questions about what you need to do?"
13. Answer any questions that come up.
14. Say, "Here are some sentence starters to help you begin writing."
15. Hold up the Report handout and point to the sentence starters.
16. Tell the students, "Okay, let's write."
17. Give the students five minutes to write the "Method" section of the report. As they work, move from student to student to check in, ask probing questions, and offer a suggestion if a student gets stuck.
18. Say, "Okay, let's move on to the last part of the report. The third part of the report is called the 'Argument'. In this section, you should explain to the reader what you figured out and how you know you are correct. All this information can be found on your Do handout and on your team's whiteboard."
19. Hold up the Do handout and point to the boards (or remind them that they have a picture of their board).
20. Ask the students, "Do you have any questions about what you need to do?"
21. Answer any questions that come up.
22. Say, "Here are some sentence starters to help you begin writing."
23. Hold up the Report handout and point to the sentence starters.
24. Tell the students, "Okay, let's write."
25. Give the students 10 minutes to write the "Argument" section of the report. As they work, move from student to student to check in, ask probing questions, and offer a suggestion if a student gets stuck.
26. Collect all the draft reports before moving on to the next activity. Tell students to mark their draft reports with a symbol that they can use to identify it later. They should NOT write their names on their draft reports.

Teaching Tip

Writing is an important component of this model. When a student writes the report on their own, it not only helps the student learn to write in the context of science (an important literacy skill) but also allows that student to write to learn (develop a better understanding of scientific concepts by writing about them). There are many writing supports (e.g., sentence starters) already embedded into this activity, so you might not need to provide much extra support for your students.

This activity can be done completely in class or started in class and then finished as homework. Do not assign the entire activity as homework unless you know that every student in your class has dedicated time, space, and resources to write outside of class and everyone understands and is comfortable with the report writing process.

If you are concerned about students feeling overwhelmed, you can develop a class "gotta-have-it" list of information that needs to be included in each section of the report. Students can then refer to this list as they write each section.

Don't worry if students do not produce a perfect report at this point. All the reports will be reviewed and revised before they are turned in to you. The reports should therefore be viewed as a starting point. Be sure to encourage students to "just get something down on paper" so "we can work on it together."

Attending to Equity

Emerging multilingual students' access to instructional tasks that require complex thinking, such as writing a report, is enhanced when linguistic complexity and instructional support match their current level of language proficiency. Therefore, it is important to scaffold the writing of the reports for the emerging multilingual students based on their current English language proficiency level. The use of sentence starters to model written language production skills can help emerging multilingual students develop English language skills because the development of academic language and academic content knowledge are interrelated processes. However, it is important that such scaffolds be used purposefully and removed when no longer needed.

ADI Go Digital

Use the **ADI Learning Hub** to go paperless. When you create a class and add students to it, students can complete all the stages right in the system.

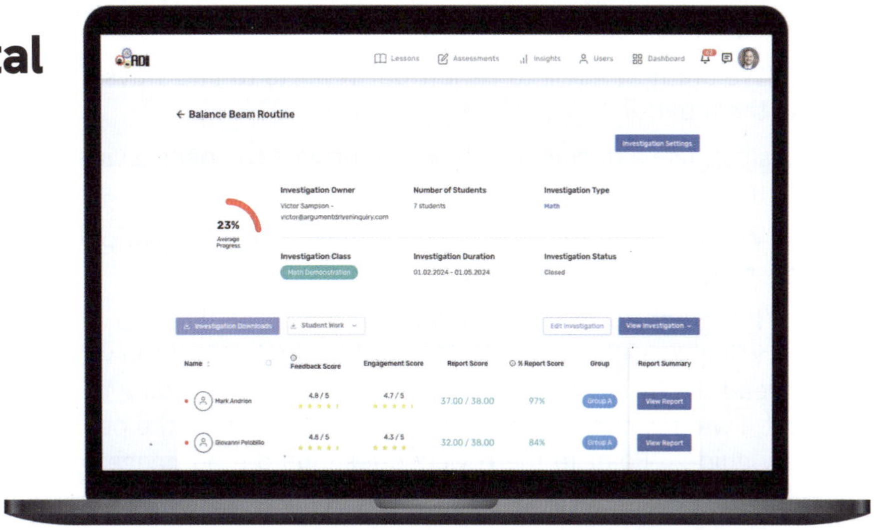

Activity B	Peer Review the Reports
Time Needed for the Activity	45 minutes
Grouping	Teams

What to Do

1. Students should work in teams of three for this activity. These can be completely different teams.
2. Show the "Peer Review Some Draft Reports" page on a screen using the Presentation mode in the Learning Hub so everyone in the classroom can see it.
3. Say, "We are now going to review our reports to find ways to make them better."
4. If possible, use a document camera to project the peer review guide (PRG) on a screen.
5. Say, "We are going to use this peer review guide to give one another feedback." Point to the image.
6. Say, "I'm going to ask you to work in teams of three to do this. I will give each team a draft report to read. You three will then read the report together. Once you are done reading the report, I want you to answer each of the questions on the peer review guide." Point to the review questions on the image of the PRG.
7. Tell the students, "You can check No, Almost, or Yes after each question." Point to the checkboxes on the image of the PRG.
8. Say, "This will be your rating for this part of the report. Make sure you agree on the rating you give the author. If you mark No or Almost, then you need to tell the author what he or she needs to do to get a Yes." Point to the space for the reviewer feedback on the image of the PRG.
9. Say, "It is really important for you to give the authors feedback that is helpful. That means you need to tell them exactly what they need to do to make their reports better."
10. Ask the students, "Do you have any questions about what you need to do?"
11. Answer any questions that come up.
12. Tell the students, "Please get into your teams of three."
13. Allow the students time to move into their triads.
14. Say, "Okay, I'm now going to give you one report to read." Give out one draft report to each triad. Make sure that the report you give a triad was not written by one of the students in that triad. Give each triad one PRG to fill out.

What to Give Out

Peer Review Guide (one per group per review, so for this investigation you need the same number of peer review guides as the number of draft reports that will be reviewed)

What to Show

The "Peer Review Some Draft Reports" page in the ADI Learning Hub

What to Look For

Make sure students are reaching consensus about what rating to give each criterion. Also, be sure that students are giving one another feedback that is specific, detailed, and helpful when they think an aspect of a report needs to be improved.

(*continued*)

15. Tell the students, "Okay, I'm going to give you 15 minutes to read the report I gave you and to fill out the peer review guide. Go ahead and get started."
16. Give the students 15 minutes to work. As they work, move around from team to team to check in and see how things are going, answer questions, and offer advice.
17. After 15 minutes pass, tell the students, "Okay, time is up. Please give me the report and the peer review guide that you filled out."
18. Collect the draft reports and the PRGs, or collect the workbooks if students are using them.
19. Say, "Okay, I am now going to give you a different report to read and a new peer review guide to fill out." Give out one more report to each triad. Make sure that the report you give a triad was not written by one of the students in that triad. Give each triad one new peer review guide to fill out as a team.
20. Say, "Okay, I'm going to give you 15 minutes to read this new report and to fill out the peer review guide. Go ahead and get started."
21. Give the students 15 minutes to work. As they work, move around from team to team to check in and see how things are going, answer questions, and offer advice.
22. After 15 minutes pass, tell the students, "Okay, time is up. Please give me the report and the peer review guide that you filled out."
23. Repeat this process for the remaining draft reports. Each triad should one have one more report to review (for a total of three reviews).
24. Collect the last draft reports and PRGs. Set all the draft reports on a table in separate stacks. Each stack should have a draft report on top (with the identification symbol facing up so students can find their reports) and the PRG on the bottom.

Teaching Tip

The peer review process is one of the best ways for students to learn how to write. When students read and review different reports, they have an opportunity to (1) see examples of texts written by others for the same purpose, (2) discuss what counts as quality and why some reports are stronger than others, (3) discuss ways to strengthen a report, and (4) pick up things that they can do to improve their reports. This helps the students learn new ways to organize and present information, which will help them write better on subsequent reports.

Attending to Equity

One helpful scaffold is to give students sentence stems for what to write in their feedback to the author. Ideal feedback is respectful, specific about what needs to be changed, and helpful in that it suggests how the author should go about making that change. You can review the writing of the emerging multilingual students in your class if they have limited English language proficiency, but make sure that all emerging multilingual students participate in the peer review process. You will see tremendous growth in their writing skills as they review their peers' writing and discuss what makes some reports stronger than others.

Activity C	Revise the Reports
Time Needed for the Activity	25 minutes
Grouping	Individuals

What to Do

1. Students should work on their own for this activity.
2. Show the "Reflect on the Feedback" page on a screen using the Presentation mode in the Learning Hub so everyone in the classroom can see it.
3. Say, "Please come pick up your reports from over here on this table." Point to the table.
4. Give the students a few minutes to pick up their reports and return to their seats.
5. Say, "Please take a few minutes to read over the peer review guide. You should use it to figure out what you need to change in your report and how you will change it."
6. Allow the students to read the PRG.
7. Show the "Create a Final Report" page on a screen using the Presentation mode in the Learning Hub so everyone in the classroom can see it.
8. Say, "Okay, let's revise our reports. I would like you to rewrite your report. You can use your draft report as a starting point but use the feedback on the peer review guide to make it better."
9. Ask the students, "Do you have any questions about what you need to do?"
10. Answer any questions that come up.
11. Tell the students, "Okay, let's write." Allow about 20 minutes for the students to revise their reports.
12. After about 20 minutes, give the following directions, "Okay, time's up. I will now come around and collect your first draft, final report, and peer review guide."
13. Collect all the first drafts, PRGs, and final reports.
14. Use the "Teacher Score" columns in the PRG to grade the final reports. Be sure to give the students feedback about their writing in the "Teacher Comments" section.

What to Show First

The "Reflect on the Feedback" page in the ADI Learning Hub

What to Show Second

The "Create a Final Report" page in the ADI Learning Hub

Attending to Equity

Students don't always use the feedback they receive from others to improve the quality of their reports. Make sure you move from student to student and encourage them to make changes based on the feedback. If students are not sure how to revise their reports, model how to use feedback. For example, you might say, "Okay, based on the peer review guide, I need to do more to describe what I did to collect data. I bet if I add information about ________, readers will have a better idea of what I did."

Teaching Tip

The peer review guide (PRG) is designed to be used with any ADI investigation, thus allowing you to use the same scoring rubric throughout the entire year. This is beneficial for several reasons. First, the criteria for what counts as a high-quality report do not change from investigation to investigation. Students, therefore, quickly learn what is expected from them when they write a report, and you do not have to spend valuable class time explaining the various components of the PRG each time students write a report. Second, the PRG makes it clear which components of a report need to be improved next time because the grade is not based on a holistic evaluation of the report. The PRG format allows students to see which aspects of their writing are strong enough and which aspects need improvement. Finally, and perhaps most importantly, the PRG provides you with a standardized measure of student performance that can be compared over multiple reports across semesters, thus allowing you to track improvement over time.

Teacher Notes

Tissue Repair

TASK

The Phenomenon

Many of us have cut our skin at one time or another. Take a few minutes to watch a timelapse video that shows how a cut heals over a period of 33 days. As you watch the video, write down some things you notice in the spaces below.

Things I **noticed**...

The Task

The video you watched showed how our bodies can replace damaged tissue. To be able to replace damaged tissue like in the video, our bodies must be able to create new skin cells to replace the ones that were damaged or destroyed because of the injury. The new skin cells come from nearby skin cells that were not damaged during the injury, such as the ones shown in the image below.

Skin cells under 400x magnification

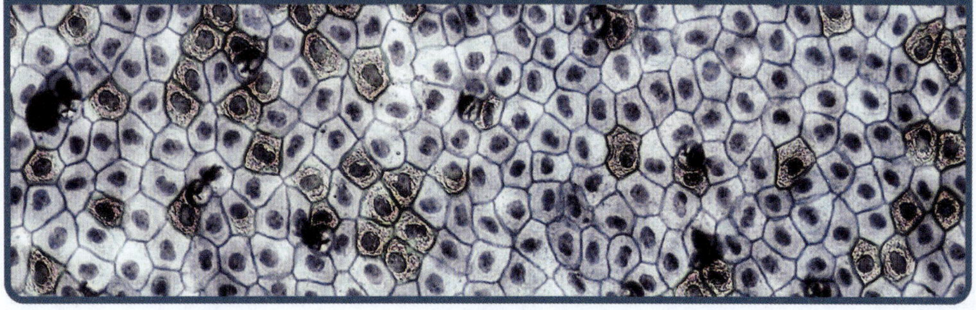

In addition, each of the new skin cells must include an exact copy of all the genes found within our existing cells to be able to function properly. Your goal in this investigation is to figure out how our bodies create new cells from existing cells without losing any important genetic information in the process. You will then create a model to explain how this important activity happens. The guiding question of this investigation is:

How do our bodies make new cells from our existing cells so our wounds can heal?

Initial Ideas

Before you start this investigation, take a few minutes to think about how our body might make new cells from existing cells. Then draw a model in the space below that shows how you think this process happens. Be sure to include labels to help explain your thinking.

Some **ideas** that I have...

Share your model with the rest of your group. Once everyone has had a chance to share their ideas, list some questions that your group has about this phenomenon.

Some **questions** we have now...

Tissue Repair

IDEAS

Some Ideas You Can Use: Cell Structures

All living things on Earth are made up of at least one cell that contains different organelles. An **organelle** is any structure found inside a cell that carries out a specific function. All cells have a cell membrane and cytoplasm (see the image below, with an illustration on the left and a photo of an actual cell on the right). The **cell membrane** separates the inside of the cell from the environment outside the cell and regulates what can pass in and out of the cell. The **cytoplasm** is all the matter within the cell except for any organelles. The cytoplasm is where most chemical processes of the cell happen.

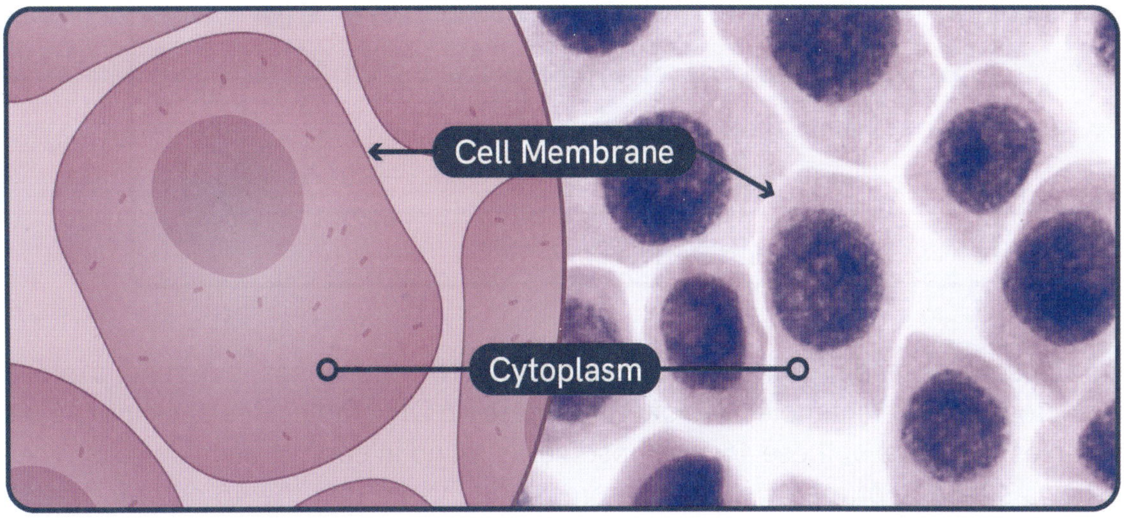

Our cells also have an organelle called a nucleus (see the image below). The **nucleus** is a large membrane-bound organelle that holds the genetic information of each cell. This genetic information is needed for a cell to function properly. It is also needed to make new organelles.

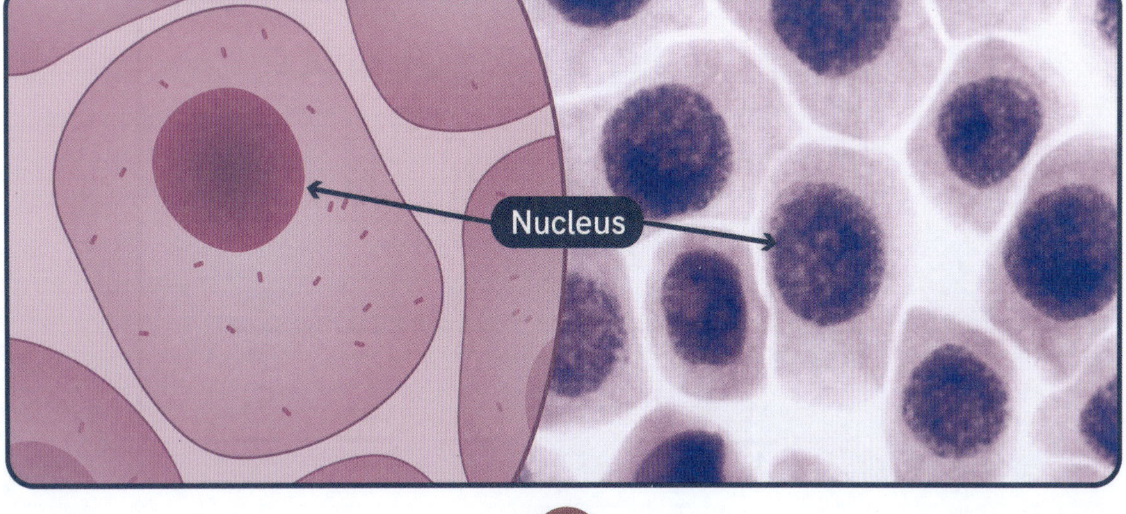

The genetic information of a cell is stored in molecules called deoxyribonucleic acid, or DNA. Each molecule of DNA is located within the nucleus of a cell. Each of these molecules is coiled around many proteins to make structures called nucleosomes. The nucleosomes, in turn, coil around each other to create structures called chromatin. The chromatin then winds up and packs together to create chromosomes. Chromatin and chromosomes are large enough to be seen under a microscope. The image below shows how DNA is packaged within the nucleus of a cell.

Organization of DNA within the nucleus of a cell

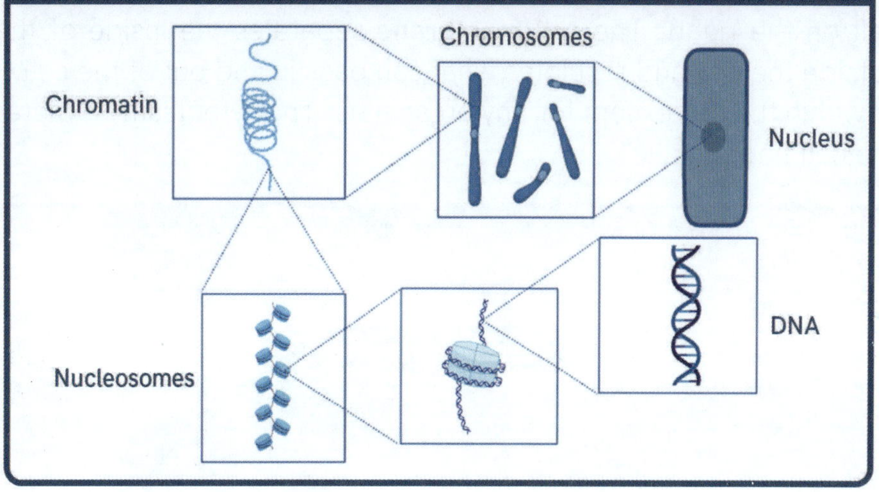

Each chromosome is made from a single molecule of DNA. In humans, our genetic information is stored across 46 different molecules of DNA. These 46 DNA molecules pack into 46 different chromosomes. A replica of these 46 chromosomes is found in each of our skin cells. The image below shows a picture of the 46 human chromosomes. Note that the 46 chromosomes in the image are not 46 different ones; instead, the chromosomes are found in pairs. Humans, as result, have two versions of 23 different chromosomes for a total of 46. Scientists call a picture of an organism's chromosomes, where the chromosomes are organized by size and appearance, a karyotype.

Human chromosomes

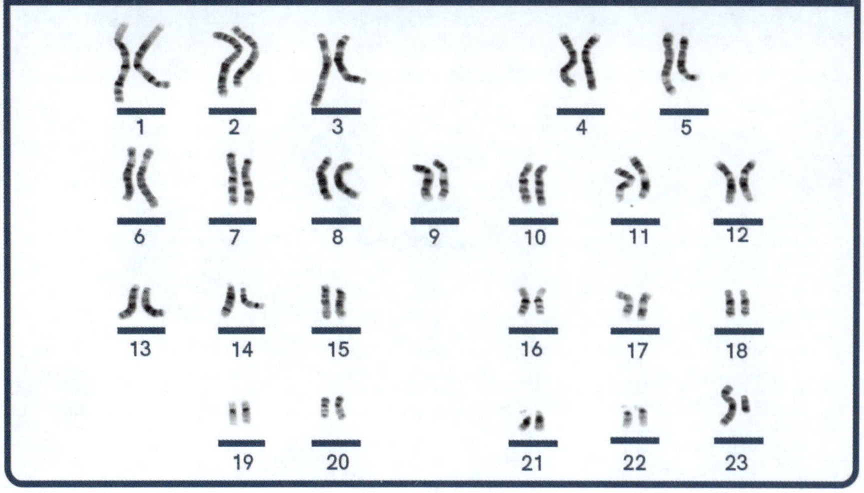

Some **useful ideas** from what I read...

Some Ideas You Can Use: Images of Cells Creating New Cells

The photo below shows new cells being created from existing cells. This photo was taken using a microscope. To create this image, a very thin slice of tissue was placed onto a microscope slide. The tissue was then stained with a dye that made the chromosomes visible. A picture was then taken of the cells that make up the tissue under 1600x magnification. As a result, the photo provides us with a clear view of the various events that happen within an existing cell as a new cell is being created. Based on these images, there seems to be five important events that happen during the process of making new cells. These important events are marked with the letters A through E. Chromosomes cannot be seen during events A and B, but chromatin can. Chromosomes can be seen during events C through E. You will be given images of cells in each of these major events during this investigation so you can use this information as you attempt to figure out what happens when new cells are created from existing cells.

Events that cells go through when creating new cells

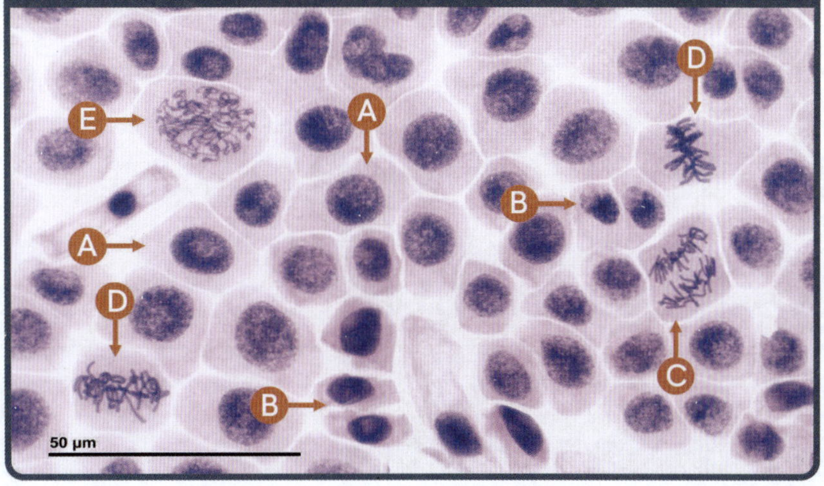

Some **useful ideas** from what I read...

Some Ideas You Can Use: Using Models to Represent a Process

Models can be used to represent an object, event, or process. A **model** can be physical, visual, or a computer-based simulation. In this investigation, you will need to use both a physical model and visual model to answer the guiding question. First, you will need to make a physical model of the chromosomes of a cell to figure out how an existing cell can produce a new cell without losing any of its genetic information in the process. The picture below shows how you can make a physical model. It is important to note that this example of physical model is just a starting point. You can add other things to your physical model to see if you make sense of what is happening during each of the five key events that happen as an existing cell makes a new cell.

A physical model of the chromosomes in a cell

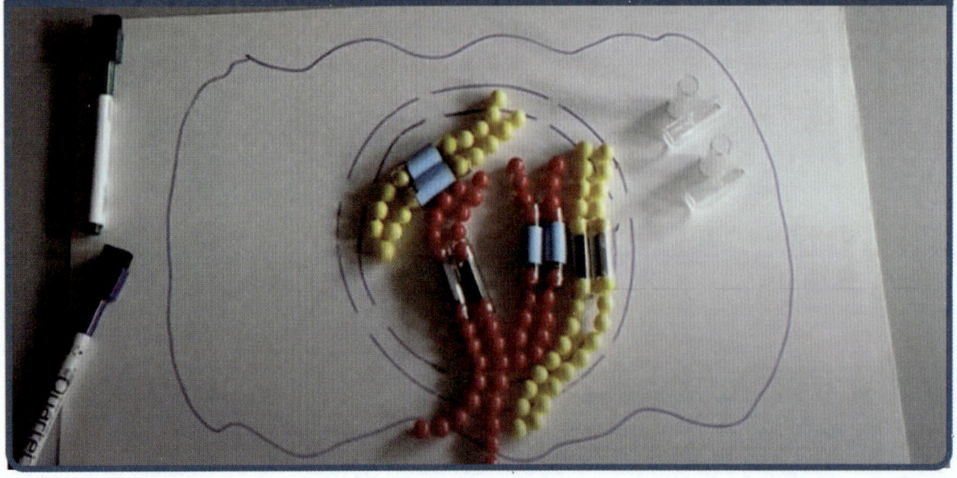

Once you are finished testing your ideas with your physical model, you can use what you figured out to create a visual model. A visual model includes images, symbols, words, and numbers to help explain a process. It does not just replicate or duplicate something that you saw like a physical model does; instead, a visual model provides a way to make predictions about what will happen in the future. You should be able to use your visual model to predict what happens when the existing cells of other organisms, which have different numbers of chromosomes, make new cells.

Some **useful ideas** from what I read...

Tissue Repair

PLAN

Plan Your Investigation

Prepare a plan for your investigation by filling out the chart below.

I am trying to answer the following **question**...

I will use the following **observations or measurements**...

I will **analyze** these observations or measurements by...

I approve of this investigation plan.

Teacher's signature

Date

Tissue Repair

DO

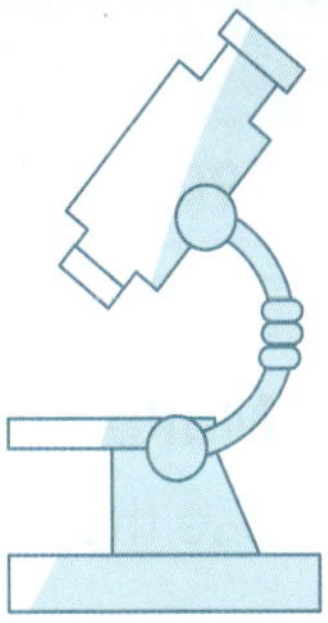

Collect Data

Keep a record of what you observe or measure during your investigation.

My **observations or measurements...**

Make Sense of Your Data

Use the space below to create your model so you can answer the guiding question.

My **model...**

Tissue Repair

SHARE

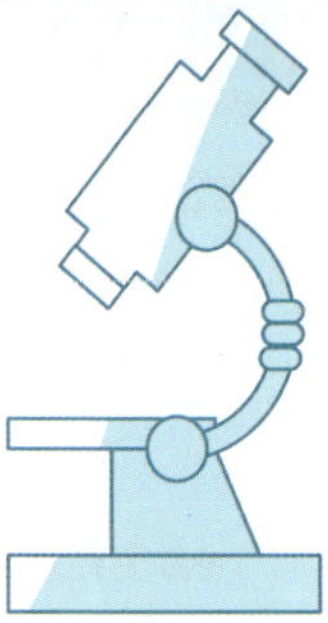

Draft Argument

Develop an argument on a whiteboard. It should include:

1. A *claim* that answers the guiding question,
2. *Evidence* in support of the claim that provides an analysis of the data and a description of what the analysis means,
3. A *justification of the evidence* that explains why your group thinks the evidence is important.

Our Guiding Question:

Claim:

Evidence:	**Justification of the Evidence:**

Argumentation Session

Share your argument with your classmates. In the space below, be sure to keep track of any ideas that you can use to revise your argument and make it better.

Some possible ways to **improve** our argument...

Tissue Repair

REFLECT

Reflective Discussion

In the space below, you can keep track of any ideas from the discussion that you think are important or will be useful in the future.

Some important **ideas**...

Tissue Repair

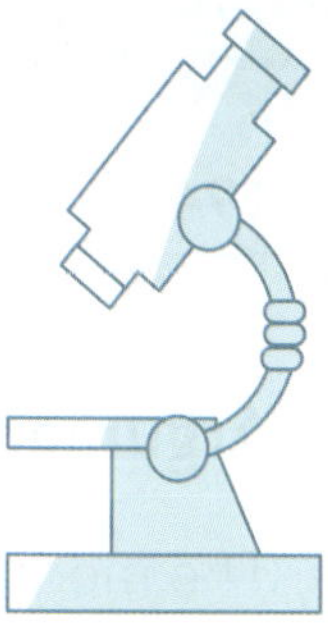

REPORT

Draft Report

Prepare a report to share what you figured out during your investigation.

Introduction

We have been studying ______________________ in class. Before we started this investigation, we explored ______________________

We noticed ______________________

Our goal for this investigation was to figure out ______________________

The guiding question was ______________________

Method

To answer this question, ______________________

Argument

The model below shows ______________________________

My evidence for this model is ______________________________

This evidence is based on several important scientific ideas. The first one is ______________________________

Elk Population Decline in Colorado Lesson Plan

OVERVIEW

Description	Students figure out what might be causing the decline of the elk population in Colorado.
Guiding Question	What potential changes in an ecosystem might be triggering the rapid decline of the Colorado elk population, and why?
What Students Figure Out	The rapid decline of the Colorado elk population could have been caused by sudden changes in the elk's ecosystem, such as an increasing population of predators or competitors or a decreasing population of possible food.

STAGES

This investigation has seven stages. Each stage includes multiple activities. The flowchart below shows what happens during each stage and how much time is needed to complete each one.

Task (Day 1)
1 — **20 minutes**
Students are introduced to a phenomenon and a guiding question to answer. They also share their initial ideas.

Ideas (Day 1)
2 — **30 minutes**
Students read about DCI, a CC, and an SEP that they can use to help them figure out the phenomenon.

Plan (Day 2)
3 — **15 minutes**
Students formulate a plan collecting and analyzing the data they need to answer the guiding question.

Do (Day 2)
4 — **30 minutes**
Students carry out their plan, make sense of the data they collected, and answer the guiding question.

Share (Day 3)
5 — **45 minutes**
Students develop, share, and revise an evidence-based argument about what they figured out.

Reflect (Day 4)
6 — **20 minutes**
Students discuss a DCI and/or a CC and reflect on the SEPs they used during the investigation.

Report (Day 4)
7 — **25 minutes**
Students write investigation reports to share what they figured out during the investigation.

Report (Day 5)
7 — **45 minutes**
Students peer review the reports written by other students and offer feedback to improve.

Report (Day 6)
7 — **30 minutes**
Students revise their reports and then submit a final draft to their teacher for evaluation.

STANDARDS ALIGNMENT

NEXT GENERATION SCIENCE STANDARDS

Performance Expectation

1. Use mathematical and/or computational representations to support explanations of factors that affect the carrying capacity of ecosystems at different scales (HS-LS2-1).
2. Evaluate claims, evidence, and reasoning that the complex interactions in ecosystems maintain relatively consistent numbers and types of organisms in stable conditions, but changing conditions may result in a new ecosystem (HS-LS2-6).

Core Ideas(s)	Crosscutting Concept(s)	Practices
Interdependent Relationships in Ecosystems (LS2.A) Ecosystem Dynamics, Functioning, and Resilience (LS2.C)	Systems and System Models (CC4) Stability and Change (CC7)	Asking Questions (SEP1) Developing and Using Models (SEP2) Planning and Carrying Out Investigations (SEP3) Analyzing and Interpreting Data (SEP4) Using Mathematics and CT (SEP5) Constructing Explanations (SEP6) Engaging in Argument from Evidence (SEP7) Obtaining, Evaluating, and Communicating Information (SEP8)

WIDA ENGLISH LANGUAGE DEVELOPMENT STANDARDS FOR SCIENCE

Explain	Argue
Interpret scientific explanations (ELD-SC.9-12.Explain.Interpretive) Construct scientific explanations (ELD-SC.9-12.Explain.Expressive)	Interpret scientific arguments (ELD-SC.9-12.Argue.Interpretive) Construct scientific arguments (ELD-SC.9-12.Argue.Expressive)

COMMON CORE ENGLISH LANGUAGE ARTS

Reading	Writing	Speaking and Listening
Key ideas and details (RST. 9-10.1 and RST.9-10.2) Craft and structure (RST.9-10.4) Integration of knowledge and ideas (RST.9-10.7, RST. 9-10.8, and RST.9-10.9) Range of reading (RST. 9-10.10)	Text types and purposes (WHST.9-10.1 and WHST.9-10.2) Production and distribution of writing (WHST.9-10.4, WHST. 9-10.5, and WHST. 9-10.6)	Comprehension and collaboration (SL.9-10.1 and SL.9-10.3) Presentation of knowledge and ideas (SL.9-10.4 and SL.9-10.6)

THINGS TO KNOW BEFORE STARTING THE LESSON

The Core Ideas

An ecosystem is a defined area that includes a community and the environment. A community, in turn, is defined as a group of two or more populations in a region, and a population is a group of organisms that belong to the same species and live in the same place at the same time. Because an ecosystem has a limited number of available resources, a population naturally settles around the ecosystem's carrying capacity, or the maximum number of organisms in a population that an ecosystem can sustainably support. A population will fluctuate around the carrying capacity, sometimes exceeding and other times falling below, but will usually correct to the limit set by the ecosystem.

Because there are almost always multiple populations in an ecosystem, these populations interact with one another as they try to survive and reproduce. These interactions include predation and herbivory, competition, and symbiosis. Predation describes when an organism called the predator kills and eats another organism, the prey, for food. Similarly, herbivory occurs when an organism feeds on another organism, usually a primary producer. Competition describes when two organisms, either from the same population or different populations, require the same resource and must try to obtain said resource before the other. Lastly, symbiosis is an interaction where two or more species purposefully live in direct contact with each other for long periods of time. There are three types of symbiotic relationships. Commensalism describes when one individual benefits from the interaction while the other individual is unaffected by it. Parasitism occurs when one individual benefits from the interaction while the other is harmed in some way. Finally, mutualism describes when both individuals benefit from the interaction.

The Crosscutting Concepts

An ecosystem is a group of living and non-living things that interact with one another and fulfill different functions. Because the nature of an ecosystem is affected by a variety of factors and is therefore very complex, we often use models to represent ecosystems. We can then use these models to test how changing different factors would affect the ecosystem as a whole, without actually changing the ecosystem itself. These models let us make predictions about and understand the world around us.

The Science and Engineering Practices

A practice of science is to **ask and refine questions** that lead to descriptions and explanations of how the natural world works. Another practice of science is to **use and construct models** as helpful tools for representing ideas and explanations. These tools include diagrams, drawings, physical replicas, mathematical representations, analogies, and computer simulations. Scientists **plan and carry out investigations** in the field or a laboratory, working collaboratively and individually. Their investigations are systematic and require clarifying what counts as data and identifying variables or

(continued)

parameters. Scientific investigations produce **data that must be analyzed and interpreted**. Because data patterns and trends are not always obvious, scientists use a range of tools and practices to identify the significant features and patterns in the data. Scientists also use **mathematics and computational thinking** for a range of tasks such as constructing simulations; statistically analyzing data; and recognizing, expressing, and applying quantitative relationships. The products of doing science are **explanations**. Scientific explanations often specify variables and are used by scientists to describe and predict different natural phenomena. **Scientific argumentation**, or the process of sharing, supporting, critiquing, and revising claims based on evidence is the process that scientists use to determine if an explanation is valid or acceptable. Finally, scientists must be able to **communicate clearly and persuasively** the ideas and methods they generate. Critiquing and communicating ideas individually and in groups is a critical professional activity.

PREPARATION

This lesson can be delivered paper-based, paperless, or remote. Use the paper-based delivery option for the investigation if students do not have 1:1 access to a computer and they are in the same classroom or physical space during the lesson. Use the paperless delivery option for the investigation if students have 1:1 access to a computer and they are in the same classroom or physical space during the lesson. Finally, if students are not in the same classroom or physical space during the lesson and have 1:1 access to a computer, use the remote delivery option.

ADI Go Digital

The digital version of this lesson is available in the **ADI Learning Hub**. Be sure to check out the built-in teacher presentation mode.

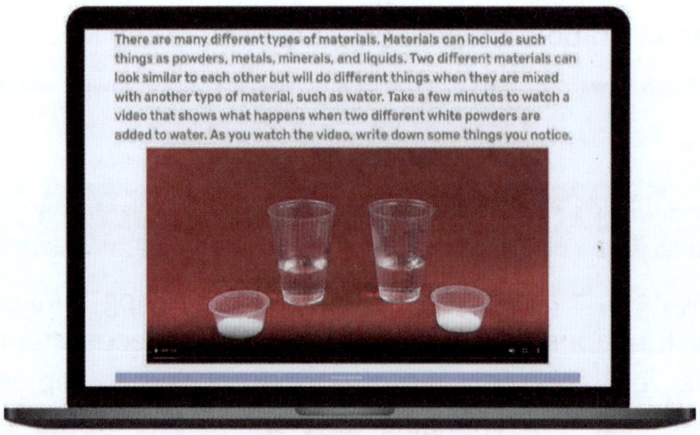

Teacher Notes

Materials Needed

The materials needed by the teacher and each student in the class for each of the three modalities are listed in the table below.

Delivery Mode	Each Student Will Need	The Teacher Will Need
Paper-Based	• Copies of each student handout for the investigation (or a workbook) • A pencil	• A teacher account in the ADI Learning Hub • Access to a computer with high-speed internet and a browser (such as Chrome, Safari, Firefox, or Edge) • A projector or smart board (for displaying a computer screen)
Paperless	• A student account in the ADI Learning Hub • Access to a computer with high-speed internet and a browser (such as Chrome, Safari, Firefox, or Edge)	• A teacher account in the ADI Learning Hub • Access to a computer with high-speed internet and a browser (such as Chrome, Safari, Firefox, or Edge)
Remote	• A student account in the ADI Learning Hub • Access to a computer with high-speed internet and a browser (such as Chrome, Safari, Firefox, or Edge)	• A teacher account in the ADI Learning Hub • Access to a computer with high-speed internet and a browser (such as Chrome, Safari, Firefox, or Edge)

Additional Materials and Preparation Needed for In-Person Delivery

This in-person investigation will require students to have access to a Chromebook, desktop, laptop, iPad, or other electronic device in order to use the simulation. The simulation is embedded in the Learning Hub, so students do not need access to external sites. If students do not have access to the Learning Hub, they can use the simulation by visiting *Carrying Capacity* by NetLogo's Modeling Commons. The teacher can determine if each student will use an electronic device or if each group will share.

How the Simulation Works

This model simulates an ecosystem of grass, elk, rabbits, and wolves. Within each 'tick', which is a time unit within the simulation, elk wander, eat, and reproduce in the habitat. They gain energy from grass and lose energy when wandering and reproducing. They die when their energy is used up. Grass is the elk's only food resource in this simulation. When a patch of grass is consumed, the square becomes black. The grass will regrow at a certain rate based on temperature. Temperature affects the grass growth rate. The habitat size determines the amount of food resources available. Students may add wolves in this simulation. Once added, a wolf will prey on any elk it encounters. Wolves do not reproduce in this simplified world. Students can also add rabbits, which eat grass. To simplify the simulation, the rabbits do not reproduce and the wolves do not prey on rabbits.

How to Use the Simulation

The simulation interface is shown in the image below.

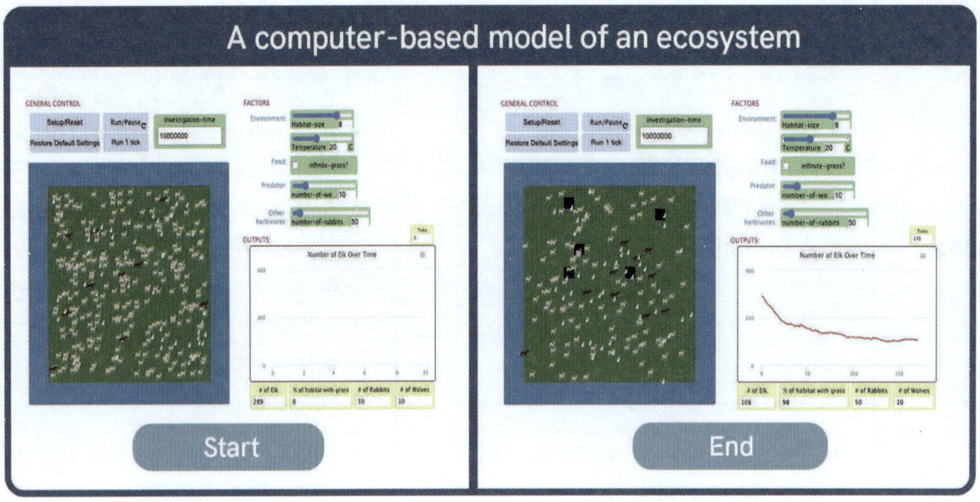

To start the simulation, press the **Setup/Reset** button to generate a population of elk in the simulation. Input the number of ticks you want the simulation to run in the **investigation-time** box. Each tick represents about 1 week of time, so a year is 52 ticks and two years is 104 ticks. Press the **Run/Pause** button to run or pause the simulation. The simulation will stop at the defined tick. The number of elk in the simulated world can be tracked over time using the output monitors and the real-time graph. Adjust the slider values for **habitat-size** and **temperature** to change the physical components of the simulated world. Click on the **infinite-grass** box to create a simulated world with unlimited resources. Adjust the slider vales for **number-of-wolves** and **number-of-rabbits** to change the biological components of the simulated world.

Safety Precautions

Remind students to follow all normal lab safety rules.

Laboratory Waste Disposal

There is no waste produced during this activity.

FACILITATION NOTES

The facilitation notes for this investigation describe how to deliver this lesson in person. They are written specifically for the paper-based mode of delivery. These same facilitation notes, however, can also be used for a paperless delivery of the lesson by simply substituting the directions that mention paper with the entry into the ADI Learning Hub. If you are teaching this lesson using the remote mode of delivery, you only need to assign due dates and prerequisites for each stage. The students can then simply follow the directions built into the ADI Learning Hub as they move through the investigation.

Elk Population Decline in Colorado Lesson Plan

Stage 1 TASK

Total Time for the Stage	45 minutes
Grouping	Individual; Teams

Learning Goals

Students should make progress on their ability to do the following by the end of this stage of the lesson:

Science

1. Ask questions that arise from careful observation of phenomena to clarify and/or seek additional information (SEP1).
2. Develop and use a model to provide mechanistic accounts of phenomena (SEP2).
3. Evaluate the merits and limitations of different models of the same phenomenon (SEP2).
4. Ask questions that arise from examining models to clarify and/or seek additional information and relationships (SEP 1).
5. Ask questions to clarify and refine a model (SEP1).

English Language Development

None

English Language Arts

1. Initiate and participate effectively in small-group collaborative discussions with diverse partners by building on others' ideas and expressing their own clearly (SL.9-10.1).
2. Present information clearly, concisely, and logically such that listeners can follow the line of reasoning and the organization, substance, and style are appropriate to the purpose, audience, and task (SL.9-10.4).
3. Adapt speech to a variety of contexts and tasks, demonstrating command of formal English when indicated or appropriate (SL.9-10.6).

Activities within the Stage

Activity A	The Phenomenon
Time Needed for the Activity	5 minutes
Grouping	Teams

What to Do

1. Have students work in their assigned teams for this activity.
2. Begin the activity by giving out a copy of the Task handout to each student.
3. Show "The Phenomenon" page on a screen using the Presentation mode in the Learning Hub so everyone in the classroom can see it.
4. Read the introduction of the phenomenon out loud and then start the video. You can also have students read the introduction of the phenomenon and watch the video on their own devices.
5. Give the students a few moments to write down their observations on their handouts before moving on to the next activity.

What to Give Out

Task handout for **Elk Population Decline in Colorado**

What to Show

"The Phenomenon" page in the ADI Learning Hub

Teaching Tip

Instead of assigning fixed roles to students by name at the beginning of the investigation, give each member of a group a different number so you can easily assign and change roles as needed during the various investigation stages. For example, you might give directions such as "1s will share out for your group" or "3s, will you please come back here and pick up these materials for your group?" or "2s and 3s, you will present the argument for your group while 1s and 4s travel to the other groups to offer feedback."

Attending to Equity

Be sure to intentionally group emerging multilingual students at times with peers who know the same languages as they do and at other times with peers whose English language development is slightly more advanced. Thoughtful grouping that varies throughout an investigation allows emerging multilingual students to benefit from the different linguistic resources of their peers.

Activity B	Share What You Noticed
Time Needed for the Activity	5 minutes
Grouping	Teams

What to Do

1. Show the "Share What You Noticed" page on a screen using the Presentation mode in the Learning Hub so everyone in the classroom can see it.
2. Say, "Okay, please share what you noticed with the others at your table."
3. Give the students about two minutes to share at their tables.
4. Say, "Okay, let's take a few minutes to share with the class. Team X (call on a team at random), what did you talk about?"
5. Allow Team X to share something they noticed. If it is related to population changes in an ecosystem, say, "That is really interesting. We are actually going to examine that in more detail during this investigation." If what is shared is not related to population changes in an ecosystem, say, "That is really interesting. Team Y (call on another team at random), what did you notice?" and allow that team to share. Follow this pattern until a team mentions something related to why the elk calves are dying off. If no team mentions this, say, "You know what I found interesting? I saw that the number of elk calves is decreasing, and I wondered how that would affect the entire elk population. Did anyone notice that as well?"
6. Move on to the next activity.

What to Show

The "Share What You Noticed" page in the ADI Learning Hub

What to Look For

As soon as a group mentions something related to why the elk calves are dying off, move on to the next activity.

Attending to Equity

Be sure to provide students with opportunities to talk in small groups during this activity. Small-group structures offer all students, including emerging multilingual students, a chance to engage in sense-making with their peers and offers them space to use their verbal and nonverbal (e.g., writing, drawing pictures, and using gestures) resources to express their ideas. Small-group structures also give students a chance to learn from other students' uses of these resources.

Activity C	What You Need to Figure Out
Time Needed for the Activity	5 minutes
Grouping	Teams

What to Do

1. Show the "What You Need to Figure Out" page on a screen using the Presentation mode in the Learning Hub so everyone in the classroom can see it.
2. Read the task out loud and have students follow along on their handouts, on their devices, or on the screen.
3. Allow students to ask clarifying questions about the task or the guiding question.
4. Once everyone understands what they will need to figure out during the investigation, move on to the next activity.

What to Show

The "What You Need to Figure Out" page in the ADI Learning Hub

Teaching Tip

Have students highlight, circle, or place a star next to the guiding question. Marking the guiding question helps students find it later during the Plan stage.

Teacher Notes

Activity D	Share Your Ideas
Time Needed for the Activity	15 minutes
Grouping	Teams

What to Do

1. Show the "Share Your ideas" page on a screen using the Presentation mode in the Learning Hub so everyone in the classroom can see it.
2. Read the directions on the handout out loud and then give students five minutes to draw their initial representations of the science.
3. Say, "Okay, will everyone share their representations with the rest of your team? Please make sure everyone has time to talk. Also, be sure to ask one another questions about the scientific ideas that are included in each representation."
4. Give the students about 8-9 minutes to talk and share. Be sure to walk around and listen to the discussions.
5. Say, "Okay, I heard some interesting ideas as I walked around the room."
6. Move on to the next stage.

What to Show

The "Share Your Ideas"page in the ADI Learning Hub

What to Look For

Each student's initial representation should include a combination of symbols, diagrams, graphs, or words. It should also clearly show how a scientific idea can be used to help complete the task and answer the guiding question.

Attending to Equity

When students are tasked with expressing their ideas, be sure to encourage them to do so through multiple modes (speaking, writing, drawings, and gestures) and to use both content-specific and everyday registers. This process helps students fully express themselves, allowing peers to evaluate, question, and build on ideas. Also, make sure that every student in every team has a chance to share their initial representation.

Activity E	More Questions
Time Needed for the Activity	15 minutes
Grouping	Teams

What to Do

1. Show the "More Questions" page on a screen using the Presentation mode in the Learning Hub so everyone in the classroom can see it.
2. Say, "Okay, I heard a bunch of interesting questions as I was walking around the room. Will you work with your team to identify two or three questions that you think are important to learn more about that came up as you all discussed the different representations you made? You can write these questions on your handout."
3. Give students about five minutes to talk and write down questions. Be sure to walk around and listen to the discussions.
4. Say, "Okay, let's take a few minutes to share. Team X (call on a team at random), will you share one question you have?"
5. Have each team share a question. Then say, "I want to share some useful scientific ideas with you. I bet this information will not only answer many of these questions but will also help you complete the task and answer the guiding question of the investigation."
6. Move on to the next stage.

What to Show

The "More Questions" page in the ADI Learning Hub

What to Look For

Each student should have 2-3 questions written on their handout.

Teaching Tip

When students are asked to create representations of the underlying scientific ideas, share these ideas with their peers, and then identify things they want to know more about, it sparks their curiosity and helps make the investigation meaningful. Also, identifying questions before reading (which students do during the Ideas stage) is an outstanding pre-reading activity that helps improve comprehension.

Elk Population Decline in Colorado Lesson Plan

Stage 2 IDEAS

Total Time for the Stage	30 minutes
Grouping	Teams

Learning Goals

Students should make progress on their ability to do the following by the end of this stage of the lesson:

Science

1. Ask questions that arise from examining theory to clarify and/or seek additional information and relationships (SEP1).
2. Critically read scientific literature adapted for classroom use to determine the central ideas or conclusions and/or to obtain scientific and/or technical information to summarize complex concepts, processes, or information presented in a text by paraphrasing them in simpler but still accurate terms (SEP8).
3. Integrate sources of information presented in different media or formats (e.g., visually and quantitatively) as well as in words to address a scientific question (SEP8).
4. Gather, read, and evaluate scientific and/or technical information from multiple authoritative sources, assessing the evidence and usefulness of each source (SEP8).
5. Explain how ecosystems have carrying capacities and how these limits result from factors such as the availability of living and non-living resources and challenges such as predation, competition, and disease (LS2.A).
6. Describe the complex set of interactions within an ecosystem that keep its numbers and types of organisms relatively constant over long periods of time under stable conditions (LS2.C).
7. Describe how extreme fluctuations in conditions and populations can challenge the functioning of ecosystems in terms of resources and habitat availability (LS2.C).
8. Use models to simulation systems and interactions between systems at different scales (CC4).
9. Construct explanations of how things change and how they remain the same (CC7).

English Language Development

1. Paraphrase central ideas in complex evidence, concepts, processes, and information to help explain how or why a phenomenon occurs (ELD-SLC.9-12.Explain.Interpretive).

English Language Arts

1. Determine the central ideas or conclusions of a text; trace the text's explanation or depiction of a complex process, phenomenon, or concept; provide an accurate summary of the text (RST.9-10.2).

(continued)

2. Determine the meaning of symbols, key terms, and other domain-specific words and phrases as they are used in a specific scientific or technical context (RST.9-10.4).
3. Read and comprehend scientific/technical texts (RST.9-10.10).
4. Initiate and participate effectively in small-group collaborative discussions with diverse partners by building on others' ideas and expressing their own clearly (SL.9-10.1).

Activities

Activity A	Discuss a Core Idea
Time Needed for the Activity	10 minutes
Grouping	Whole class/Teams

What to Do

1. Have students work in their assigned teams during this stage. Give out a copy of the Ideas handout to each student.
2. The first activity provides students with information about carrying capacity.
3. Show the "Read About a Core Idea You Can Use" page on a screen using the Presentation mode in the Learning Hub so everyone in the classroom can see it. You can also click on the images to make them full screen and easier for students to see.
4. Have students read the text on their own, either on their handouts or on the screen.
5. Give the students about three minutes to discuss the text with the others in their teams and develop a shared list of ideas that they think are important or useful.
6. Ask a team to share an idea they thought was useful. Have the remaining teams share a different idea until there are no more ideas to share.
7. Move on to the next activity.

What to Give Out

Ideas handout for **Elk Population Decline in Colorado**

What to Show

The "Read About a Core Idea You Can Use" page in the ADI Learning Hub

What to Look For

Students should write down some of the following ideas in the box labeled, "Some things I know from what I read" on their handouts:

- An ecosystem is a defined area that includes a community and the environment.
- A community consists of two or more populations, or a group of organisms that belong to the same species and live in the same place at the same time.
- The carrying capacity of an ecosystem is the maximum population size that the ecosystem can sustain.
- Populations usually fluctuate between being slightly over and slightly under the ecosystem's carrying capacity.
- When a population is greater than the ecosystem's carrying capacity, it is described as having "overshot".

Teaching Tip

Students do not need to do a close read of this text or completely understand all the ideas in it before moving on to the next activity. They will be encouraged to revisit this text later in the investigation when creating their arguments and writing their reports. All students need to be able to do during this activity is to work with the other members of their group to identify one or two ideas that they think are important to keep in mind or are potentially helpful.

Activity B	Read About Another Core Idea
Time Needed for the Activity	10 minutes
Grouping	Teams

What to Do

1. The second activity provides students with information about interactions between organisms.
2. Show the "Read About Another Core Idea You Can Use" page on a screen using the Presentation mode in the Learning Hub so everyone in the classroom can see it. You can also click on the images to make them full screen and easier for students to see.
3. Have students read the text on their own, either on their handouts or on the screen.
4. Give the students about three minutes to discuss the text with the others in their teams and develop a shared list of ideas that they think are important or useful.
5. Ask a team to share an idea they thought was useful. Have the remaining teams share a different idea until there are no more ideas to share.
6. Move on to the next activity.

What to Show

The "Read About Another Core Idea You Can Use" page in the ADI Learning Hub

What to Look For

Students should write down some of the following ideas in the box labeled, "Some things I know from what I read" on their handouts:

- Individual organisms interact with one another within an ecosystem in three ways: predation and herbivory, competition, and symbiosis.
- Predation happens when an animal, or predator, kills and eats another animal, its prey.
- Herbivory happens when an organism feeds on a plant or other type of producer.
- Competition occurs when two or more organisms have similar needs and try to obtain the same limited resource.

(*continued*)

- Symbiosis describes two or more species purposefully living in direct contact with each other for long periods of time.
- There are three main types of symbiosis: commensalism, parasitism, and mutualism.
- Commensalism is an interaction in which one organism benefits from the interaction while the other is neither helped nor harmed.
- Parasitism occurs when one individual benefits from the interaction while harming the other individual.
- Mutualism describes when all individuals benefit from the interaction.

Attending to Equity

There are many supports for helping students comprehend what they read already embedded into this activity (e.g., activating prior knowledge, providing a shared experience, making connections, synthesizing, and talking with peers). So, you might not need to provide much extra support. If you are concerned about some students not understanding what they are reading because of their scores on past reading comprehension tests, you can read it out loud as they follow along. As you read the text out loud, be sure to stop at each important idea and ask the students to put a star (or other annotation) next to it in the margin of their handouts. They can then discuss these ideas in their small groups.

Teacher Notes

Activity C	Read About a Final Core Idea
Time Needed for the Activity	10 minutes
Grouping	Teams

What to Do

1. The third activity provides students with information about using models to represent a process.
2. Show the "Read about a Final Core Idea You Can Use" page on a screen using the Presentation mode in the Learning Hub so everyone in the classroom can see it. You can also click on the images to make them full screen and easier for students to see.
3. Have students read the text on their own, either on their handouts or on the screen.
4. Give the students about three minutes to discuss the text with the others in their teams and develop a shared list of ideas that they think are important or useful.
5. Ask a team to share an idea they thought was useful. Have the remaining teams share a different idea until there are no more ideas to share.
6. Move on to the next activity.

What to Show

The "Read About a Final Core Idea You Can Use" page in the ADI Learning Hub

What to Look For

Students should write down some of the following ideas in the box labeled, "Some things I know from what I read" on their handouts:

- Models are used to represent an object, event, or process.
- Models can be physical, visual, or a computer-based simulation.
- Models have limited precision and reliability because they make simplifying assumptions and approximations about a system.

Attending to Equity

To support emerging multilingual students during this activity, be sure to create a word bank on the wall of high-frequency, high-need concrete vocabulary that is needed to communicate scientific ideas. The word bank highlights terms needed for basic communication in an academic context. It also helps with the spelling. You can also encourage your emerging multilingual students to:

1. Use strategic learning techniques such as concept mapping, drawing, memorizing, comparing, contrasting, and reviewing.
2. Decode (sound out) words using skills such as recognizing sound-letter relationships and identifying cognates, affixes, roots, and base words.
3. Use visual and contextual support and support from peers and teachers to enhance and confirm understanding of the texts.

Elk Population Decline in Colorado Lesson Plan

Stage 3 PLAN

Total Time for the Stage	15 minutes
Grouping	Teams

Learning Goals

Students should make progress on their ability to do the following by the end of this stage of the lesson:

Science

1. Evaluate a question to determine if it is testable and relevant (SEP1).
2. Plan an investigation individually and collaboratively to produce data to serve as the basis for evidence as part of building and revising models (SEP3).
3. Consider possible variables or effects and evaluate the confounding investigation's design to ensure variables are controlled (SEP3).
4. Ask questions to determine relationships, including quantitative relationships, between independent and dependent variables (SEP1).
5. Decide on types, quantity, and accuracy of data needed and consider limitations on the precision of the data (e.g., number of trials, cost, risk, and time), and refine the design accordingly (SEP3).
6. Select appropriate tools to collect, record, and analyze data (SEP3).

English Language Development

1. Define investigable questions on observations, information, and/or data about a phenomenon (ELD-SLC.9-12.Explain.Interpretive).
2. Develop reasoning to illustrate and/or predict the relationships between variables (ELD-SLC.9-12.Explain.Expressive).

English Language Arts

1. Initiate and participate effectively in small-group collaborative discussions with diverse partners by building on others' ideas and expressing their own clearly (SL.9-10.1).
2. Present information clearly, concisely, and logically such that listeners can follow the line of reasoning and the organization, substance, and style are appropriate to the purpose, audience, and task (SL.9-10.4).
3. Adapt speech to a variety of contexts and tasks, demonstrating command of formal English when indicated or appropriate (SL.9-10.6).

Activities

Activity A	Some Materials You Can Use
Time Needed for the Activity	5 minutes
Grouping	Teams

What to Do

1. Show the "Some Materials You Can Use" page on a screen using the Presentation mode in the Learning Hub so everyone in the classroom can see it. You can also click on the images to make them full screen and easier for students to see.
2. Say, "Okay, it sounds like we now have enough information to plan an investigation that will help us figure out what is causing the decline in the elk population in Colorado."
3. Say, "Before we get started, I want to show you some materials that we are going to use during our investigation. Will you all please look at the screen so I can show you what we have?"
4. Direct students to look at the screen where the simulation has been pulled up.
5. Ask, "What do you think we could use this simulation for?"
6. Allow students to share ideas. Then say, "We can use this simulation to test how different changes in an ecosystem affect populations within that ecosystem."
7. Say, "Okay, that is all the stuff we can use. Will you all head back to your teams so we can plan our investigation?"
8. Move on to the next activity.

What to Set Up

Pull up the simulation in the Learning Hub on a screen so that all students can see it. If you do not have a Learning Hub subscription, you can access the simulation by visiting *Carrying Capacity* by NetLogo's Media Commons.

What to Show

The "Some Materials You Can Use" page in the ADI Learning Hub

Teaching Tip

This activity helps students see for themselves all the materials they will be able to use during the hands-on part of the investigation instead of just reading about them or looking at pictures. It is important to show each material to the students and then point out how they can use it during the investigation so they can participate in developing a plan during the next activity.

Activity B	Make a Plan
Time Needed for the Activity	10 minutes
Grouping	Teams

What to Do

1. Have students work in their assigned teams during this activity. Give a copy of the Plan handout to each student.
2. Show the "Make a Plan" page on a screen using the Presentation mode in the Learning Hub so everyone in the classroom can see it.
3. Say, "Work as a team to create a plan for your investigation by filling out the chart on your handout. As part of your plan, be sure to include which tools and techniques you will use to solve problems."
4. Give the students at least 10 minutes to complete this task. As students work in small teams to create their investigation proposals, be sure to move from team to team to answer questions and offer advice.
5. When a team thinks that they have a good plan, read it over. If you agree, sign off at the bottom of their handouts to let them know that their plan has been approved. If you do not agree, tell them what they will need to add or change before you will approve it. Once they have made the changes, you can sign their handouts.
6. Once all the teams are done, you can move on to the next activity. You can also use an approved investigation plan as an "exit ticket" out of class.

What to Give Out

Plan handout for **Elk Population Decline in Colorado**

What to Show

The "Make a Plan" page in the ADI Learning Hub

What to Look For

The investigation plan should, at a minimum, include:

- The question they are trying to answer.
- What conditions they will test using the simulation.
- How they will analyze the data they collect from the simulation.
- How they will use their analysis to answer the guiding question.

Attending to Equity

To support emerging multilingual students during this stage, be sure to create a word bank on the wall of high-frequency, high-need concrete vocabulary that is needed to communicate scientific processes and ideas. The word bank highlights terms needed for basic communication in an academic context. It also helps with the spelling. Also, you can encourage your emerging multilingual students to use modes (speaking, writing, drawings, and gestures) and to use both content-specific and everyday registers to discuss and create their plans. This process helps students fully express themselves, allowing peers to evaluate, question, and build on ideas.

Teaching Tip

One way to do a quick check-in with a group during this activity is to use the "two questions, a nudge, and a goal" technique. This technique includes the following steps:

1. Sit down with a group of students.
2. Ask, "Would one of you tell me a little about what you all have been doing or thinking about?" (Question 1)
3. A student answers on behalf of the group.
4. Ask, "Would anyone else like to add anything to that?" (Question 2)
5. A different student provides additional information on behalf of the group.
6. Ask, "Have you thought about [a hint or something you want to highlight]?" (A nudge)
7. Say, "Okay, I'm going to check on the other groups. When I get back, I want to see [something that you want the group to accomplish]." (A goal)

When you have completed the check-in with one group, move on to the next group.

Teacher Notes

Elk Population Decline in Colorado Lesson Plan

Stage 4 DO

Total Time for the Stage	45 minutes
Grouping	Teams

Learning Goals

Students should make progress on their ability to do the following by the end of this stage of the lesson:

Science

1. Conduct an investigation individually and collaboratively to produce data to serve as the basis for evidence (SEP3).
2. Analyze data using tools and/or technologies to make valid and reliable scientific claims (SEP4).
3. Consider limitations of data analysis (e.g., measurement error and sample selection) when analyzing and interpreting data (SEP4).
4. Evaluate the impact of new data on a working explanation and/or model of a proposed process or system (SEP4).
5. Apply ratios, rates, percentages, and unit conversions in the context of complicated measurement problems involving quantities with derived or compound units (SEP5).
6. Make a quantitative and/or qualitative claim regarding the relationship between dependent and independent variables (SEP6).
7. Construct and revise an explanation based on valid and reliable evidence obtained from a variety of sources and the assumption that theories and laws that describe the natural world operate today as they did in the past and will continue to do so in the future (SEP6).
8. Apply scientific ideas, principles, and/or evidence to provide an explanation of phenomena taking into account possible unanticipated effects (SEP6).

English Language Development

1. Paraphrase central ideas in complex evidence, concepts, processes, and information to help explain how or why a phenomenon occurs (ELD-SLC.9-12.Explain.Interpretive).
2. Evaluate the extent to which reasoning, theory, and/or models link evidence to claims and support conclusions (ELD-SLC.9-12.Explain.Interpretive).
3. Develop reasoning to illustrate the relationships between variables (ELD-SLC.9-12.Explain.Expressive).

English Language Arts

1. Initiate and participate effectively in small-group collaborative discussions with diverse partners by building on others' ideas and expressing their own clearly (SL.9-10.1).
2. Present information clearly, concisely, and logically such that listeners can follow the line of reasoning and the organization, substance, and style are appropriate to the purpose, audience, and task (SL.9-10.4).
3. Adapt speech to a variety of contexts and tasks, demonstrating command of formal English when indicated or appropriate (SL.9-10.6).

Activities

Activity A	Carry Out Your Plan
Time Needed for the Activity	25 minutes
Grouping	Teams

What to Do

1. Students should work in their assigned teams for this activity. Begin this activity by giving out a copy of the Do handout to each student.
2. Show the "Carry Out Your Plan" page on a screen using the Presentation mode in the Learning Hub so everyone in the classroom can see it.
3. Say, "Okay, it sounds like we are now ready to carry out our plans and collect the data or observations we need to answer the guiding question. You can get the materials and tools you need to use from this table." Point to the table where you have placed them.
4. Ask, "Does anyone have any questions about these materials or tools?"
5. Answer any questions that come up.
6. Say, "As you work, be sure to record your observations and measurements on your handout so you can keep track of everything you do. Does anyone have any final questions?"
7. Answer any questions that come up.
8. Say, "Okay, go ahead and get started. You have the next 25 minutes to work. I'll be available to help if any team needs me."
9. Give the students at least 25 minutes to carry out their plans and solve the problem. As students work in small teams, be sure to move from team to team to answer questions and offer advice.

What to Set Up

Have each group access the simulation through the Learning Hub by using a personal or shared electronic device. If students do not have a Learning Hub subscription, they can access the simulation by visiting *Carrying Capacity* by NetLogo's Media Commons.

What to Give Out

Do handout for **Elk Population Decline in Colorado**

What to Show

The "Carry Out Your Plan" page in the ADI Learning Hub

Teaching Tip

One way to do a quick check-in with a group during this activity is to use the "two questions, a nudge, and a goal" technique. This technique includes the following steps:

1. Sit down with a group of students.
2. Ask, "Would one of you tell me a little about what you all have been doing or thinking about?" (Question 1)
3. A student answers on behalf of the group.
4. Ask, "Would anyone else like to add anything to that?" (Question 2)
5. A different student provides additional information on behalf of the group.
6. Ask, "Have you thought about [a hint or something you want to highlight]?" (A nudge)
7. Say, "Okay, I'm going to check on the other groups. When I get back, I want to see [something that you want the group to accomplish]." (A goal)

When you have completed the check-in with one group, move on to the next group.

Attending to Equity

Be sure to intentionally group emerging multilingual students at times with peers who know the same languages as they do and at other times with peers whose English language development is slightly more advanced. Thoughtful grouping that varies during different investigation (or activities within an investigation) allows emerging multilingual students to benefit from the different linguistic resources of their peers. We also recommend assessing emerging multilingual students based on how much they contribute to the group during this activity, rather than solely on what they write or draw on their handouts.

Teacher Notes

Activity B	Make Sense of Your Data
Time Needed for the Activity	20 minutes
Grouping	Teams

What to Do

1. Students should work in their assigned teams for this activity. Begin this activity by showing the "Make Sense of Your Data" page on a screen using the Presentation mode in the Learning Hub so everyone in the classroom can see it.
2. Say, "Okay, it sounds like we are now ready to make sense of the data we collected. One way to do this is to create a table, graph, or other visualization that shows a trend or change over time, a comparison between teams, or a relationship between things you tested. You can create a table, graph, or other visualization here on your handout."
3. Point to the box on the handout.
4. Say, "Please work together in your teams to figure out the best way to make sense of your data. Remember, you are looking for any trends, changes, differences, or relationships in the data that will help you answer the guiding question. As you work, be sure to use the space on your handout to perform any calculations that you need to do and create any representations that you need. Does anyone have any last questions?"
5. Answer any questions that come up.
6. Say, "Okay, go ahead and get started. You have the rest of the period to work. I'll be available to help if any team needs me."
7. Give the students at least 15 minutes to make sense of their data. As students work together in small teams, be sure to move from team to team to answer questions and offer advice.
8. Make sure students return the materials and tools to the table where they found them before dismissing class. Cleanup should take no longer than five minutes.

What to Show

The "Make Sense of Your Data" page in the ADI Learning Hub

What to Look For

One of the most useful ways to make sense of the data during this activity is to make a table that compares the initial conditions of the ecosystem with their effect on the elk population.

Teaching Tip

Collecting and returning materials or tools to a central table works best if you assign the job to a specific number within each team. For example, you can say, "2s, please come up and grab the materials and tools that your team needs" at the start of the activity. Then you can say, "3s, please return the materials and tools your team used back to this table," at the end of the activity. Assigning jobs like this not only cuts down on how many students need to be up and moving around, but also allows you to hold specific students accountable for the materials and tools.

Attending to Equity

Encourage your emerging multilingual students to speak using learning strategies such as requesting assistance, employing nonverbal cues, and using synonyms and circumlocution (conveying ideas by defining or describing when exact English words are not known) during this activity. It is also helpful to point out and/or draw emerging multilingual learners' attention to elements of the English sound system in newly acquired vocabulary such as long and short vowels, silent letters, and consonant clusters as they come up naturally during this activity.

Teacher Notes

Elk Population Decline in Colorado Lesson Plan

Stage 5 SHARE

Total Time Needed for the Stage	45 minutes
Grouping	Teams

Learning Goals

Students should make progress on their ability to do the following by the end of this stage of the lesson:

Science

1. Make and defend a claim based on evidence about the natural world that reflects scientific knowledge and student-generated evidence (SEP7).
2. Present an oral argument based on evidence (SEP7).
3. Evaluate the claims, evidence, and/or reasoning to determine the merits of arguments (SEP7).
4. Ask questions to clarify relationships, including quantitative relationships, between independent and dependent variables (SEP1).
5. Ask questions that challenge the interpretation of a dataset or the suitability of a method (SEP1).
6. Respectfully give critiques on scientific arguments by probing reasoning and evidence and challenging ideas and conclusions (SEP7).
7. Respectfully receive critiques on scientific arguments by responding thoughtfully to diverse perspectives and determining what additional information is required to resolve contradictions. (SEP7).
8. Communicate scientific information or ideas in multiple formats including orally, graphically, textually, and mathematically (SEP8).

English Language Development

1. Identify appropriate and sufficient evidence from data, models, and/or information from investigations of a phenomenon (ELD-SLC.9-12.Argue.Interpretive).
2. Compare reasoning and claims based on evidence from competing arguments (ELD-SLC.9-12.Argue.Interpretive).
3. Defend or refute a claim based on data and evidence (ELD-SLC.9-12.Argue.Expressive).
4. Establish and maintain an appropriate tone and stance (ELD-SLC.9-12.Argue.Expressive).
5. Signal logical relationships among reasoning, evidence, data, and/or models when making and defending a claim, counterclaim, and/or rebuttal (ELD-SLC.9-12.Argue.Expressive).

English Language Arts

1. Initiate and participate effectively in small-group collaborative discussions with diverse partners by building on others' ideas and expressing their own clearly (SL.9-10.1).
2. Integrate multiple sources of information presented in diverse media or formats (e.g., visually, quantitatively, and orally) evaluating the credibility and accuracy of each source (SL.9-10.2).
3. Evaluate a speaker's reasoning and use of evidence and rhetoric, identifying any fallacious reasoning or exaggerated or distorted evidence (SL.9-10.3).

(continued)

4. Present information clearly, concisely, and logically such that listeners can follow the line of reasoning and the organization, substance, and style are appropriate to the purpose, audience, and task (SL.9-10.4).
5. Make strategic use of media (e.g., textual, graphical, and visual) in presentations to enhance understanding of findings, reasoning, and evidence and to add interest (SL.9.10.5).
6. Adapt speech to a variety of contexts and tasks, demonstrating command of formal English when indicated or appropriate (SL.9-10.6).

Activities

Activity A	Create a Draft Argument
Time Needed for the Activity	15 minutes
Grouping	Teams

What to Do

1. Students should work in their assigned teams for this activity. Begin this activity by giving out a copy of the Share handout to each student.
2. Show the "Create a Draft Argument" page on a screen using the Presentation mode in the Learning Hub so everyone in the classroom can see it.
3. Say, "Okay, it sounds like we have all solved the task. I would now like you to create a draft argument to share what you figured out with the other teams. Your argument should include everything in this image here." Point to the image of a science-based argument projected on the screen. Be sure to go over what should be included in each component of the argument and where students can find the information they need.
4. Say, "I would like you to make your draft argument on a whiteboard. You can use the materials on this table to make your argument." Point to the table where you have placed the materials they need.
5. Ask, "Does anyone have any questions?"
6. Answer any questions that come up.
7. Say, "I'm going to give you exactly 15 minutes to get this done. Your draft argument doesn't need to be perfect. We just need something down on the whiteboards so we can share our ideas."
8. Give each team of students 15 minutes to create a draft argument. As students work in small teams, be sure to move from team to team to answer questions and offer advice.

What to Set Up

Place the following items on a table so students can use them during this stage:

- 24″ x 36″ whiteboard
- 3-4 dry-erase pens in different colors
- Eraser or cloth towel

What to Give Out

Share handout for **Elk Population Decline in Colorado**

What to Show

The "Create a Draft Argument" page in the ADI Learning Hub

What to Look for

Students should include core ideas that they read about during the Ideas stage in their justification.

ARGUMENT-DRIVEN INQUIRY

Teaching Tip

Be sure to plant some different ideas with each group about how to make a strong argument as you move from group to group to check in, answer questions, and offer advice. This strategy helps to ensure that there will be some really strong examples of the evidence and justification components of the argument for students to see and discuss during the next activity. For example, you could help two groups have a really strong evidence component by telling one group, "I think a graph would be really helpful here" and telling another group, "I think your evidence would be stronger if you explained, in words, what this graph shows." You could help a third group have a really strong justification by saying, "Let's brainstorm some ideas that will help you defend your choice of evidence," or "Let's go back to our Ideas handout and see if there are any concepts we can include."

It is important you make sure that all the arguments are different at this point in the investigation by planting different ideas with different groups because the students will have nothing to critique or discuss during the next activity if they all have the same argument. The goal during this activity is to set the stage for the next activity by ensuring that the various components of a strong argument can be found on different whiteboards within the class. These strong components will then be taken up by all the groups because they reflect good ideas. Be sure to give students a voice and choice during this activity, so they can make mistakes and learn from them.

Attending to Equity

It is often helpful to assign and change roles at least once during this activity. For example, you might assign the 1s to be "in charge of writing" at the beginning of the activity. You can then easily switch roles about five minutes into the activity by saying, "1s, please hand the pens to the 2s. 2s, you are now in charge of writing for the group." By changing roles, you can ensure that all students have a chance to participate and contribute their ideas to the creation of a draft argument.

ADI Go Digital

The **ADI Learning Hub** has tons of features that allow students to share their thinking with each other when they are doing the investigation paperless or remote.

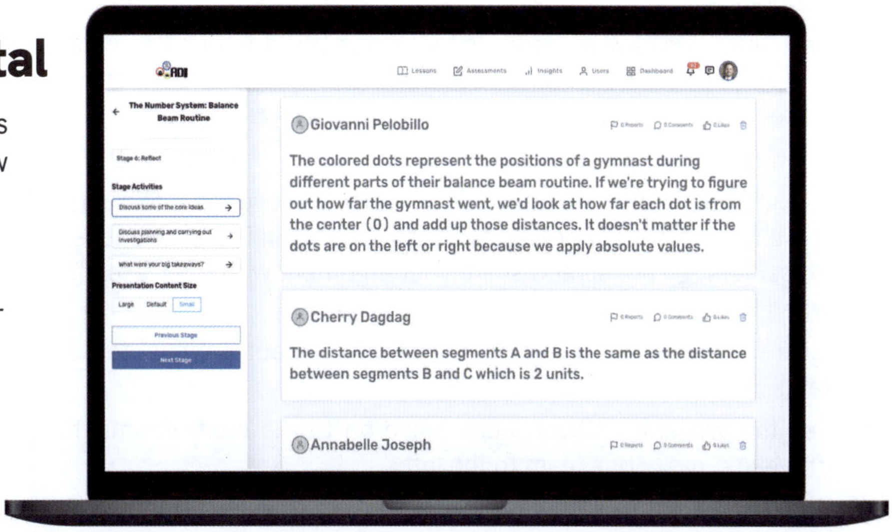

ARGUMENT-DRIVEN INQUIRY

Activity B	Present Your Draft Argument
Time Needed for the Activity	15 minutes
Grouping	Teams

What to Do

1. Students should work in their assigned teams for this activity.
2. Show the "Present Your Draft Argument" page on a screen using the Presentation mode in the Learning Hub so everyone in the classroom can see it.
3. Say, "We are now going to share our arguments. Please set up your whiteboards so everyone can see them." Allow students time to set up their whiteboards. They should be set up on stands or chairs, so they are facing toward the outside of the room.
4. Say, "Okay, before we get started, I want to explain what we are going to do next. I'm going to ask some of you to present your arguments to your classmates. If you are presenting your argument, your job is to share your team's claim, evidence, and justification of the evidence. The rest of you will be travelers. If you are a traveler, your job is to listen to the presenters, ask them questions if you do not understand something, and then offer them some suggestions about ways to make their argument better. Do you have any questions about these instructions?"
5. Answer any questions that come up.
6. Say, "As you talk to one another, make sure you write down any good ideas or suggestions that are important in the space on your handout, so you don't forget about them." Point to the box labeled, "Some possible ways to improve our argument" on the handout.
7. Say, "After we have a chance to learn from one another, I'm going to give you some time to revise your arguments and make them better. Do you have any questions about what we are doing?"
8. Answer any questions that come up.
9. Say, "Okay. Let's get started. Travelers, move one team to the left. Don't move to the next team until I tell you to move. Presenters, go ahead and share your argument with the travelers when they get there."
10. Give each team of presenters and travelers about 3-4 minutes to talk.
11. Tell the students, "Okay. Let's move on to the next argument. Travelers, move one team to the left."
12. Again, give each team of presenters and travelers about 3-4 minutes to talk.

What to Show

The "Present Your Draft Argument" page in the ADI Learning Hub

What to Look for

The critiques and feedback that students give one another during this activity should:

- Be shared in a way that is kind (respectful),
- Provide information about what, exactly, needs to be changed or improved (specific), and
- Offer a suggestion about how to make something better rather than just saying something is wrong (helpful).

In addition, students should be focusing their critiques on the science included in the arguments, not the appearance of the arguments.

(continued)

13. Repeat this process until each team has had their argument read and critiqued three times. Then tell the students to return to their workstations.
14. Be sure to move between workstations during this activity to keep everyone involved and model what it means to communicate and critique scientific ideas and reasoning.

Attending to Equity

Be sure to encourage students to provide feedback that is respectful, specific, and helpful during this activity. It is important to encourage student-to-student talk because it helps students develop speaking and listening skills and enables them to make sense of what and how they know. It also provides a context for emerging multilingual students to use both productive (speaking and writing) and receptive (listening and reading) language and learn science through meaningful interactions and scientific sense-making.

Teaching Tip

There are two types of formats for sharing and critiquing arguments. The first format is called a *gallery walk.* In this format, each group sets up their argument at their workstation so others can see it. The entire group then moves to a different workstation. While at a different workstation, the group is expected to read the argument at that workstation (which was created by a different group) and offer critiques and suggestions for improvement. The students should move to a different workstation after a few minutes to read and critique an argument written by a different group. Students should also be encouraged to keep a record of good ideas or potential ways to improve their own arguments as they move from workstation to workstation to offer critiques and feedback. After repeating this process in the gallery walk 3-4 times, every group will have had an opportunity to read and critique three or four different arguments and receive feedback from three or four different groups.

The second format is called a *modified gallery walk.* In the modified gallery walk format, one or two members of the group stay at their workstation to share their group's ideas (these students are called presenters) while the other members of their group go to different groups, one group at a time, to listen to and critique the arguments developed by their classmates (these students are called travelers). The travelers should visit at least three different workstations during the argumentation session. We also recommend that the presenters keep a record of the critiques made by their classmates and any suggestions for improvement. The travelers should also be encouraged to keep a record of good ideas or potential ways to improve their own arguments as they travel from group to group.

Attending to Equity

Another good way to have students, especially emerging multilingual learners, give one another feedback is to have them use four different color sticky notes to make comments. Give each group a single pink, blue, yellow, and green sticky note. Then direct the students to use the pink sticky note to suggest "something to add" to the argument, the blue sticky note to suggest "something to change" in the argument, the yellow sticky note to highlight "something they like" in the argument, and the green sticky note to ask a question about "something they don't understand". This strategy is useful because everyone in the classroom will be actively engaged during this activity, and everyone will have a chance to see different arguments. This format, however, does not give students a chance to support, critique, and challenge the ideas, solutions, scientific reasoning of other groups through talk because two different groups are never at the same workstation at the same time.

Activity C	Revise Your Argument
Time Needed for the Activity	15 minutes
Grouping	Teams

What to Do

1. Students should work in their assigned teams for this activity.
2. Show the "Revise Your Argument" page on a screen using the Presentation mode in the Learning Hub so everyone in the classroom can see it.
3. Say, "I'm now going to give you all about 10 minutes to revise your argument. Please return to your original teams."
4. Wait for the students to move back into their original teams.
5. Say, "Okay, take a few minutes to talk in your teams and determine what you want to change to make your argument better. Once you have decided what to change, go ahead and make the changes to your whiteboard."
6. Ask, "Do you have any questions about what you need to do?"
7. Answer any questions that come up.
8. Tell the students, "Okay. Let's get started."
9. Give the students 10 minutes to work in their teams on their arguments. As they work, move from team to team to check in, ask probing questions, and offer a suggestion if a team gets stuck.
10. When students are finished with their final arguments, have them take a picture of it or have them put the boards in a safe spot so they can refer to the picture or the boards during the Report stage.

What to Show

The "Revise Your Argument" page in the ADI Learning Hub

What to Look For

A strong argument includes the following information:

- A claim that relates the decline of the elk calf population to factors like predation, competition, and resource availability.
- A table that compares the initial conditions of the ecosystem with their effect on the elk population.
- An explanation of the table.
- An explanation of the chosen evidence that uses core ideas from the investigation.

Teaching Tip

It is important to give students time to meet with their original groups so they can discuss what they learned by interacting with individuals from the other groups. At this point in the process, students often realize that their initial problem-solving model was flawed in some way. It is therefore important to encourage students to "redo the science" as needed. At the end of this activity, each group should have a final argument that is much better than their original one.

Elk Population Decline in Colorado Lesson Plan

Stage 6 REFLECT

Total Time for the Stage	25 minutes
Grouping	Whole class, Teams

Learning Goals

Students should make progress on their ability to do the following by the end of this stage of the lesson:

Science

1. Ask questions that arise from careful observation of phenomena, or unexpected results, to clarify and/or seek additional information as needed (SEP1).
2. Explain how ecosystems have carrying capacities and how these limits result from factors such as the availability of living and non-living resources and challenges such as predation, competition, and disease (LS2.A).
3. Describe the complex set of interactions within an ecosystem that keep its numbers and types of organisms relatively constant over long periods of time under stable conditions (LS2.C).
4. Describe how extreme fluctuations in conditions and populations can challenge the functioning of ecosystems in terms of resources and habitat availability (LS2.C).
5. Use models to simulation systems and interactions between systems at different scales (CC4).
6. Construct explanations of how things change and how they remain the same (CC7).

English Language Development

1. Paraphrase central ideas in complex evidence, concepts, processes, and information to help explain how or why a phenomenon occurs (ELD-SLC.9-12.Explain.Interpretive).
2. Use scientific explanations about a phenomenon in a different context or situation (ELD-SLC.9-12.Explain.Expressive).

English Language Arts

1. Participate effectively in small-group collaborative discussions with diverse partners on texts by building on others' ideas and expressing their own clearly and persuasively (SL.9-10.1)

Activities

Activity A	Discuss a Core Idea
Time Needed for the Activity	10 minutes
Grouping	Whole class/Teams

What to Do

1. Students should be in small teams for this activity. However, they do not need to be in the same teams as they were during previous stages.
2. Begin this activity by giving out a copy of the Reflect handout to each student.
3. Show the "Discuss a Core Idea" page on a screen using the Presentation mode in the Learning Hub so everyone in the classroom can see it.
4. Say, "Okay. Now that we have all revised our arguments and made them better, let's take some time to discuss some of the core ideas you used during this investigation and to justify the evidence you used in your arguments, so we are all on the same page. Look at the picture on the screen."
5. Pause for about 30 seconds.
6. Say, "Take a few minutes to talk in your teams and see if you can use what you know about carrying capacity and interactions between organisms to explain what is happening in the picture."
7. Give the students at least three minutes to talk in their small teams.
8. Say, "Okay. Let's share with the whole class. Team A (or whichever team you want to share first), what did you talk about?"
9. Allow the first team to share.
10. Say, "Interesting. Would anyone like to add to, challenge, or refine the ideas shared by Team A?"

(continued)

What to Give Out

Reflect handout for **Elk Population Decline in Colorado**

What to Show

The "Discuss a Core Idea" page in the ADI Learning Hub

Explanation of the Image

The image that the students discuss during this activity shows three different samples of human blood cells.

What to Look For During the Discussion

Students who understand the content will likely mention some, if not all, of the following ideas when providing an explanation for the image:

- An ecosystem is a defined area that includes a community and the environment.
- A community consists of two or more populations, or a group of organisms that belong to the same species and live in the same place at the same time.
- The carrying capacity of an ecosystem is the maximum population size that the ecosystem can sustain.
- Populations usually fluctuate between being slightly over and slightly under the ecosystem's carrying capacity.
- When a population is greater than the ecosystem's carrying capacity, it is described as having "overshot."
- Individual organisms interact with one another within an ecosystem in three ways: predation and herbivory, competition, and symbiosis.

(continued)

11. Allow students to respond.
12. Say, "Okay. Team B (or whichever team you want to share second), what did you talk about?"
13. Allow the second team to share.
14. Say, "Interesting. Would anyone like to add to, challenge, or refine the ideas shared by Team B?"
15. Allow students to respond.
16. Keep up this pattern of questioning until all teams have shared. Be sure to keep encouraging students to talk about the content until they reach a correct and consensual explanation for what is shown in the image. If they are unable to reach a consensus or reach consensus using an inaccurate idea, be sure to provide an accurate explanation for the image before moving on to the next activity.

- Predation happens when an animal, or predator, kills and eats another animal, its prey.
- Herbivory happens when an organism feeds on a plant or other type of producer.
- Competition occurs when two or more organisms have similar needs and try to obtain the same limited resource.
- Symbiosis is when two or more species purposefully live in direct contact with each other for long periods of time.
- There are three main types of symbiosis: commensalism, parasitism, and mutualism.
- Commensalism is an interaction in which one organism benefits from the interaction while the other is neither helped nor harmed.
- Parasitism occurs when one individual benefits from the interaction while harming the other individual.
- Mutualism describes when all individuals benefit from the interaction.

Attending to Equity

Be sure to intentionally group emerging multilingual students at times with peers who know the same languages as they do and at other times with peers whose English language development is slightly more advanced. Also, be sure to ask students how they have seen the core idea in the everyday lives; this will make the topic of discussion more engaging and relatable.

Teacher Notes

Teaching Tip

This activity is a discussion, not a presentation about what the students "should have seen" or "should have learned" during the investigation. The more students talk during this activity, the more meaningful the experience will be for them and the more you can learn about their thinking. Be sure to use talk goals and moves to help facilitate the discussion. Below are some examples of talk goals that you can use to help facilitate the discussion and some questions that you can ask the students to help accomplish each goal. These talk goals and questions are only suggestions.

Goal 1: Encourage students to share, expand on, or clarify an idea.

- What do you see going on here?
- What do you think is happening?
- How might we explain that?
- Can you tell me more about that?
- What do you mean by that?
- Can you give me an example of that?
- Can you say that in another way?

Goal 2: Encourage students to support an idea.

- Why do you think that?
- What is your evidence for that?
- How did you arrive at that conclusion?
- What makes you say that?
- How do you know that?

Goal 3: Encourage students to think with others.

- Would anyone like to add to that idea?
- What should we keep or change about that idea?
- Can anyone else give me some examples of that?
- Can anyone take that idea and push it a little further?
- Does anyone want to respond to that idea?
- Do you agree or disagree with that idea, and why?
- What do we think about that idea?
- Does that fit with what we have figured out so far?
- Is there another way to do it?

Goal 4: Encourage students to think about what they know and don't know yet.

- I'm confused now. What have we figured out so far?
- What do we need to figure out now?
- So, what do we agree on right now, and what are we still trying to figure out?
- What are we sure and not so sure about?
- What questions do you have?

Teacher Notes

Activity B	Discuss the Practices
Time Needed for the Activity	10 minutes
Grouping	Whole class/Teams

What to Do

1. Show the first page of "Discuss the Practices" on a screen using the Presentation mode in the Learning Hub so everyone in the classroom can see it.
2. Say, "Okay. Now that we have talked about some of the core ideas that we used during our investigation, let's take a few minutes to talk about the science practices that we used during the investigation. An example of a scientific practice used was analyzing and interpreting data. On the screen there is a question that I want you all to think about: What other things did you do during your investigation that you feel were good examples of a scientific practice or attempting to figure out something in a scientific way?"
3. Pause for about 30 seconds.
4. Say, "Take a few minutes to talk about this question in your teams."
5. Give the students at least three minutes to talk in their small teams.
6. Say, "Okay. Let's share with the whole class. Team A (or whichever team you want to share first), what did you talk about?"
7. Allow the first team to share.
8. Say, "Okay. Team B (or whichever team you want to share second), what did you talk about?"
9. Allow the second team to share.
10. Continue asking teams until every team has a chance to share.
11. Show the second page of "Discuss the Practices" on a screen using the Presentation mode in the Learning Hub so everyone in the classroom can see it.
12. Say, "Here is another question that I want you all to think about: What things did you do during your investigation that were not good examples of a scientific practice or not very scientific?"
13. Pause for about 30 seconds.

What to Show First

The first page of "Discuss the Practices" in the ADI Learning Hub

What to Show Second

The second page of "Discuss the Practices" in the ADI Learning Hub

What to Show Third

The third page of "Discuss the Practices" in the ADI Learning Hub

(continued)

14. Say, "Take a few minutes to talk about this question in your team."
15. Give the students at least three minutes to talk in their small teams.
16. Say, "Okay. Let's share with the whole class. Team A (or whichever team you want to share first), what did you talk about?"
17. Allow the first team to share.
18. Say, "Okay. Team B (or whichever team you want to share second), what did you talk about?"
19. Allow the second team to share.
20. Continue asking teams until every team has a chance to share.
21. Show the third page of "Discuss the Practices" on a screen using the Presentation mode in the Learning Hub so everyone in the classroom can see it.
22. Say, "Here is a final question that I want you all to think about: What should we all do during our next investigation to make sure we are using scientific practices and being scientific in the way we try to figure something out?"
23. Pause for about 30 seconds.
24. Say, "Take a few minutes to talk about this question in your team."
25. Give the students at least three minutes to talk in their small teams.
26. Say, "Okay. Let's share with the whole class. Team A (or whichever team you want to share first), what did you talk about?"
27. Allow the first team to share.
28. Say, "Okay. Team B (or whichever team you want to share second), what did you talk about?"
29. Allow the second team to share.
30. Continue asking teams until every team has a chance to share.
31. Once everyone has shared, offer any suggestions or advice about what they can do to improve their next investigation and then move on to the next stage.

Teaching Tip

A great way to wrap up this activity is for the students to contribute to a list of class norms or "gotta-haves" to keep in mind when planning and carrying out an investigation in science. To create this list, each group can suggest for a potential norm or "gotta-have" to add to the list at the end of this activity, then the class votes on the one that they think is most important. The most important one gets added to the list. The list then serves as reminder for students during the Plan and Do stages of subsequent investigations. Over time, the list will grow and each addition to the list will be more nuanced than the last as students develop expertise with scientific processes.

Attending to Equity

Be sure to give all students, especially emerging multilingual students, enough time to talk in small groups before asking anyone to share with the entire class during this activity. Small-group structures offer all students, including emerging multilingual students, a chance to engage in sense-making with their peers as well as space to use their verbal and nonverbal (e.g., writing, drawing pictures, and using gestures) resources to express their ideas. Small-group structures also give students a chance to learn from other students' uses of these resources. When students are tasked with expressing their ideas, be sure that they know it is okay for them to do so in a mode or combination of modes (speaking, writing, drawings, and gestures).

Elk Population Decline in Colorado Lesson Plan

Stage 7 REPORT

Total Time Needed for the Stage	100 minutes (over three days)
Grouping	Individual, Teams

Learning Goals

Students should make progress on their ability to do the following by the end of this stage of the lesson:

Science

1. Use mathematical and/or computational representations to support explanations of factors that affect the carrying capacity of ecosystems at different scales (HS-LS2-1).
2. Evaluate claims, evidence, and reasoning that the complex interactions in ecosystems maintain relatively consistent numbers and types of organisms in stable conditions, but changing conditions may result in a new ecosystem (HS-LS2-6).
3. Make and defend a claim based on evidence about the natural world that reflects scientific knowledge and student-generated evidence (SEP7).
4. Construct a written argument based on evidence (SEP7).
5. Evaluate the claims, evidence, and/or reasoning to determine the merits of arguments (SEP7).
6. Respectfully give critiques on scientific arguments by and challenging ideas and conclusions (SEP7).
7. Communicate scientific information or ideas in multiple formats including graphically and textually (SEP8).

English Language Development

1. Identify appropriate and sufficient evidence from data, models, and/or information from investigations of a phenomenon (ELD-SLC.9-12.Argue.Interpretive).
2. Compare reasoning and claims based on evidence from competing arguments (ELD-SLC.9-12.Argue.Interpretive).
3. Defend or refute a claim based on data and evidence (ELD-SLC.9-12.Argue.Expressive).
4. Establish and maintain an appropriate tone and stance (ELD-SLC.9-12.Argue.Expressive).
5. Signal logical relationships among reasoning, evidence, data, and/or models when making and defending a claim (ELD-SLC.9-12.Argue.Expressive).

English Language Arts

1. Write routinely over extended time frames (time for reflection and revision) and shorter time frames for a range of discipline-specific tasks, purposes, and audiences (WHST.9-10.10).
2. Write arguments focused on discipline-specific content (WHST.9-10.1).
3. Write informative/explanatory texts, including the narration of scientific procedures/experiments or technical process (WHST.9-10.2).
4. Produce clear and coherent writing in which the development, organization, and style are appropriate to task, purpose, and audience (WHST.9-10.4).
5. Strengthen writing as needed by revising, editing, rewriting, and focusing on addressing what is most significant for a specific purpose and audience (WHST.9-10.5).
6. Determine the central ideas or conclusions of a text; trace the text's explanation or depiction of a complex process, phenomenon, or concept; provide an accurate summary of the text (RST.9-10.2).
7. Determine the meaning of symbols, key terms, and other domain-specific words and phrases as they are used in a specific scientific or technical context (RST.9-10.4).
8. Analyze the structure of the relationships among concepts in a text, including relationships among key terms (RST.9-10.5).
9. Assess the extent to which the reasoning and evidence in a text support the author's claim (RST.9-10.8).
10. Compare and contrast findings presented in a text to those from other sources (including their own experiments), noting when the findings support or contradict other accounts (RST.9-10.9).

Teacher Notes

Activities

Activity B	Draft Report
Time Needed for the Activity	20 minutes
Grouping	Individual

What to Do

1. Have students work on their own during this activity. Give a copy of the Report handout to each student.
2. Show the "Create a Draft Report" page on a screen using the Presentation mode in the Learning Hub so everyone in the classroom can see it.
3. Say, "You are now going to write an investigation report to share what you figured out. The first part of the report is called the 'Introduction'. In this section, you should explain to the reader what you were investigating, why you were investigating it, and what question you were trying to answer. All this information can be found on your Task handout."
4. Hold up the Task handout.
5. Ask the students, "Do you have any questions about what you need to do?"
6. Answer any questions that come up.
7. Say, "Here are some sentence starters to help you begin writing."
8. Hold up the Report handout and point to the sentence starters.
9. Tell the students, "Okay, let's write."
10. Give the students five minutes to write the "Introduction" section of the report. As they work, move from student to student to check in, ask probing questions, and offer a suggestion if a student gets stuck.
11. Say, "Okay, let's move on to the second part of the report. The second part of the report is called the 'Method'. In this section, you should explain to the reader what you did during the investigation to complete the task and answer the guiding question.

(*continued*)

What to Give Out

Report handout for **Elk Population Decline in Colorado**

What to Show

The "Create a Draft Report" page in the ADI Learning Hub

All this information can be found in the Plan handout. Remember that you all planned and carried out different investigations, so do not assume that the reader will know what you did."

12. Ask the students, "Do you have any questions about what you need to do?"
13. Answer any questions that come up.
14. Say, "Here are some sentence starters to help you begin writing."
15. Hold up the Report handout and point to the sentence starters.
16. Tell the students, "Okay, let's write."
17. Give the students five minutes to write the "Method" section of the report. As they work, move from student to student to check in, ask probing questions, and offer a suggestion if a student gets stuck.
18. Say, "Okay, let's move on to the last part of the report. The third part of the report is called the 'Argument'. In this section, you should explain to the reader what you figured out and how you know you are correct. All this information can be found on your Do handout and on your team's whiteboard."
19. Hold up the Do handout and point to the boards (or remind them that they have a picture of their board).
20. Ask the students, "Do you have any questions about what you need to do?"
21. Answer any questions that come up.
22. Say, "Here are some sentence starters to help you begin writing."
23. Hold up the Report handout and point to the sentence starters.
24. Tell the students, "Okay, let's write."
25. Give the students 10 minutes to write the "Argument" section of the report. As they work, move from student to student to check in, ask probing questions, and offer a suggestion if a student gets stuck.
26. Collect all the draft reports before moving on to the next activity. Tell students to mark their draft reports with a symbol that they can use to identify it later. They should NOT write their names on their draft reports.

Teaching Tip

Writing is an important component of this model. When a student writes the report on their own, it not only helps the student learn to write in the context of science (an important literacy skill) but also allows that student to write to learn (develop a better understanding of scientific concepts by writing about them). There are many writing supports (e.g., sentence starters) already embedded into this activity, so you might not need to provide much extra support for your students.

This activity can be done completely in class or started in class and then finished as homework. Do not assign the entire activity as homework unless you know that every student in your class has dedicated time, space, and resources to write outside of class and everyone understands and is comfortable with the report writing process.

If you are concerned about students feeling overwhelmed, you can develop a class "gotta-have-it" list of information that needs to be included in each section of the report. Students can then refer to this list as they write each section.

Don't worry if students do not produce a perfect report at this point. All the reports will be reviewed and revised before they are turned in to you. The reports should therefore be viewed as a starting point. Be sure to encourage students to "just get something down on paper" so "we can work on it together."

Attending to Equity

Emerging multilingual students' access to instructional tasks that require complex thinking, such as writing a report, is enhanced when linguistic complexity and instructional support match their current level of language proficiency. Therefore, it is important to scaffold the writing of the reports for the emerging multilingual students based on their current English language proficiency level. The use of sentence starters to model written language production skills can help emerging multilingual students develop English language skills because the development of academic language and academic content knowledge are interrelated processes. However, it is important that such scaffolds be used purposefully and removed when no longer needed.

ADI Go Digital

Use the **ADI Learning Hub** to go paperless. When you create a class and add students to it, students can complete all the stages right in the system.

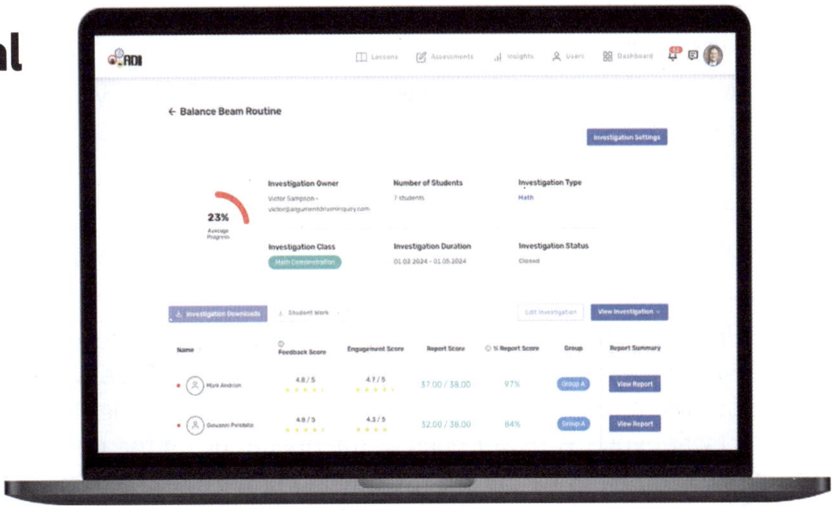

Activity B	Peer Review the Reports
Time Needed for the Activity	45 minutes
Grouping	Teams

What to Do

1. Students should work in teams of three for this activity. These can be completely different teams.
2. Show the "Peer Review Some Draft Reports" page on a screen using the Presentation mode in the Learning Hub so everyone in the classroom can see it.
3. Say, "We are now going to review our reports to find ways to make them better."
4. If possible, use a document camera to project the peer review guide (PRG) on a screen.
5. Say, "We are going to use this peer review guide to give one another feedback." Point to the image.
6. Say, "I'm going to ask you to work in teams of three to do this. I will give each team a draft report to read. You three will then read the report together. Once you are done reading the report, I want you to answer each of the questions on the peer review guide." Point to the review questions on the image of the PRG.
7. Tell the students, "You can check No, Almost, or Yes after each question." Point to the checkboxes on the image of the PRG.
8. Say, "This will be your rating for this part of the report. Make sure you agree on the rating you give the author. If you mark No or Almost, then you need to tell the author what he or she needs to do to get a Yes." Point to the space for the reviewer feedback on the image of the PRG.
9. Say, "It is really important for you to give the authors feedback that is helpful. That means you need to tell them exactly what they need to do to make their reports better."
10. Ask the students, "Do you have any questions about what you need to do?"
11. Answer any questions that come up.
12. Tell the students, "Please get into your teams of three."
13. Allow the students time to move into their triads.
14. Say, "Okay, I'm now going to give you one report to read." Give out one draft report to each triad. Make sure that the report you give a triad was not written by one of the students in that triad. Give each triad one PRG to fill out.

What to Give Out

Peer Review Guide (one per group per review, so for this investigation you need the same number of peer review guides as the number of draft reports that will be reviewed)

What to Show

The "Peer Review Some Draft Reports" page in the ADI Learning Hub

What to Look For

Make sure students are reaching consensus about what rating to give each criterion. Also, be sure that students are giving one another feedback that is specific, detailed, and helpful when they think an aspect of a report needs to be improved.

(continued)

15. Tell the students, "Okay, I'm going to give you 15 minutes to read the report I gave you and to fill out the peer review guide. Go ahead and get started."
16. Give the students 15 minutes to work. As they work, move around from team to team to check in and see how things are going, answer questions, and offer advice.
17. After 15 minutes pass, tell the students, "Okay, time is up. Please give me the report and the peer review guide that you filled out."
18. Collect the draft reports and the PRGs, or collect the workbooks if students are using them.
19. Say, "Okay, I am now going to give you a different report to read and a new peer review guide to fill out." Give out one more report to each triad. Make sure that the report you give a triad was not written by one of the students in that triad. Give each triad one new peer review guide to fill out as a team.
20. Say, "Okay, I'm going to give you 15 minutes to read this new report and to fill out the peer review guide. Go ahead and get started."
21. Give the students 15 minutes to work. As they work, move around from team to team to check in and see how things are going, answer questions, and offer advice.
22. After 15 minutes pass, tell the students, "Okay, time is up. Please give me the report and the peer review guide that you filled out."
23. Repeat this process for the remaining draft reports. Each triad should one have one more report to review (for a total of three reviews).
24. Collect the last draft reports and PRGs. Set all the draft reports on a table in separate stacks. Each stack should have a draft report on top (with the identification symbol facing up so students can find their reports) and the PRG on the bottom.

Teaching Tip

The peer review process is one of the best ways for students to learn how to write. When students read and review different reports, they have an opportunity to (1) see examples of texts written by others for the same purpose, (2) discuss what counts as quality and why some reports are stronger than others, (3) discuss ways to strengthen a report, and (4) pick up things that they can do to improve their reports. This helps the students learn new ways to organize and present information, which will help them write better on subsequent reports.

Attending to Equity

One helpful scaffold is to give students sentence stems for what to write in their feedback to the author. Ideal feedback is respectful, specific about what needs to be changed, and helpful in that it suggests how the author should go about making that change. You can review the writing of the emerging multilingual students in your class if they have limited English language proficiency, but make sure that all emerging multilingual students participate in the peer review process. You will see tremendous growth in their writing skills as they review their peers' writing and discuss what makes some reports stronger than others.

Activity C	Revise the Reports
Time Needed for the Activity	25 minutes
Grouping	Individuals

What to Do

1. Students should work on their own for this activity.
2. Show the "Reflect on the Feedback" page on a screen using the Presentation mode in the Learning Hub so everyone in the classroom can see it.
3. Say, "Please come pick up your reports from over here on this table." Point to the table.
4. Give the students a few minutes to pick up their reports and return to their seats.
5. Say, "Please take a few minutes to read over the peer review guide. You should use it to figure out what you need to change in your report and how you will change it."
6. Allow the students to read the PRG.
7. Show the "Create a Final Report" page on a screen using the Presentation mode in the Learning Hub so everyone in the classroom can see it.
8. Say, "Okay, let's revise our reports. I would like you to rewrite your report. You can use your draft report as a starting point but use the feedback on the peer review guide to make it better."
9. Ask the students, "Do you have any questions about what you need to do?"
10. Answer any questions that come up.
11. Tell the students, "Okay, let's write." Allow about 20 minutes for the students to revise their reports.
12. After about 20 minutes, give the following directions, "Okay, time's up. I will now come around and collect your first draft, final report, and peer review guide."
13. Collect all the first drafts, PRGs, and final reports.
14. Use the "Teacher Score" columns in the PRG to grade the final reports. Be sure to give the students feedback about their writing in the "Teacher Comments" section.

What to Show First

The "Reflect on the Feedback" page in the ADI Learning Hub

What to Show Second

The "Create a Final Report" page in the ADI Learning Hub

Attending to Equity

Students don't always use the feedback they receive from others to improve the quality of their reports. Make sure you move from student to student and encourage them to make changes based on the feedback. If students are not sure how to revise their reports, model how to use feedback. For example, you might say, "Okay, based on the peer review guide, I need to do more to describe what I did to collect data. I bet if I add information about ________, readers will have a better idea of what I did."

Teaching Tip

The peer review guide (PRG) is designed to be used with any ADI investigation, thus allowing you to use the same scoring rubric throughout the entire year. This is beneficial for several reasons. First, the criteria for what counts as a high-quality report do not change from investigation to investigation. Students, therefore, quickly learn what is expected from them when they write a report, and you do not have to spend valuable class time explaining the various components of the PRG each time students write a report. Second, the PRG makes it clear which components of a report need to be improved next time because the grade is not based on a holistic evaluation of the report. The PRG format allows students to see which aspects of their writing are strong enough and which aspects need improvement. Finally, and perhaps most importantly, the PRG provides you with a standardized measure of student performance that can be compared over multiple reports across semesters, thus allowing you to track improvement over time.

Teacher Notes

Elk Population Decline in Colorado

TASK

The Phenomenon

Elk are the second largest species within the deer family, Cervidae, and one of the largest terrestrial mammals in North America. Elk weigh between 500 and 700 pounds, are 5 feel tall at the shoulder, and are about eight feet long from nose to tail. An elk eats an average of 3 pounds of grass per day for every 100 pounds of body weight; therefore, a large elk that weighs 700 pounds needs to eat about 21 pounds of grass each day. The pictures below show some examples of elk.

Examples of Elk

Elk are native to Colorado. However, the number of elk in Colorado has been rapidly declining over the past few years and scientists are not sure why. Take a few minutes to watch a video about what is happening to the size of the Colorado elk population. As you watch the video, write down some things you notice in the space below.

Things **I noticed...**

ARGUMENT-DRIVEN INQUIRY

The Task

The video highlighted the fact that the elk calves in Colorado are dying more often than usual. This observation is a cause for concern because elk breed only once a year, and each female only gives birth to one calf at a time. Your goal in this investigation is to figure out what changes to an ecosystem could cause elk calves to start dying off, and then develop a model to explain why the entire elk population in Colorado is in decline as a result. The guiding question of this investigation is:

What potential changes in an ecosystem might be triggering the rapid decline of the Colorado elk population, and why?

Initial Ideas

Before you start this investigation, take a few minutes to think about all the different components of an ecosystem. Then, draw a model in the space below that shows how you think a change in one or more of these components might cause the size of an elk population to start to decrease. Be sure to include labels to help explain your thinking.

Some **ideas** that I have...

Share your model with the other members of your group. Once everyone has had a chance to share their idea, list some questions that your group has about this phenomenon.

Some questions my group has now...

Elk Population Decline in Colorado

IDEAS

Some Ideas You Can Use: Carrying Capacity

An **ecosystem** is a defined area that includes a community and the environment. A **community** is defined as two or more populations. A **population** is a group of organisms that belong to the same species and live in the same place at the same time. A **species** is a type of organism. The image below shows the difference between a species, a population, a community, and an ecosystem. In this image, there are three species: pine trees, deer, and beetle. Since there is more than one individual in each species, there are also three populations. The three populations create a single community. The community along with the sunlight, water, and space is an ecosystem.

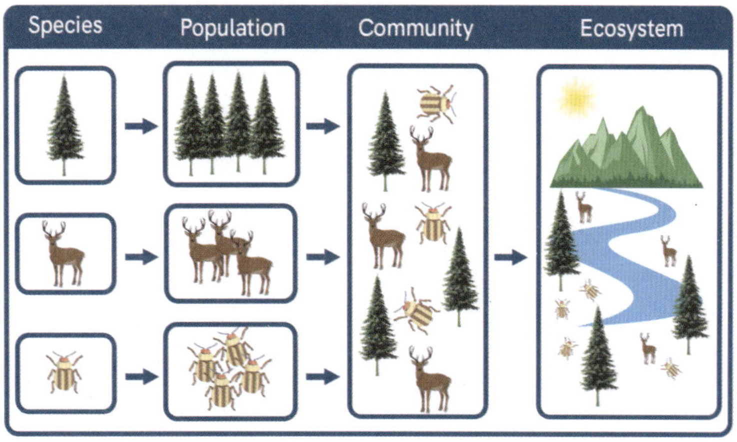

The graph below shows how the size of a population will change over time based on the carrying capacity of an ecosystem. The **carrying capacity** of an ecosystem is the maximum population size that the ecosystem can sustain. A small population will grow exponentially while there are plenty of resources available. The population will then "overshoot" the carrying capacity of the ecosystem. Overshoot occurs when the size of the population is greater than the carrying capacity, so some individuals in the population die because they cannot find enough resources to survive. At this point, the death rate of the population is greater than the birth rate. Once the population size is less than the carrying capacity, the birth rate will be greater than the death rate again.

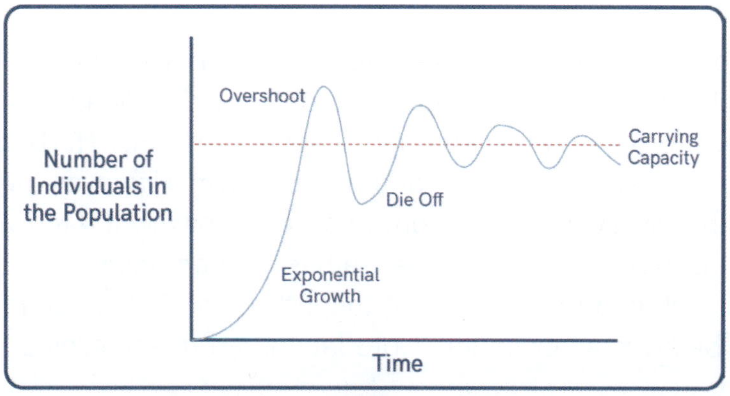

As the birth rate and the death rate of the population change, the number of individuals in the population will fluctuate above and below the carrying capacity. Eventually, the population reaches a stable equilibrium point at or near the carrying capacity of the ecosystem. Ecosystems, however, are dynamic in nature and the carrying capacity for any population can decrease or increase when a component of an ecosystem or some interaction between two or more components of an ecosystem change in some way.

Some **useful ideas** from what I read...

Some Ideas You Can Use: Interactions Between Organisms

There are several different ways that individual organisms can interact with one another within an ecosystem. The different interactions are organized into three categories based on the outcome of the interaction: predation and herbivory, competition, and symbiosis. However, the exact nature of these interactions differs by organism or population and can change over time depending on environmental conditions.

Predation involves one individual, the predator, killing and eating another individual, the prey. In most examples of this interaction, the predator and prey are both animals; however, protozoans are known to prey on bacteria and other protozoans. **Herbivory** is like predation because it involves one organism feeding on another organism. However, there are two important differences between herbivory and predation. First, in herbivory the individual that is eaten is a primary producer, such as a plant or algae, rather than an animal or a protozoan. Second, herbivory often does not lead to the death of the individual that is eaten. The bird eating the insect in the picture below is an example of predation and the caterpillar eating the leaf is an example of herbivory.

Competition is an interaction that occurs when two or more organisms with similar needs are trying to obtain the same limited resource. This type of interaction is harmful to at least one individual because limited access to a resource constrains their growth, reproduction, or survival. Competition between individuals can occur within the same population or across different populations. The pictures below show competition between trees and smaller plants for light and between male turkeys for a mate.

Symbiosis is an interaction characterized by two or more species living purposefully in direct contact with each other for long periods of time. There are three main types of symbiosis. The first type of symbiosis is called commensalism. **Commensalism** is an interaction in which one individual benefits while the other is neither helped nor harmed. The relationship between the bird and tree in the first picture below is an example of commensalism since the bird can nest in the tree without harming or helping the tree in the process. The second type of symbiosis is called parasitism. **Parasitism** occurs when one individual, the parasite, benefits from another individual, the host, while harming the host in the process. Parasites feed on host tissue or fluids and can be found within or outside of the host body. The bark beetle is an example of a parasite. The third type of symbiosis is called mutualism. **Mutualism** is an interaction where all individuals involved benefit from living in direct contact. The relationship between bark beetles and the fungus they carry is an example of mutualism because both the beetle and fungi benefit from the relationship. The fungi benefit from the beetles because the beetles move them to new trees and the beetles benefit from the fungi because the fungi help break down tree tissues, which makes it easier for the beetles to feed.

Some **useful ideas** from what I read...

Some Ideas You Can Use: System Models

Models can be used to represent a system and the way the parts that make up that system interact with one another. A **model** can be physical, mathematical, or a computer-based simulation. You will use a computer-based simulation as your model in this investigation (see image below). This simulation has been coded so it represents both the biological and physical components of an ecosystem. The **biological components** include populations of grass, elk, rabbits, and wolves. The elk and rabbits feed on the grass, which makes them both primary consumers (or herbivores). The wolves are secondary consumers (or predators) that feed on the elk and rabbit. The **physical components** of the ecosystem in the model include the amount of space and temperature. This model, as a result, will enable you to figure out what changes to an ecosystem cause an elk population to increase or decrease in size over time.

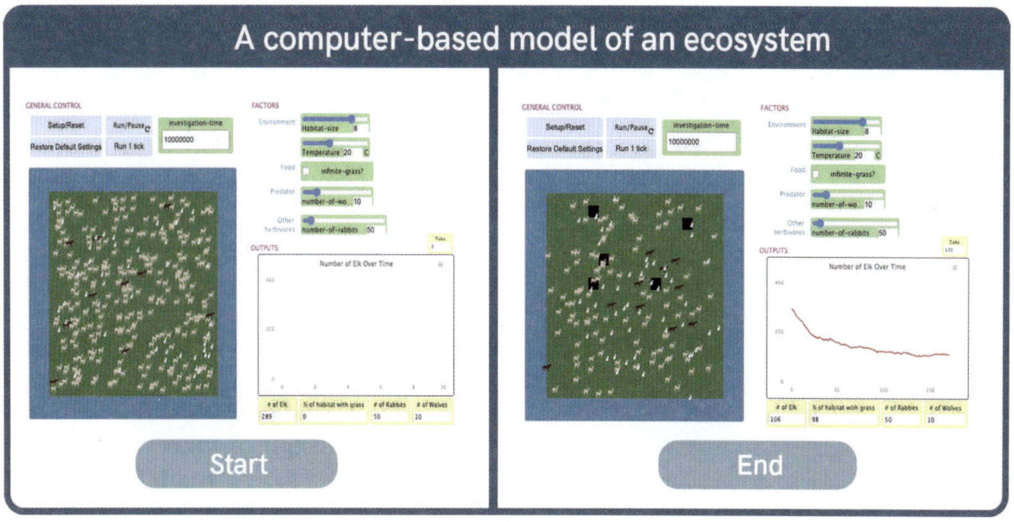

You can change a biological component of the ecosystem, such as the "number of wolves", "numbers of rabbits", and the amount of grass (finite or infinite) at the start of the simulation. These biological components of the ecosystem will be the **inputs** of the system model. You can also change the physical components of the ecosystem in the model by changing the "temperature", which affects how quickly the grass grows, or the "habitat size", which affects how much space there is for grass to grow and the initial size of the elk population at the start of the simulation. These physical components will affect the **interactions** that take place in the system model. How the size of the elk population changes over time or its final size after a period time (such as 200 years) will be the **output of interest** for this investigation. Models like this can be used to predict the behavior of a system under different conditions, such as when the biological and physical components of an ecosystem are altered in some way, but it is important to keep in mind that **these predictions have limited precision and reliability** due to the assumptions and approximation inherent in this model.

Some **useful ideas** from what I read...

Elk Population Decline in Colorado

PLAN

Plan Your Investigation

Prepare a plan for your investigation by filling out the chart below.

I am trying to answer the following **question**...

I will use the following **observations or measurements**...

I will **collect** these observations or measurements by...

I will **analyze** these observations or measurements by...

I approve of this investigation plan.

Teacher's signature

Date

Elk Population Decline in Colorado

DO

Collect Data

Keep a record of what you observe or measure during your investigation.

My **observations or measurements...**

Make Sense of Your Data

Use the space below to create your model so you can answer the guiding question.

My **model...**

Elk Population Decline in Colorado

SHARE

Draft Argument

Develop an argument on a whiteboard. It should include:

1. A *claim* that answers the guiding question,
2. *Evidence* in support of the claim that provides an analysis of the data and a description of what the analysis means,
3. A *justification of the evidence* that explains why your group thinks the evidence is important.

Our Guiding Question:

Claim:

Evidence:	Justification of the Evidence:

Argumentation Session

Share your argument with your classmates. In the space below, be sure to keep track of any ideas that you can use to revise your argument and make it better.

Some possible ways to **improve** our argument...

Elk Population Decline in Colorado

REFLECT

Reflective Discussion

In the space below, you can keep track of any ideas from the discussion that you think are important or will be useful in the future.

Some important **ideas**...

Elk Population Decline in Colorado

REPORT

Draft Report

Prepare a report to share what you figured out during your investigation.

Introduction

We have been studying ______________________________ in class. Before we started this investigation, we explored ______________________________

____

____

We noticed ______________________________

____

____

Our goal for this investigation was to figure out ______________________________

____

____

The guiding question was ______________________________

____

Method

To answer this question, ______________________________

____

____

____

____

____

____

____

____

____

Argument

The model below shows ____

My evidence for this model is____

This evidence is based on several important scientific ideas. The first one is ____________________

Bark Beetles and Climate Change in Washington Lesson Plan

OVERVIEW

Description	Students figure out how to reduce bark beetle outbreaks and prevent resulting tree loss in Washington state.
Guiding Question	What can people do to reduce bark beetle outbreak occurrences and tree loss in Washington state?
What Students Figure Out	We can limit bark beetle outbreaks and tree loss by reducing the number of susceptible host trees by removing some trees from a forest before they are infested by bark beetles, making it harder for beetles to find new trees to infest.

STAGES

This investigation has seven stages. Each stage includes multiple activities. The flowchart below shows what happens during each stage and how much time is needed to complete each one.

Task (Day 1)
1 — **20 minutes**
Students are introduced to a phenomenon and a guiding question to answer. They also share their initial ideas.

Ideas (Day 1)
2 — **30 minutes**
Students read about DCI, a CC, and an SEP that they can use to help them figure out the phenomenon.

Plan (Day 2)
3 — **15 minutes**
Students formulate a plan collecting and analyzing the data they need to answer the guiding question.

Do (Day 2)
4 — **30 minutes**
Students carry out their plan, make sense of the data they collected, and answer the guiding question.

Share (Day 3)
5 — **45 minutes**
Students develop, share, and revise an evidence-based argument about what they figured out.

Reflect (Day 4)
6 — **20 minutes**
Students discuss a DCI and/or a CC and reflect on the SEPs they used during the investigation.

Report (Day 4)
7 — **25 minutes**
Students write investigation reports to share what they figured out during the investigation.

Report (Day 5)
7 — **45 minutes**
Students peer review the reports written by other students and offer feedback to improve.

Report (Day 6)
7 — **30 minutes**
Students revise their reports and then submit a final draft to their teacher for evaluation.

STANDARDS ALIGNMENT

NEXT GENERATION SCIENCE STANDARDS

Performance Expectation

1. Use mathematical representations to support and revise explanations based on evidence about factors affecting biodiversity and populations in ecosystems of different scales (HS-LS2-2).
2. Evaluate claims, evidence, and reasoning that the complex interactions in ecosystems maintain relatively consistent numbers and types of organisms in stable conditions, but changing conditions may result in a new ecosystem (HS-LS2-6).
3. Design, evaluate, and refine a solution for reducing the impacts of human activities on the environment and biodiversity (HS-LS2-7).

Core Ideas(s)

Interdependent Relationships in Ecosystems (LS2.A)
Ecosystem Dynamics, Functioning, and Resilience (LS2.C)
Biodiversity and Humans (LS4.D)
Developing Possible Solutions (ETS1.B)

Crosscutting Concept(s)

Systems and System Models (CC4)
Stability and Change (CC7)

Practices

Asking Questions (SEP1)
Developing and Using Models (SEP2)
Planning and Carrying Out Investigations (SEP3)
Analyzing and Interpreting Data (SEP4)
Using Mathematics and CT (SEP5)
Constructing Explanations (SEP6)
Engaging in Argument from Evidence (SEP7)
Obtaining, Evaluating, and Communicating Information (SEP8)

WIDA ENGLISH LANGUAGE DEVELOPMENT STANDARDS FOR SCIENCE

Explain

Interpret scientific explanations (ELD-SC.9-12.Explain.Interpretive)
Construct scientific explanations (ELD-SC.9-12.Explain.Expressive)

Argue

Interpret scientific arguments (ELD-SC.9-12.Argue.Interpretive)
Construct scientific arguments (ELD-SC.9-12.Argue.Expressive)

COMMON CORE ENGLISH LANGUAGE ARTS

Reading	Writing	Speaking and Listening
Key ideas and details (RST.9-10.1 and RST.9-10.2) Craft and structure (RST.9-10.4) Integration of knowledge and ideas (RST.9-10.7, RST.9-10.8, and RST.9-10.9) Range of reading (RST.9-10.10)	Text types and purposes (WHST.9-10.1 and WHST.9-10.2) Production and distribution of writing (WHST.9-10.4, WHST.9-10.5, and WHST.9-10.6)	Comprehension and collaboration (SL.9-10.1 and SL.9-10.3) Presentation of knowledge and ideas (SL.9-10.4 and SL.9-10.6)

THINGS TO KNOW BEFORE STARTING THE LESSON

The Core Ideas

An ecosystem is a defined area that includes a community and the environment. A community, in turn, is defined as a group of two or more populations in a region, and a population is a group of organisms that belong to the same species and live in the same place at the same time. Because an ecosystem has a limited number of available resources, a population naturally settles around the ecosystem's carrying capacity, or the maximum number of organisms in a population that an ecosystem can sustainably support. A population will fluctuate around the carrying capacity, sometimes exceeding and other times falling below, but will usually correct to the limit set by the ecosystem. Therefore, ecosystems remain relatively consistent and stable over long periods of time.

However, extreme changes to an ecosystem can cause the ecosystem to lose the ability to stabilize itself and to instead transform into a new ecosystem, often causing the loss of biodiversity in the region. These extreme changes can be the result of human activity—such as habitat destruction, pollution, introduction of an invasive species, and climate change—damaging the ecosystem. Humans depend on ecosystems and their associated biodiversity, so we must generate solutions to repair ecosystems damaged by these activities.

The Crosscutting Concepts

An ecosystem is a group of living and non-living things that interact with one another and fulfill different functions. Because the nature of an ecosystem is affected by a variety of factors and is therefore very complex, we often use models to represent ecosystems. We can then use these models to test how changing different factors would affect the ecosystem as a whole, without actually changing the ecosystem itself. These models let us make predictions about and understand the world around us.

The Science and Engineering Practices

A practice of science is to **ask and refine questions** that lead to descriptions and explanations of how the natural world works. Another practice of science is to **use and construct models** as helpful tools for representing ideas and explanations. These tools include diagrams, drawings, physical replicas, mathematical representations, analogies, and computer simulations. Scientists **plan and carry out investigations** in the field or a laboratory, working collaboratively and individually. Their investigations are systematic and require clarifying what counts as data and identifying variables or parameters. Scientific investigations produce **data that must be analyzed and interpreted**. Because data patterns and trends are not always obvious, scientists use a range of tools and practices to identify the significant features and patterns in the data. Scientists also use **mathematics and computational thinking** for a range of tasks such as constructing simulations; statistically analyzing data; and recognizing, expressing, and applying quantitative relationships. The products of doing science are **explanations**. Scientific explanations often specify variables and are used by scientists to describe and predict different natural phenomena. **Scientific argumentation**, or the process of sharing, supporting, critiquing, and revising claims based on evidence is the process that scientists use to determine if an explanation is valid or acceptable. Finally, scientists must be able to **communicate clearly and persuasively** the ideas and methods they generate. Critiquing and communicating ideas individually and in groups is a critical professional activity.

PREPARATION

This lesson can be delivered paper-based, paperless, or remote. Use the paper-based delivery option of the investigation if students do not have 1:1 access to a computer and they are in the same classroom or physical space during the lesson. Use the paperless delivery option of the investigation if students have 1:1 access to a computer and they are in the same classroom or physical space during the lesson. Finally, if students are not in the same classroom or physical space during the lesson and have 1:1 access to a computer, use the remote delivery option.

Teacher Notes

Materials Needed

The materials needed by the teacher and each student in the class for each of the three modalities are listed in the table below.

Delivery Mode	Each Student Will Need	The Teacher Will Need
Paper-Based	• Copies of each student handout for the investigation (or a workbook) • A pencil	• A teacher account in the ADI Learning Hub • Access to a computer with high-speed internet and a browser (such as Chrome, Safari, Firefox, or Edge) • A projector or smart board (for displaying a computer screen)
Paperless	• A student account in the ADI Learning Hub • Access to a computer with high-speed internet and a browser (such as Chrome, Safari, Firefox, or Edge)	• A teacher account in the ADI Learning Hub • Access to a computer with high-speed internet and a browser (such as Chrome, Safari, Firefox, or Edge)
Remote	• A student account in the ADI Learning Hub • Access to a computer with high-speed internet and a browser (such as Chrome, Safari, Firefox, or Edge)	• A teacher account in the ADI Learning Hub • Access to a computer with high-speed internet and a browser (such as Chrome, Safari, Firefox, or Edge)

Additional Materials and Preparation Needed for In-Person Delivery

This in-person investigation will require students to have access to a Chromebook, desktop, laptop, iPad, or other electronic device in order to use the simulation. The simulation is embedded in the Learning Hub, so students do not need access to external sites. If students do not have access to the Learning Hub, they can use the simulation by visiting *Climate-Driven Bark Beetle Outbreaks* by NetLogo. The teacher can determine if each student will use an electronic device or if each group will share.

How the Simulation Works

This model simulates an ecosystem of bark beetles and pine trees. The model has 289 patches (light brown squares). Each patch may grow one pine tree. Bark beetles follow a pheromone to find their host tree. They first infest the tree with the strongest pheromone which is not fully occupied. If no pheromone is sensed, bark beetles attack a mature tree nearby and release pheromones to attract other dispersing beetles. Each bark beetle will produce two offspring when it infests a tree. When the average temperature rises during summer, some beetles will produce two more offspring beetles. Beetles can live up to two hypothetical years but die after reproduction. Each year, 49% of the beetles will die during winter. When the average temperature rises during winter, fewer beetles die. Trees become reddish after receiving 35 attacks and die after getting 50 attacks. Droughts will lower the tree defense threshold from 50 to 45, 40, or 35. Every empty patch grows a new pine tree at a chance of 5% each year.

How the Simulation Works

The simulation interface is shown in the image below.

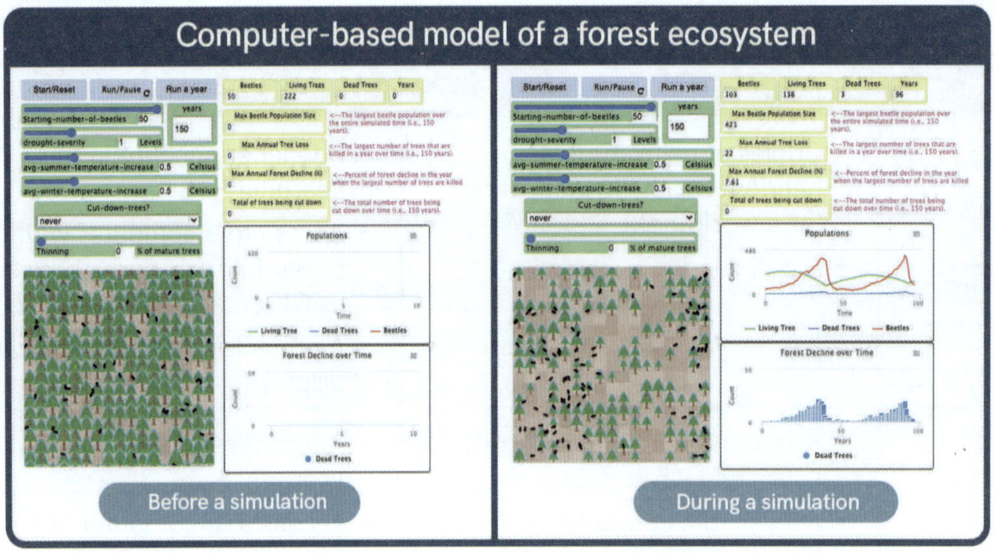

To start the simulation, set up the forest by using the slider **Starting-number-of-beetles** to set up the starting beetle population. Click **Start/Reset** to confirm the settings. Put a number in **years** to define when the model ends. Click **Run/Pause** to run or pause the model or click **Run a year** to run the model for a hypothetical year. Use the sliders to define the temperature increases and the severity of the drought. Use the switch **Cut-down-trees?** to choose whether and how to cut down trees to mitigate the outbreak. Use the slider **Thinning** to set up how many percent of mature trees are to be cut down for thinning the forest. These trees include the infected and the uninfected.

FACILITATION NOTES

The facilitation notes for this investigation describe how to deliver this lesson in person. They are written specifically for the paper-based mode of delivery. These same facilitation notes, however, can also be used for a paperless delivery of the lesson by simply substituting the directions that mention paper with the entry into the ADI Learning Hub. If you are teaching this lesson using the remote mode of delivery, you only need to assign due dates and prerequisites for each stage. The students can then simply follow the directions built into the ADI Learning Hub as they move through the investigation.

ADI Go Digital

The digital version of this lesson is available in the **ADI Learning Hub**. Be sure to check out the built-in teacher presentation mode.

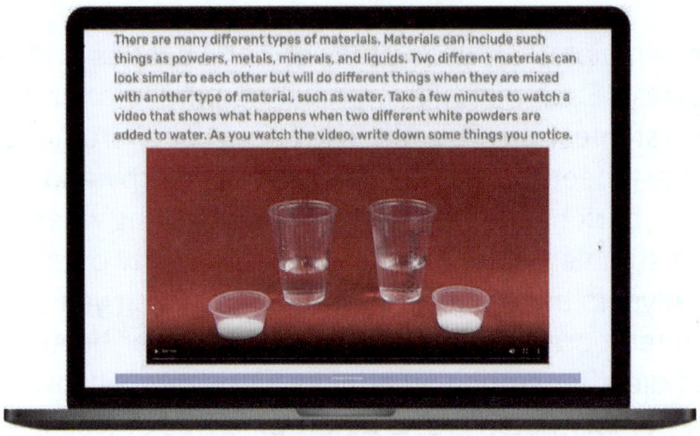

Bark Beetles and Climate Change in Washington Lesson Plan

Stage 1 TASK

Total Time for the Stage	45 minutes
Grouping	Individual; Teams

Learning Goals

Students should make progress on their ability to do the following by the end of this stage of the lesson:

Science

1. Ask questions that arise from careful observation of phenomena to clarify and/or seek additional information (SEP1).
2. Develop and use a model to provide mechanistic accounts of phenomena (SEP2).
3. Evaluate the merits and limitations of different models of the same phenomenon (SEP2).
4. Ask questions that arise from examining models to clarify and/or seek additional information and relationships (SEP 1).
5. Ask questions to clarify and refine a model (SEP1).

English Language Development

None

English Language Arts

1. Initiate and participate effectively in small-group collaborative discussions with diverse partners by building on others' ideas and expressing their own clearly (SL.9-10.1).
2. Present information clearly, concisely, and logically such that listeners can follow the line of reasoning and the organization, substance, and style are appropriate to the purpose, audience, and task (SL.9-10.4).
3. Adapt speech to a variety of contexts and tasks, demonstrating command of formal English when indicated or appropriate (SL.9-10.6).

Activities within the Stage

Activity A	The Phenomenon
Time Needed for the Activity	5 minutes
Grouping	Teams

What to Do

1. Have students work in their assigned teams for this activity.
2. Begin the activity by giving out a copy of the Task handout to each student.
3. Show "The Phenomenon" page on a screen using the Presentation mode in the Learning Hub so everyone in the classroom can see it.
4. Read the introduction of the phenomenon out loud and then start the video. You can also have students read the introduction of the phenomenon and watch the video on their own devices.
5. Give the students a few moments to write down their observations on their handouts before moving on to the next activity.

What to Give Out

Task handout for **Bark Beetles and Climate Change in Washington**

What to Show

"The Phenomenon" page in the ADI Learning Hub

Teaching Tip

Instead of assigning fixed roles to students by name at the beginning of the investigation, give each member of a group a different number so you can easily assign and change roles as needed during the various investigation stages. For example, you might give directions such as "1s will share out for your group" or "3s, will you please come back here and pick up these materials for your group?" or "2s and 3s, you will present the argument for your group while 1s and 4s travel to the other groups to offer feedback."

Attending to Equity

Be sure to intentionally group emerging multilingual students at times with peers who know the same languages as they do and at other times with peers whose English language development is slightly more advanced. Thoughtful grouping that varies throughout an investigation allows emerging multilingual students to benefit from the different linguistic resources of their peers.

Activity B	Share What You Noticed
Time Needed for the Activity	5 minutes
Grouping	Teams

What to Do

1. Show the "Share What You Noticed" page on a screen using the Presentation mode in the Learning Hub so everyone in the classroom can see it.
2. Say, "Okay, please share what you noticed with the others at your table."
3. Give the students about two minutes to share at their tables.
4. Say, "Okay, let's take a few minutes to share with the class. Team X (call on a team at random), what did you talk about?"
5. Allow Team X to share something they noticed. If it is related to ecosystems and human activity, say, "That is really interesting. We are actually going to examine that in more detail during this investigation." If what is shared is not related to ecosystems and human activity, say, "That is really interesting. Team Y (call on another team at random), what did you notice?" and allow that team to share. Follow this pattern until a team mentions something related to how we could stop bark beetle outbreaks from damaging trees. If no team mentions this, say, "You know what I found interesting? I saw that bark beetles are harmful to trees in Washington, and I wondered how we might prevent future bark beetle outbreaks. Did anyone think about that as well?"
6. Move on to the next activity.

What to Show

The "Share What You Noticed" page in the ADI Learning Hub

What to Look For

As soon as a group mentions something related to how we could stop bark beetle outbreaks from damaging trees, move on to the next activity.

Attending to Equity

Be sure to provide students with opportunities to talk in small groups during this activity. Small-group structures offer all students, including emerging multilingual students, a chance to engage in sense-making with their peers and offers them space to use their verbal and nonverbal (e.g., writing, drawing pictures, and using gestures) resources to express their ideas. Small-group structures also give students a chance to learn from other students' uses of these resources.

Activity C	What You Need to Figure Out
Time Needed for the Activity	5 minutes
Grouping	Teams

What to Do

1. Show the "What You Need to Figure Out" page on a screen using the Presentation mode in the Learning Hub so everyone in the classroom can see it.
2. Read the task out loud and have students follow along on their handout, on their devices, or on the screen.
3. Allow students to ask clarifying questions about the task or the guiding question.
4. Once everyone understands what they will need to figure out during the investigation, move on to the next activity.

What to Show

The "What You Need to Figure Out" page in the ADI Learning Hub

Teaching Tip

Have students highlight, circle, or place a star next to the guiding question. Marking the guiding question helps students find it later during the Plan stage.

Teacher Notes

Activity D	Share Your Ideas
Time Needed for the Activity	15 minutes
Grouping	Teams

What to Do

1. Show the "Share Your ideas" page on a screen using the Presentation mode in the Learning Hub so everyone in the classroom can see it.
2. Read the directions on the handout out loud and then give students five minutes to draw their initial representations of the science.
3. Say, "Okay, will everyone share their representations with the rest of your team? Please make sure everyone has time to talk. Also, be sure to ask one another questions about the scientific ideas that are included in each representation."
4. Give the students about 8-9 minutes to talk and share. Be sure to walk around and listen to the discussions.
5. Say, "Okay, I heard some interesting ideas as I walked around the room."
6. Move on to the next stage.

What to Show

The "Share Your Ideas"page in the ADI Learning Hub

What to Look For

Each student's initial representation should include a combination of symbols, diagrams, graphs, or words. It should also clearly show how a scientific idea can be used to help complete the task and answer the guiding question.

Attending to Equity

When students are tasked with expressing their ideas, be sure to encourage them to do so through multiple modes (speaking, writing, drawings, and gestures) and to use both content-specific and everyday registers. This process helps students fully express themselves, allowing peers to evaluate, question, and build on ideas. Also, make sure that every student in every team has a chance to share their initial representation.

Activity E	More Questions
Time Needed for the Activity	15 minutes
Grouping	Teams

What to Do

1. Show the "More Questions" page on a screen using the Presentation mode in the Learning Hub so everyone in the classroom can see it.
2. Say, "Okay, I heard a bunch of interesting questions as I was walking around the room. Will you work with your team to identify two or three questions that you think are important to learn more about that came up as you all discussed the different representations you made? You can write these questions on your handout."
3. Give students about five minutes to talk and write down questions. Be sure to walk around and listen to the discussions.
4. Say, "Okay, let's take a few minutes to share. Team X (call on a team at random), will you share one question you all have?"
5. Have each team share a question. Then say, "I want to share information about some useful scientific ideas with you. I bet this information will not only answer many of these questions but will also help you complete the task and answer the guiding question of the investigation."
6. Move on to the next stage.

What to Show

The "More Questions" page in the ADI Learning Hub

What to Look For

Each student should have 2-3 questions written on their handout.

Teaching Tip

When students are asked to create representations of the underlying scientific ideas, share these ideas with their peers, and then identify things they want to know more about, it sparks their curiosity and helps make the investigation meaningful. Also, identifying questions before reading (which students do during the Ideas stage) is an outstanding pre-reading activity that helps improve comprehension.

Bark Beetles and Climate Change in Washington Lesson Plan

Stage 2 IDEAS

Total Time for the Stage	30 minutes
Grouping	Teams

Learning Goals

Students should make progress on their ability to do the following by the end of this stage of the lesson:

Science

1. Ask questions that arise from examining theory to clarify and/or seek additional information and relationships (SEP1).
2. Critically read scientific literature adapted for classroom use to determine the central ideas or conclusions and/or to obtain scientific and/or technical information to summarize complex concepts, processes, or information presented in a text by paraphrasing them in simpler but still accurate terms (SEP8).
3. Integrate sources of information presented in different media or formats (e.g., visually and quantitatively) as well as in words to address a scientific question (SEP8).
4. Gather, read, and evaluate scientific and/or technical information from multiple authoritative sources, assessing the evidence and usefulness of each source (SEP8).
5. Explain how ecosystems have carrying capacities and how these limits result from factors such as the availability of living and non-living resources and challenges such as predation, competition, and disease (LS2.A).
6. Describe the complex set of interactions within an ecosystem that keep its numbers and types of organisms relatively constant over long periods of time under stable conditions (LS2.C).
7. Describe how extreme fluctuations in conditions and populations can challenge the functioning of ecosystems in terms of resources and habitat availability (LS2.C).
8. Describe how anthropogenic changes in an environment—such as habitat destruction, pollution, and overexploitation—can disrupt an ecosystem and threaten the survival of a species (LS2.C).
9. Identify how biodiversity is increased by the formation of new species in speciation and decreased by the loss of old species in extinction (LS4.D).
10. Explain how humans depend on the biodiversity of the living world for its resources, but human activity is having adverse impacts on biodiversity (LS4.D).
11. Evaluate solutions by taking into account a range of constraints such as cost, safety, reliability, and aesthetics as well as social, cultural, and environmental impacts (ETS1.B).
12. Use models to simulation systems and interactions between systems at different scales (CC4).
13. Construct explanations of how things change and how they remain the same (CC7).

English Language Development

1. Paraphrase central ideas in complex evidence, concepts, processes, and information to help explain how or why a phenomenon occurs (ELD-SLC.9-12.Explain.Interpretive).

English Language Arts

1. Determine the central ideas or conclusions of a text; trace the text's explanation or depiction of a complex process, phenomenon, or concept; provide an accurate summary of the text (RST.9-10.2).
2. Determine the meaning of symbols, key terms, and other domain-specific words and phrases as they are used in a specific scientific or technical context (RST.9-10.4).
3. Read and comprehend scientific/technical texts (RST.9-10.10).
4. Initiate and participate effectively in small-group collaborative discussions with diverse partners by building on others' ideas and expressing their own clearly (SL.9-10.1).

Teacher Notes

Activities

Activity A	Read About a Core Idea
Time Needed for the Activity	10 minutes
Grouping	Teams

What to Do

1. Have students work in their assigned teams during this stage. Give out a copy of the Ideas handout to each student.
2. The first activity provides students with information about ecosystem dynamics, functioning, and resilience.
3. Show the "Read About a Core Idea You Can Use" page on a screen using the Presentation mode in the Learning Hub so everyone in the classroom can see it. You can also click on the images to make them full screen and easier for students to see.
4. Have students read the text on their own, either on their handouts or on the screen.
5. Give the students about three minutes to discuss the text with the others in their teams and develop a shared list of ideas that they think are important or useful.
6. Ask a team to share an idea they thought was useful. Have the remaining teams share a different idea until there are no more ideas to share.
7. Move on to the next activity.

Teaching Tip

Students do not need to do a close read of this text or completely understand all the ideas in it before moving on to the next activity. They will be encouraged to revisit this text later in the investigation when creating their arguments and writing their reports. All students need to be able to do during this activity is to work with the other members of their group to identify one or two ideas that they think are important to keep in mind or are potentially helpful.

What to Give Out

Ideas handout for **Bark Beetles and Climate Change in Washington State**

What to Show

The "Read About a Core Idea You Can Use" page in the ADI Learning Hub

What to Look For

Students should write down some of the following ideas in the box labeled, "Some things I know from what I read" on their handouts:

- An ecosystem is a defined area that includes a community and the environment.
- A community consists of two or more populations, or a group of organisms that belong to the same species and live in the same place at the same time.
- The carrying capacity of an ecosystem is the maximum population size that the ecosystem can sustain.
- Populations usually fluctuate between being slightly over and slightly under the ecosystem's carrying capacity.
- While ecosystems can remain stable despite occasional changes, extreme changes can damage an ecosystem and cause it to transform into a different ecosystem.
- Extreme changes to an ecosystem often result in a loss of biodiversity.
- Extreme changes due to human activity include habitat destruction, pollution, introduction of invasive species, and climate change.

Activity B	Read About a Core Idea
Time Needed for the Activity	10 minutes
Grouping	Teams

What to Do

1. The second activity provides students with information about developing solutions to address decreasing biodiversity.
2. Show the "Read About Another Core Idea You Can Use" page on a screen using the Presentation mode in the Learning Hub so everyone in the classroom can see it. You can also click on the images to make them full screen and easier for students to see.
3. Have students read the text on their own, either on their handouts or on the screen.
4. Give the students about three minutes to discuss the text with the others in their teams and develop a shared list of ideas that they think are important or useful.
5. Ask a team to share an idea they thought was useful. Have the remaining teams share a different idea until there are no more ideas to share.
6. Move on to the next activity.

What to Show

The "Read About Another Core Idea You Can Use" page in the ADI Learning Hub

What to Look For

Students should write down some of the following ideas in the box labeled, "Some things I know from what I read" on their handouts:

- Humans depend on the biodiversity of ecosystems for survival, but human activity can negatively impact biodiversity.
- The bark beetle outbreaks in Washington are caused by climate change, which is driven by human activity.
- Bark beetle outbreaks can be reduced in the long term through policies that address climate change, such as limiting greenhouse gas emissions.
- Bark beetle outbreaks can be reduced in the short term through slowing bark beetle reproduction, such as by removing infested trees and cutting away infested branches.

Attending to Equity

There are many supports for helping students comprehend what they read already embedded into this activity (e.g., activating prior knowledge, providing a shared experience, making connections, synthesizing, and talking with peers). So, you might not need to provide much extra support. If you are concerned about some students not understanding what they are reading because of their scores on past reading comprehension tests, you can read it out loud as they follow along. As you read the text out loud, be sure to stop at each important idea and ask the students to put a star (or other annotation) next to it in the margin of their handouts. They can then discuss these ideas in their small groups.

Activity C	Read About a Final Core Idea
Time Needed for the Activity	10 minutes
Grouping	Teams

What to Do

1. The third activity provides students with information about using models to represent a process.
2. Show the "Read About a Final Core Idea You Can Use" page on a screen using the Presentation mode in the Learning Hub so everyone in the classroom can see it. You can also click on the images to make them full screen and easier for students to see.
3. Have students read the text on their own, either on their handouts or on the screen.
4. Give the students about three minutes to discuss the text with the others in their teams and develop a shared list of ideas that they think are important or useful.
5. Ask a team to share an idea they thought was useful. Have the remaining teams share a different idea until there are no more ideas to share.
6. Move on to the next activity

What to Show

The "Read About a Final Core Idea You Can Use" page in the ADI Learning Hub

What to Look For

Students should write down some of the following ideas in the box labeled, "Some things I know from what I read" on their handouts:

- Models are used to represent an object, event, or process.
- Models can be physical, visual, or a computer-based simulation.
- Models have limited precision and reliability because they make simplifying assumptions and approximations about a system.

Attending to Equity

To support emerging multilingual students during this activity, be sure to create a word bank on the wall of high-frequency, high-need concrete vocabulary that is needed to communicate scientific ideas. The word bank highlights terms needed for basic communication in an academic context. It also helps with the spelling. You can also encourage your emerging multilingual students to:

1. Use strategic learning techniques such as concept mapping, drawing, memorizing, comparing, contrasting, and reviewing.
2. Decode (sound out) words using skills such as recognizing sound-letter relationships and identifying cognates, affixes, roots, and base words.
3. Use visual and contextual support and support from peers and teachers to enhance and confirm understanding of the texts.

Stage 3 PLAN

Total Time for the Stage	15 minutes
Grouping	Teams

Learning Goals

Students should make progress on their ability to do the following by the end of this stage of the lesson:

Science

1. Evaluate a question to determine if it is testable and relevant (SEP1).
2. Plan an investigation individually and collaboratively to produce data to serve as the basis for evidence as part of building and revising models (SEP3).
3. Consider possible variables or effects and evaluate the confounding investigation's design to ensure variables are controlled (SEP3).
4. Ask questions to determine relationships, including quantitative relationships, between independent and dependent variables (SEP1).
5. Decide on types, quantity, and accuracy of data needed and consider limitations on the precision of the data (e.g., number of trials, cost, risk, and time), and refine the design accordingly (SEP3).
6. Select appropriate tools to collect, record, and analyze data (SEP3).

English Language Development

1. Define investigable questions on observations, information, and/or data about a phenomenon (ELD-SLC.9-12.Explain.Interpretive).
2. Develop reasoning to illustrate and/or predict the relationships between variables (ELD-SLC.9-12.Explain.Expressive).

English Language Arts

1. Initiate and participate effectively in small-group collaborative discussions with diverse partners by building on others' ideas and expressing their own clearly (SL.9-10.1).
2. Present information clearly, concisely, and logically such that listeners can follow the line of reasoning and the organization, substance, and style are appropriate to the purpose, audience, and task (SL.9-10.4).
3. Adapt speech to a variety of contexts and tasks, demonstrating command of formal English when indicated or appropriate (SL.9-10.6).

Activities

Activity A	Some Materials You Can Use
Time Needed for the Activity	5 minutes
Grouping	Teams

What to Do

1. Show the "Some Materials You Can Use" page on a screen using the Presentation mode in the Learning Hub so everyone in the classroom can see it. You can also click on the images to make them full screen and easier for students to see.
2. Say, "Okay, it sounds like we now have enough information to plan an investigation that will help us figure out how to reduce tree loss from bark beetle outbreaks."
3. Say, "Before we get started, I want to show you some materials that we are going to use during our investigation. Will you all please look at the screen so I can show you what we have?"
4. Direct students to look at the screen where the simulation has been pulled up.
5. Ask, "What do you think we could use this simulation for?"
6. Allow students to share ideas. Then say, "We can use this simulation to determine what factors affect bark beetle reproduction."
7. Say, "Okay, that is all the stuff we can use. Will you all head back to your teams so we can plan our investigation?"
8. Move on to the next activity.

What to Set Up

Pull up the simulation in the Learning Hub on a screen so that all students can see it. If you do not have a Learning Hub subscription, you can access the simulation by visiting *Climate-Driven Bark Beetle Outbreaks* by NetLogo.

What to Show

The "Some Materials You Can Use" page in the ADI Learning Hub

Teaching Tip

This activity helps students see for themselves all the materials they will be able to use during the hands-on part of the investigation instead of just reading about them or looking at pictures. It is important to show each material to the students and then point out how they can use it during the investigation so they can participate in developing a plan for the investigation during the next activity.

Activity B	Make a Plan
Time Needed for the Activity	10 minutes
Grouping	Teams

What to Do

1. Have students work in their assigned teams during this activity. Give a copy of the Plan handout to each student.
2. Show the "Make a Plan" page on a screen using the Presentation mode in the Learning Hub so everyone in the classroom can see it.
3. Say, "Work as a team to create a plan for your investigation by filling out the chart on your handout. As part of your plan, be sure to include which tools and techniques you will use to solve problems."
4. Give the students at least 10 minutes to complete this task. As students work in small teams to create their investigation proposals, be sure to move from team to team to answer questions and offer advice.
5. When a team thinks that they have a good plan, read it over. If you agree, sign off at the bottom of their handouts to let them know that their plan has been approved. If you do not agree, tell them what they will need to add or change before you will approve it. Once they have made the changes you can sign their handouts.
6. Once all the teams are done, you can move on to the next activity. You can also use an approved investigation plan as an "exit ticket" out of class.

What to Give Out

Plan handout for **Bark Beetles and Climate Change in Washington**

What to Show

The "Make a Plan" page in the ADI Learning Hub

What to Look For

The investigation plan should, at a minimum, include:

- The question they are trying to answer.
- What conditions they will test using the simulation.
- How they will analyze the data they collect from the simulation.
- How they will use their analysis to answer the guiding question.

Attending to Equity

To support emerging multilingual students during this stage, be sure to create a word bank on the wall of high-frequency, high-need concrete vocabulary that is needed to communicate scientific processes and ideas. The word bank highlights terms needed for basic communication in an academic context. It also helps with the spelling. Also, you can encourage your emerging multilingual students to use modes (speaking, writing, drawings, and gestures) and to use both content-specific and everyday registers to discuss and create their plans. This process helps students fully express themselves, allowing peers to evaluate, question, and build on ideas.

Teaching Tip

One way to do a quick check-in with a group during this activity is to use the "two questions, a nudge, and a goal" technique. This technique includes the following steps:

1. Sit down with a group of students.
2. Ask, "Would one of you tell me a little about what you all have been doing or thinking about?" (Question 1)
3. A student answers on behalf of the group.
4. Ask, "Would anyone else like to add anything to that?" (Question 2)
5. A different student provides additional information on behalf of the group.
6. Ask, "Have you thought about [a hint or something you want to highlight]?" (A nudge)
7. Say, "Okay, I'm going to check on the other groups. When I get back, I want to see [something that you want the group to accomplish]." (A goal)

When you have completed the check-in with one group, move on to the next group.

Teacher Notes

Stage 4 DO

Total Time for the Stage	45 minutes
Grouping	Teams

Learning Goals

Students should make progress on their ability to do the following by the end of this stage of the lesson:

Science

1. Conduct an investigation individually and collaboratively to produce data to serve as the basis for evidence (SEP3).
2. Analyze data using tools and/or technologies to make valid and reliable scientific claims (SEP4).
3. Consider limitations of data analysis (e.g., measurement error and sample selection) when analyzing and interpreting data (SEP4).
4. Evaluate the impact of new data on a working explanation and/or model of a proposed process or system (SEP4).
5. Apply ratios, rates, percentages, and unit conversions in the context of complicated measurement problems involving quantities with derived or compound units (SEP5).
6. Make a quantitative and/or qualitative claim regarding the relationship between dependent and independent variables (SEP6).
7. Construct and revise an explanation based on valid and reliable evidence obtained from a variety of sources and the assumption that theories and laws that describe the natural world operate today as they did in the past and will continue to do so in the future (SEP6).
8. Apply scientific ideas, principles, and/or evidence to provide an explanation of phenomena taking into account possible unanticipated effects (SEP6).

English Language Development

1. Paraphrase central ideas in complex evidence, concepts, processes, and information to help explain how or why a phenomenon occurs (ELD-SLC.9-12.Explain.Interpretive).
2. Evaluate the extent to which reasoning, theory, and/or models link evidence to claims and support conclusions (ELD-SLC.9-12.Explain.Interpretive).
3. Develop reasoning to illustrate the relationships between variables (ELD-SLC.9-12.Explain.Expressive).

English Language Arts

1. Initiate and participate effectively in small-group collaborative discussions with diverse partners by building on others' ideas and expressing their own clearly (SL.9-10.1).
2. Present information clearly, concisely, and logically such that listeners can follow the line of reasoning and the organization, substance, and style are appropriate to the purpose, audience, and task (SL.9-10.4).
3. Adapt speech to a variety of contexts and tasks, demonstrating command of formal English when indicated or appropriate (SL.9-10.6).

Activities

Activity A	Carry Out Your Plan
Time Needed for the Activity	25 minutes
Grouping	Teams

What to Do

1. Students should work in their assigned teams for this activity. Begin this activity by giving out a copy of the Do handout to each student.
2. Show the "Carry Out Your Plan" page on a screen using the Presentation mode in the Learning Hub so everyone in the classroom can see it.
3. Say, "Okay, it sounds like we are now ready to carry out our plans and collect the data or observations we need to answer the guiding question. You can get the materials and tools you need to use from this table." Point to the table where you have placed them.
4. Ask, "Does anyone have any questions about these materials or tools?"
5. Answer any questions that come up.
6. Say, "As you work, be sure to record your observations and measurements on your handout so you can keep track of everything you do. Does anyone have any final questions?"
7. Answer any questions that come up.
8. Say, "Okay, go ahead and get started. You have the next 25 minutes to work. I'll be available to help if any team needs me."
9. Give the students at least 25 minutes to carry out their plans and solve the problem. As students work in small teams, be sure to move from team to team to answer questions and offer advice.

What to Set Up

Have each group access the simulation through the Learning Hub by using a personal or shared electronic device. If students do not have a Learning Hub subscription, they can access the simulation by visiting *Climate-Driven Bark Beetle Outbreaks* by NetLogo.

What to Give Out

Do handout for **Bark Beetles and Climate Change in Washington**

What to Show

The "Carry Out Your Plan" page in the ADI Learning Hub

Teaching Tip

One way to do a quick check-in with a group during this activity is to use the "two questions, a nudge, and a goal" technique. This technique includes the following steps:

1. Sit down with a group of students.
2. Ask, "Would one of you tell me a little about what you all have been doing or thinking about?" (Question 1)
3. A student answers on behalf of the group.
4. Ask, "Would anyone else like to add anything to that?" (Question 2)
5. A different student provides additional information on behalf of the group.
6. Ask, "Have you thought about [a hint or something you want to highlight]?" (A nudge)
7. Say, "Okay, I'm going to check on the other groups. When I get back, I want to see [something that you want the group to accomplish]." (A goal)

When you have completed the check-in with one group, move on to the next group.

Attending to Equity

Be sure to intentionally group emerging multilingual students at times with peers who know the same languages as they do and at other times with peers whose English language development is slightly more advanced. Thoughtful grouping that varies during different investigation (or activities within an investigation) allows emerging multilingual students to benefit from the different linguistic resources of their peers. We also recommend assessing emerging multilingual students based on how much they contribute to the group during this activity, rather than solely on what they write or draw on their handouts.

Teacher Notes

Activity B	Make Sense of Your Data
Time Needed for the Activity	20 minutes
Grouping	Teams

What to Do

1. Students should work in their assigned teams for this activity. Begin this activity by showing the "Make Sense of Your Data" page on a screen using the Presentation mode in the Learning Hub so everyone in the classroom can see it.
2. Say, "Okay, it sounds like we are now ready to make sense of the data we collected. One way to do this is to create a table, graph, or other visualization that shows a trend or change over time, a comparison between teams, or a relationship between things you tested. You can create a table, graph, or other visualization here on your handout."
3. Point to the box on the handout.
4. Say, "Please work together in your teams to figure out the best way to make sense of your data. Remember, you are looking for any trends, changes, differences, or relationships in the data that will help you answer the guiding question. As you work, be sure to use the space on your handout to perform any calculations that you need to do and create any representations that you need. Does anyone have any last questions?"
5. Answer any questions that come up.
6. Say, "Okay, go ahead and get started. You have the rest of the period to work. I'll be available to help if any team needs me."
7. Give the students at least 15 minutes to make sense of their data. As students work together in small teams, be sure to move from team to team to answer questions and offer advice.
8. Make sure students return the materials and tools to the table where they found them before dismissing class. Cleanup should take no longer than five minutes.

What to Show

The "Make Sense of Your Data" page in the ADI Learning Hub

What to Look For

One of the most useful ways to make sense of the data during this activity is to first make a table that compares the initial conditions of the ecosystems, the bark beetle population over time, and the amount of tree loss over time.

Teaching Tip

Collecting and returning materials or tools to a central table works best if you assign the job to a specific number within each team. For example, you can say, "2s, please come up and grab the materials and tools that your team needs" at the start of the activity. Then you can say, "3s, please return the materials and tools your team used back to this table," at the end of the activity. Assigning jobs like this not only cuts down on how many students need to be up and moving around, but also allows you to hold specific students accountable for the materials and tools.

Attending to Equity

Encourage your emerging multilingual students to speak using learning strategies such as requesting assistance, employing nonverbal cues, and using synonyms and circumlocution (conveying ideas by defining or describing when exact English words are not known) during this activity. It is also helpful to point out and/or draw emerging multilingual learners' attention to elements of the English sound system in newly acquired vocabulary such as long and short vowels, silent letters, and consonant clusters as they come up naturally during this activity.

Teacher Notes

Stage 5 SHARE

Total Time Needed for the Stage	45 minutes
Grouping	Teams

Learning Goals

Students should make progress on their ability to do the following by the end of this stage of the lesson:

Science

1. Make and defend a claim based on evidence about the natural world that reflects scientific knowledge and student-generated evidence (SEP7).
2. Present an oral argument based on evidence (SEP7).
3. Evaluate the claims, evidence, and/or reasoning to determine the merits of arguments (SEP7).
4. Ask questions to clarify relationships, including quantitative relationships, between independent and dependent variables (SEP1).
5. Ask questions that challenge the interpretation of a dataset or the suitability of a method (SEP1).
6. Respectfully give critiques on scientific arguments by probing reasoning and evidence and challenging ideas and conclusions (SEP7).
7. Respectfully receive critiques on scientific arguments by responding thoughtfully to diverse perspectives and determining what additional information is required to resolve contradictions. (SEP7).
8. Communicate scientific information or ideas in multiple formats including orally, graphically, textually, and mathematically (SEP8).

English Language Development

1. Identify appropriate and sufficient evidence from data, models, and/or information from investigations of a phenomenon (ELD-SLC.9-12.Argue.Interpretive).
2. Compare reasoning and claims based on evidence from competing arguments (ELD-SLC.9-12. Argue.Interpretive).
3. Defend or refute a claim based on data and evidence (ELD-SLC.9-12.Argue.Expressive).
4. Establish and maintain an appropriate tone and stance (ELD-SLC.9-12.Argue.Expressive).
5. Signal logical relationships among reasoning, evidence, data, and/or models when making and defending a claim, counterclaim, and/or rebuttal (ELD-SLC.9-12.Argue.Expressive).

English Language Arts

1. Initiate and participate effectively in small-group collaborative discussions with diverse partners by building on others' ideas and expressing their own clearly (SL.9-10.1).
2. Integrate multiple sources of information presented in diverse media or formats (e.g., visually, quantitatively, and orally) evaluating the credibility and accuracy of each source (SL.9-10.2).
3. Evaluate a speaker's reasoning and use of evidence and rhetoric, identifying any fallacious reasoning or exaggerated or distorted evidence (SL.9-10.3).

(continued)

4. Present information clearly, concisely, and logically such that listeners can follow the line of reasoning and the organization, substance, and style are appropriate to the purpose, audience, and task (SL.9-10.4).
5. Make strategic use of media (e.g., textual, graphical, and visual) in presentations to enhance understanding of findings, reasoning, and evidence and to add interest (SL.9.10.5).
6. Adapt speech to a variety of contexts and tasks, demonstrating command of formal English when indicated or appropriate (SL.9-10.6).

Activities

Activity A	Create a Draft Argument
Time Needed for the Activity	15 minutes
Grouping	Teams

What to Do

1. Students should work in their assigned teams for this activity. Begin this activity by giving out a copy of the Share handout to each student.
2. Show the "Create a Draft Argument" page on a screen using the Presentation mode in the Learning Hub so everyone in the classroom can see it.
3. Say, "Okay, it sounds like we have all solved the task. I would now like you to create a draft argument to share what you figured out with the other teams. Your argument should include everything in this image here." Point to the image of a science-based argument projected on the screen. Be sure to go over what should be included in each component of the argument and where students can find the information they need.
4. Say, "I would like you to make your draft argument on a whiteboard. You can use the materials on this table to make your argument." Point to the table where you have placed the materials they need.
5. Ask, "Does anyone have any questions?"
6. Answer any questions that come up.
7. Say, "I'm going to give you exactly 15 minutes to get this done. Your draft argument doesn't need to be perfect. We just need something down on the whiteboards so we can share our ideas."
8. Give each team of students 15 minutes to create a draft argument. As students work in small teams, be sure to move from team to team to answer questions and offer advice.

What to Set Up

Place the following items on a table so students can use them during this stage:

- 24″ x 36″ whiteboard
- 3-4 dry-erase pens in different colors
- Eraser or cloth towel

What to Give Out

Share handout for **Bark Beetles and Climate Change in Washington**

What to Show

The "Create a Draft Argument" page in the ADI Learning Hub

What to Look for

Students should include core ideas that they read about during the Ideas stage in their justification.

Teaching Tip

Be sure to plant some different ideas with each group about how to make a strong argument as you move from group to group to check in, answer questions, and offer advice. This strategy helps to ensure that there will be some really strong examples of the evidence and justification components of the argument for students to see and discuss during the next activity. For example, you could help two groups have a really strong evidence component by telling one group, "I think a graph would be really helpful here" and telling another group, "I think your evidence would be stronger if you explained, in words, what this graph shows." You could help a third group have a really strong justification by saying, "Let's brainstorm some ideas that will help you defend your choice of evidence," or "Let's go back to our Ideas handout and see if there are any concepts we can include."

It is important you make sure that all the arguments are different at this point in the investigation by planting different ideas with different groups, because the students will have nothing to critique or discuss during the next activity if they all have the same argument. The goal during this activity is to set the stage for the next activity by ensuring that the various components of a strong argument can be found on different whiteboards within the class. These strong components will then be taken up by all the groups because they reflect good ideas. Be sure to give students a voice and choice during this activity, so they can make mistakes and learn from them.

Attending to Equity

It is often helpful to assign and change roles at least once during this activity. For example, you might assign the 1s to be "in charge of writing" at the beginning of the activity. You can then easily switch roles about five minutes into the activity by saying, "1s, please hand the pens to the 2s. 2s, you are now in charge of writing for the group." By changing roles, you can ensure that all students have a chance to participate and contribute their ideas to the creation of a draft argument.

ADI Go Digital

The **ADI Learning Hub** has tons of features that allow students to share their thinking with each other when they are doing the investigation paperless or remote.

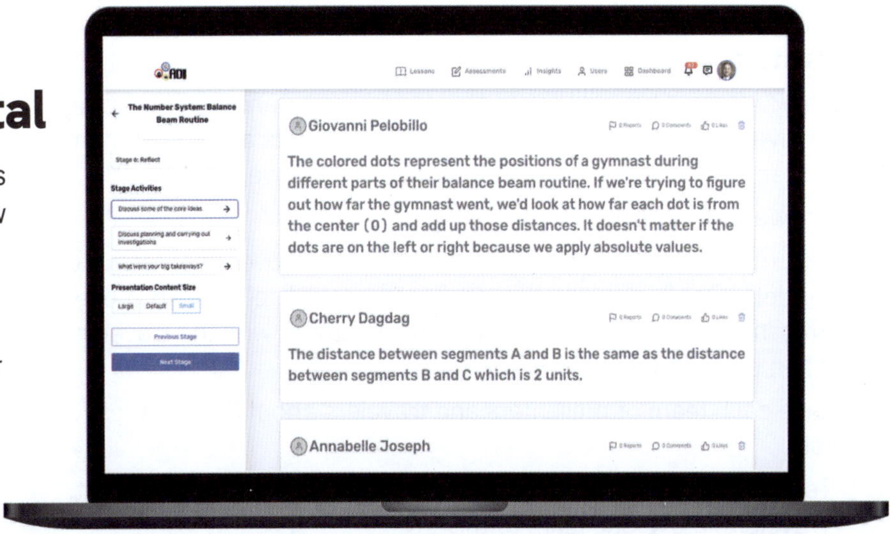

Activity B	Present Your Draft Argument
Time Needed for the Activity	15 minutes
Grouping	Teams

What to Do

1. Students should work in their assigned teams for this activity.
2. Show the "Present Your Draft Argument" page on a screen using the Presentation mode in the Learning Hub so everyone in the classroom can see it.
3. Say, "We are now going to share our arguments. Please set up your whiteboards so everyone can see them." Allow students time to set up their whiteboards. They should be set up on stands or chairs, so they are facing toward the outside of the room.
4. Say, "Okay, before we get started, I want to explain what we are going to do next. I'm going to ask some of you to present your arguments to your classmates. If you are presenting your argument, your job is to share your team's claim, evidence, and justification of the evidence. The rest of you will be travelers. If you are a traveler, your job is to listen to the presenters, ask them questions if you do not understand something, and then offer them some suggestions about ways to make their argument better. Do you have any questions about these instructions?"
5. Answer any questions that come up.
6. Say, "As you talk to one another, make sure you write down any good ideas or suggestions that are important in the space on your handout, so you don't forget about them." Point to the box labeled, "Some possible ways to improve our argument" on the handout.
7. Say, "After we have a chance to learn from one another, I'm going to give you some time to revise your arguments and make them better. Do you have any questions about what we are doing?"
8. Answer any questions that come up.
9. Say, "Okay. Let's get started. Travelers, move one team to the left. Don't move to the next team until I tell you to move. Presenters, go ahead and share your argument with the travelers when they get there."
10. Give each team of presenters and travelers about 3-4 minutes to talk.
11. Tell the students, "Okay. Let's move on to the next argument. Travelers, move one team to the left."
12. Again, give each team of presenters and travelers about 3-4 minutes to talk.

What to Show

The "Present Your Draft Argument" page in the ADI Learning Hub

What to Look for

The critiques and feedback that students give one another during this activity should:

- Be shared in a way that is kind (respectful),
- Provide information about what, exactly, needs to be changed or improved (specific), and
- Offer a suggestion about how to make something better rather than just saying something is wrong (helpful).

In addition, students should be focusing their critiques on the science included in the arguments, not the appearance of the arguments.

(continued)

13. Repeat this process until each team has had their argument read and critiqued three times. Then tell the students to return to their workstations.
14. Be sure to move between workstations during this activity to keep everyone involved and model what it means to communicate and critique scientific ideas and reasoning.

Attending to Equity

Be sure to encourage students to provide feedback that is respectful, specific, and helpful during this activity. It is important to encourage student-to-student talk because it helps students develop speaking and listening skills and enables them to make sense of what and how they know. It also provides a context for emerging multilingual students to use both productive (speaking and writing) and receptive (listening and reading) language and learn science through meaningful interactions and scientific sense-making.

Teaching Tip

There are two types of formats for sharing and critiquing arguments. The first format is called a *gallery walk.* In this format, each group sets up their argument at their workstation so others can see it. The entire group then moves to a different workstation. While at a different workstation, the group is expected to read the argument at that workstation (which was created by a different group) and offer critiques and suggestions for improvement. The students should move to a different workstation after a few minutes to read and critique an argument written by a different group. Students should also be encouraged to keep a record of good ideas or potential ways to improve their own arguments as they move from workstation to workstation to offer critiques and feedback. After repeating this process in the gallery walk 3-4 times, every group will have had an opportunity to read and critique three or four different arguments and receive feedback from three or four different groups.

The second format is called a *modified gallery walk.* In the modified gallery walk format, one or two members of the group stay at their workstation to share their group's ideas (these students are called presenters) while the other members of their group go to different groups, one group at a time, to listen to and critique the arguments developed by their classmates (these students are called travelers). The travelers should visit at least three different workstations during the argumentation session. We also recommend that the presenters keep a record of the critiques made by their classmates and any suggestions for improvement. The travelers should also be encouraged to keep a record of good ideas or potential ways to improve their own arguments as they travel from group to group.

Attending to Equity

Another good way to have students, especially emerging multilingual learners, give one another feedback is to have them use four different color sticky notes to make comments. Give each group a single pink, blue, yellow, and green sticky note. Then direct the students to use the pink sticky note to suggest "something to add" to the argument, the blue sticky note to suggest "something to change" in the argument, the yellow sticky note to highlight "something they like" in the argument, and the green sticky note to ask a question about "something they don't understand". This strategy is useful because everyone in the classroom will be actively engaged during this activity, and everyone will have a chance to see different arguments. This format, however, does not give students a chance to support, critique, and challenge the ideas, solutions, scientific reasoning of other groups through talk because two different groups are never at the same workstation at the same time.

Activity C	Revise Your Argument
Time Needed for the Activity	15 minutes
Grouping	Teams

What to Show

The "Revise Your Argument" page in the ADI Learning Hub

What to Do

1. Students should work in their assigned teams for this activity.
2. Show the "Revise Your Argument" page on a screen using the Presentation mode in the Learning Hub so everyone in the classroom can see it.
3. Say, "I'm now going to give you all about 10 minutes to revise your argument. Please return to your original teams."
4. Wait for the students to move back into their original teams.
5. Say, "Okay, take a few minutes to talk in your teams and determine what you want to change to make your argument better. Once you have decided what to change, go ahead and make the changes to your whiteboard."
6. Ask, "Do you have any questions about what you need to do?"
7. Answer any questions that come up.
8. Tell the students, "Okay. Let's get started."
9. Give the students 10 minutes to work in their teams on their arguments. As they work, move from team to team to check in, ask probing questions, and offer a suggestion if a team gets stuck.
10. When students are finished with their final arguments, have them take a picture of it or have them put the boards in a safe spot so they can refer to the picture or the boards during the Report stage.

What to Look For

A strong argument includes the following information:

- A claim that states something similar to, "We can limit bark beetle outbreaks and tree loss by reducing the number of susceptible host trees by removing some trees from a forest before they are infested by bark beetles, making it harder for beetles to find a new tree to infest."
- A table that compares the initial conditions of the ecosystems, the bark beetle population over time, and the amount of tree loss over time.
- An explanation of the table.
- An explanation of the chosen evidence that uses core ideas from the investigation.

Teaching Tip

It is important to give students time to meet with their original groups so they can discuss what they learned by interacting with individuals from the other groups. At this point in the process, students often realize that their initial problem-solving model was flawed in some way. It is therefore important to encourage students to "redo the science" as needed. At the end of this activity, each group should have a final argument that is much better than their original one.

Bark Beetles and Climate Change in Washington Lesson Plan

Stage 6 REFLECT

Total Time for the Stage	25 minutes
Grouping	Whole class, Teams

Learning Goals

Students should make progress on their ability to do the following by the end of this stage of the lesson:

Science

1. Ask questions that arise from careful observation of phenomena, or unexpected results, to clarify and/or seek additional information as needed (SEP1).
2. Explain how ecosystems have carrying capacities and how these limit results from factors such as the availability of living and non-living resources and challenges such as predation, competition, and disease (LS2.A).
3. Describe the complex set of interactions within an ecosystem that keep its numbers and types of organisms relatively constant over long periods of time under stable conditions (LS2.C).
4. Describe how extreme fluctuations in conditions and populations can challenge the functioning of ecosystems in terms of resources and habitat availability (LS2.C).
5. Describe how anthropogenic changes in an environment—such as habitat destruction, pollution, and overexploitation—can disrupt an ecosystem and threaten the survival of a species (LS2.C).
6. Identify how biodiversity is increased by the formation of new species in speciation and decreased by the loss of old species in extinction (LS4.D).
7. Explain how humans depend on the biodiversity of the living world for its resources, but human activity is having adverse impacts on biodiversity (LS4.D).
8. Evaluate solutions by taking into account a range of constraints such as cost, safety, reliability, and aesthetics as well as social, cultural, and environmental impacts (ETS1.B).
9. Use models to simulation systems and interactions between systems at different scales (CC4).
10. Construct explanations of how things change and how they remain the same (CC7).

English Language Development

1. Paraphrase central ideas in complex evidence, concepts, processes, and information to help explain how or why a phenomenon occurs (ELD-SLC.9-12.Explain.Interpretive).
2. Use scientific explanations about a phenomenon in a different context or situation (ELD-SLC.9-12.Explain.Expressive).

English Language Arts

1. Participate effectively in small-group collaborative discussions with diverse partners on texts by building on others' ideas and expressing their own clearly and persuasively (SL.9-10.1).

Activities

Activity A	Discuss a Core Idea
Time Needed for the Activity	10 minutes
Grouping	Whole class/Teams

What to Give Out

Reflect handout for **Bark Beetles and Climate Change in Washington**

What to Do

1. Students should be in small teams for this activity. However, they do not need to be in the same teams as they were during previous stages.
2. Begin this activity by giving out a copy of the Reflect handout to each student.
3. Show the "Discuss a Core Idea" page on a screen using the Presentation mode in the Learning Hub so everyone in the classroom can see it.
4. Say, "Okay. Now that we have all revised our arguments and made them better, let's take some time to discuss some of the core ideas you used during this investigation and to justify the evidence in your arguments so we are all on the same page. Look at the picture on the screen."
5. Pause for about 30 seconds.
6. Say, "Take a few minutes to talk in your teams and see if you can use what you know about ecosystem dynamics, functioning, resilience, and developing solutions to problems of decreasing biodiversity to explain what is happening in the picture."
7. Give the students at least three minutes to talk in their small teams.
8. Say, "Okay. Let's share with the whole class. Team A (or whichever team you want to share first), what did you talk about?"
9. Allow the first team to share.
10. Say, "Interesting. Would anyone like to add to, challenge, or refine the ideas shared by Team A?"

(continued)

What to Show

The "Discuss a Core Idea" page in the ADI Learning Hub

Explanation of the Image

The image that the students discuss during this activity shows different populations within an ecosystem.

What to Look For During the Discussion

Students who understand the content will likely mention some, if not all, of the following ideas when providing an explanation for the image:

- An ecosystem is a defined area that includes a community and the environment.
- A community consists of two or more populations, or a group of organisms that belong to the same species and live in the same place at the same time.
- The carrying capacity of an ecosystem is the maximum population size that the ecosystem can sustain.
- Populations usually fluctuate between being slightly over and slightly under the ecosystem's carrying capacity.

(continued)

11. Allow students to respond.
12. Say, "Okay. Team B (or whichever team you want to share second), what did you talk about?"
13. Allow the second team to share.
14. Say, "Interesting. Would anyone like to add to, challenge, or refine the ideas shared by Team B?"
15. Allow students to respond.
16. Keep up this pattern of questioning until all teams have shared. Be sure to keep encouraging students to talk about the content until they reach a correct and consensual explanation for what is shown in the image. If they are unable to reach a consensus or reach consensus using an inaccurate idea, be sure to provide an accurate explanation for the image before moving on to the next activity.

- While ecosystems can remain stable despite occasional changes, extreme changes can damage an ecosystem and cause it to transform into a different ecosystem.
- Extreme changes to an ecosystem often result in a loss of biodiversity.
- Extreme changes due to human activity include habitat destruction, pollution, introduction of invasive species, and climate change.

Teaching Tip

This activity is a discussion, not a presentation about what the students "should have seen" or "should have learned" during the investigation. The more students talk during this activity, the more meaningful the experience will be for them and the more you can learn about their thinking. Be sure to use talk goals and moves to help facilitate the discussion. Below are some examples of talk goals that you can use to help facilitate the discussion and some questions that you can ask the students to help accomplish each goal. These talk goals and questions are only suggestions.

Goal 1: Encourage students to share, expand on, or clarify an idea.
- What do you see going on here?
- What do you think is happening?
- How might we explain that?
- Can you tell me more about that?
- What do you mean by that?
- Can you give me an example of that?
- Can you say that in another way?

Goal 2: Encourage students to support an idea.
- Why do you think that?
- What is your evidence for that?
- How did you arrive at that conclusion?
- What makes you say that?
- How do you know that?

Goal 3: Encourage students to think with others.
- Would anyone like to add to that idea?
- What should we keep or change about that idea?
- Can anyone else give me some examples of that?
- Can anyone take that idea and push it a little further?
- Does anyone want to respond to that idea?
- Do you agree or disagree with that idea, and why?
- What do we think about that idea?
- Does that fit with what we have figured out so far?
- Is there another way to do it?

Goal 4: Encourage students to think about what they know and don't know yet.
- I'm confused now. What have we figured out so far?
- What do we need to figure out now?
- So, what do we agree on right now, and what are we still trying to figure out?
- What are we sure and not so sure about?
- What questions do you have?

Attending to Equity

Be sure to intentionally group emerging multilingual students at times with peers who know the same languages as they do and at other times with peers whose English language development is slightly more advanced. Also, be sure to ask students how they have seen the core idea in the everyday lives; this will make the topic of discussion more engaging and relatable.

Teacher Notes

Activity B	Discuss the Practices
Time Needed for the Activity	10 minutes
Grouping	Whole class/Teams

What to Do

1. Show the first page of "Discuss the Practices" on a screen using the Presentation mode in the Learning Hub so everyone in the classroom can see it.
2. Say, "Okay. Now that we have talked about some of the core ideas that we used during our investigation, let's take a few minutes to talk about the science practices that we used during the investigation. An example of a scientific practice used was analyzing and interpreting data. On the screen there is a question that I want you all to think about: What other things did you do during your investigation that you feel were good examples of a scientific practice or attempting to figure out something in a scientific way?"
3. Pause for about 30 seconds.
4. Say, "Take a few minutes to talk about this question in your teams."
5. Give the students at least three minutes to talk in their small teams.
6. Say, "Okay. Let's share with the whole class. Team A (or whichever team you want to share first), what did you talk about?"
7. Allow the first team to share.
8. Say, "Okay. Team B (or whichever team you want to share second), what did you talk about?"
9. Allow the second team to share.
10. Continue asking teams until every team has a chance to share.
11. Show the second page of "Discuss the Practices" on a screen using the Presentation mode in the Learning Hub so everyone in the classroom can see it.
12. Say, "Here is another question that I want you all to think about: What things did you do during your investigation that were not good examples of a scientific practice or not very scientific?"
13. Pause for about 30 seconds.

What to Show First

The first page of "Discuss the Practices" in the ADI Learning Hub

What to Show Second

The second page of "Discuss the Practices" in the ADI Learning Hub

What to Show Third

The third page of "Discuss the Practices" in the ADI Learning Hub

(*continued*)

14. Say, "Take a few minutes to talk about this question in your team."
15. Give the students at least three minutes to talk in their small teams.
16. Say, "Okay. Let's share with the whole class. Team A (or whichever team you want to share first), what did you talk about?"
17. Allow the first team to share.
18. Say, "Okay. Team B (or whichever team you want to share second), what did you talk about?"
19. Allow the second team to share.
20. Continue asking teams until every team has a chance to share.
21. Show the third page of "Discuss the Practices" on a screen using the Presentation mode in the Learning Hub so everyone in the classroom can see it.
22. Say, "Here is a final question that I want you all to think about: What should we all do during our next investigation to make sure we are using scientific practices and being scientific in the way we try to figure something out?"
23. Pause for about 30 seconds.
24. Say, "Take a few minutes to talk about this question in your team."
25. Give the students at least three minutes to talk in their small teams.
26. Say, "Okay. Let's share with the whole class. Team A (or whichever team you want to share first), what did you talk about?"
27. Allow the first team to share.
28. Say, "Okay. Team B (or whichever team you want to share second), what did you talk about?"
29. Allow the second team to share.
30. Continue asking teams until every team has a chance to share.
31. Once everyone has shared, offer any suggestions or advice about what they can do to improve their next investigation and then move on to the next stage.

Teaching Tip

A great way to wrap up this activity is for the students to contribute to a list of class norms or "gotta-haves" to keep in mind when planning and carrying out an investigation in science. To create this list, each group can suggest for a potential norm or "gotta-have" to add to the list at the end of this activity, then the class votes on the one that they think is most important. The most important one gets added to the list. The list then serves as reminder for students during the Plan and Do stages of subsequent investigations. Over time, the list will grow and each addition to the list will be more nuanced than the last as students develop expertise with scientific processes.

Attending to Equity

Be sure to give all students, especially emerging multilingual students, enough time to talk in small groups before asking anyone to share with the entire class during this activity. Small-group structures offer all students, including emerging multilingual students, a chance to engage in sense-making with their peers as well as space to use their verbal and nonverbal (e.g., writing, drawing pictures, and using gestures) resources to express their ideas. Small-group structures also give students a chance to learn from other students' uses of these resources. When students are tasked with expressing their ideas, be sure that they know it is okay for them to do so in a mode or combination of modes (speaking, writing, drawings, and gestures).

Bark Beetles and Climate Change in Washington Lesson Plan

Stage 7 REPORT

Total Time Needed for the Stage	100 minutes (over three days)
Grouping	Individual, Teams

Learning Goals

Students should make progress on their ability to do the following by the end of this stage of the lesson:

Science

1. Use mathematical representations to support and revise explanations based on evidence about factors affecting biodiversity and populations in ecosystems of different scales (HS-LS2-2).
2. Evaluate claims, evidence, and reasoning that the complex interactions in ecosystems maintain relatively consistent numbers and types of organisms in stable conditions, but changing conditions may result in a new ecosystem (HS-LS2-6).
3. Design, evaluate, and refine a solution for reducing the impacts of human activities on the environment and biodiversity (HS-LS2-7).
4. Make and defend a claim based on evidence about the natural world that reflects scientific knowledge and student-generated evidence (SEP7).
5. Construct a written argument based on evidence (SEP7).
6. Evaluate the claims, evidence, and/or reasoning to determine the merits of arguments (SEP7).
7. Respectfully give critiques by probing reasoning and evidence and challenging ideas and conclusions (SEP7).
8. Communicate scientific information or ideas in multiple formats including graphically and textually (SEP8).

English Language Development

1. Identify appropriate and sufficient evidence from data, models, and/or information from investigations of a phenomenon (ELD-SLC.9-12.Argue.Interpretive).
2. Compare reasoning and claims based on evidence from competing arguments (ELD-SLC.9-12.Argue.Interpretive).
3. Defend or refute a claim based on data and evidence (ELD-SLC.9-12.Argue.Expressive).
4. Establish and maintain an appropriate tone and stance (ELD-SLC.9-12.Argue.Expressive).
5. Signal logical relationships among reasoning, evidence, data, and/or models when making and defending a claim (ELD-SLC.9-12.Argue.Expressive).

English Language Arts

1. Write routinely over extended time frames (time for reflection and revision) and shorter time frames for a range of discipline-specific tasks, purposes, and audiences (WHST.9-10.10).
2. Write arguments focused on discipline-specific content (WHST.9-10.1).
3. Write informative/explanatory texts, including the narration of scientific procedures/experiments or technical process (WHST.9-10.2).
4. Produce clear and coherent writing in which the development, organization, and style are appropriate to task, purpose, and audience (WHST.9-10.4).

(continued)

5. Strengthen writing as needed by revising, editing, rewriting, and focusing on addressing what is most significant for a specific purpose and audience (WHST.9-10.5).
6. Determine the central ideas or conclusions of a text; trace the text's explanation or depiction of a complex process, phenomenon, or concept; provide an accurate summary of the text (RST.9-10.2).
7. Determine the meaning of symbols, key terms, and other domain-specific words and phrases as they are used in a specific scientific or technical context (RST.9-10.4).
8. Analyze the structure of the relationships among concepts in a text, including relationships among key terms (RST.9-10.5).
9. Assess the extent to which the reasoning and evidence in a text support the author's claim (RST.9-10.8).
10. Compare and contrast findings presented in a text to those from other sources (including their own experiments), noting when the findings support or contradict other accounts (RST.9-10.9).

Activities

Activity A	Draft Report
Time Needed for the Activity	20 minutes
Grouping	Individual

What to Do

1. Have students work on their own during this activity. Give a copy of the Report handout to each student.
2. Show the "Create a Draft Report" page on a screen using the Presentation mode in the Learning Hub so everyone in the classroom can see it.
3. Say, "You are now going to write an investigation report to share what you figured out. The first part of the report is called the 'Introduction'. In this section, you should explain to the reader what you were investigating, why you were investigating it, and what question you were trying to answer. All this information can be found on your Task handout."
4. Hold up the Task handout.
5. Ask the students, "Do you have any questions about what you need to do?"
6. Answer any questions that come up.
7. Say, "Here are some sentence starters to help you begin writing."
8. Hold up the Report handout and point to the sentence starters.
9. Tell the students, "Okay, let's write."
10. Give the students five minutes to write the "Introduction" section of the report. As they work, move from student to student to check in, ask probing questions, and offer a suggestion if a student gets stuck.

What to Give Out

Report handout for **Bark Beetles and Climate Change in Washington**

What to Show

The "Create a Draft Report" page in the ADI Learning Hub

(continued)

11. Say, "Okay, let's move on to the second part of the report. The second part of the report is called the 'Method'. In this section, you should explain to the reader what you did during the investigation to complete the task and answer the guiding question. All this information can be found in the Plan handout. Remember that you all planned and carried out different investigations, so do not assume that the reader will know what you did."
12. Ask the students, "Do you have any questions about what you need to do?"
13. Answer any questions that come up.
14. Say, "Here are some sentence starters to help you begin writing."
15. Hold up the Report handout and point to the sentence starters.
16. Tell the students, "Okay, let's write."
17. Give the students five minutes to write the "Method" section of the report. As they work, move from student to student to check in, ask probing questions, and offer a suggestion if a student gets stuck.
18. Say, "Okay, let's move on to the last part of the report. The third part of the report is called the 'Argument'. In this section, you should explain to the reader what you figured out and how you know you are correct. All this information can be found on your Do handout and on your team's whiteboard."
19. Hold up the Do handout and point to the boards (or remind them that they have a picture of their board).
20. Ask the students, "Do you have any questions about what you need to do?"
21. Answer any questions that come up.
22. Say, "Here are some sentence starters to help you begin writing."
23. Hold up the Report handout and point to the sentence starters.
24. Tell the students, "Okay, let's write."
25. Give the students 10 minutes to write the "Argument" section of the report. As they work, move from student to student to check in, ask probing questions, and offer a suggestion if a student gets stuck.
26. Collect all the draft reports before moving on to the next activity. Tell students to mark their draft reports with a symbol that they can use to identify it later. They should NOT write their names on their draft reports.

Teaching Tip

Writing is an important component of this model. When a student writes the report on their own, it not only helps the student learn to write in the context of science (an important literacy skill) but also allows that student to write to learn (develop a better understanding of scientific concepts by writing about them). There are many writing supports (e.g., sentence starters) already embedded into this activity, so you might not need to provide much extra support for your students.

This activity can be done completely in class or started in class and then finished as homework. Do not assign the entire activity as homework unless you know that every student in your class has dedicated time, space, and resources to write outside of class and everyone understands and is comfortable with the report writing process.

If you are concerned about students feeling overwhelmed, you can develop a class "gotta-have-it" list of information that needs to be included in each section of the report. Students can then refer to this list as they write each section.

Don't worry if students do not produce a perfect report at this point. All the reports will be reviewed and revised before they are turned in to you. The reports should therefore be viewed as a starting point. Be sure to encourage students to "just get something down on paper" so "we can work on it together."

Attending to Equity

Emerging multilingual students' access to instructional tasks that require complex thinking, such as writing a report, is enhanced when linguistic complexity and instructional support match their current level of language proficiency. Therefore, it is important to scaffold the writing of the reports for the emerging multilingual students based on their current English language proficiency level. The use of sentence starters to model written language production skills can help emerging multilingual students develop English language skills because the development of academic language and academic content knowledge are interrelated processes. However, it is important that such scaffolds be used purposefully and removed when no longer needed.

ADI Go Digital

Use the **ADI Learning Hub** to go paperless. When you create a class and add students to it, students can complete all the stages right in the system.

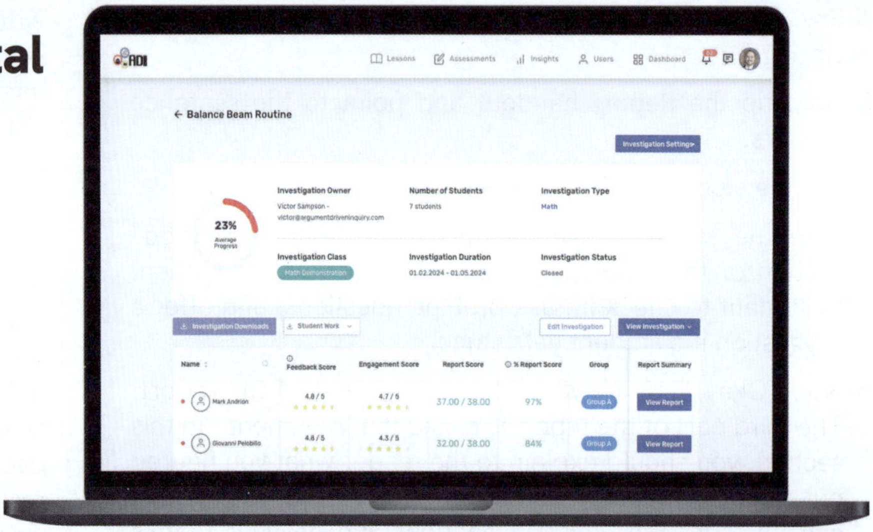

Activity B	Peer Review the Reports
Time Needed for the Activity	45 minutes
Grouping	Teams

What to Do

1. Students should work in teams of three for this activity. These can be completely different teams.
2. Show the "Peer Review Some Draft Reports" page on a screen using the Presentation mode in the Learning Hub so everyone in the classroom can see it.
3. Say, "We are now going to review our reports to find ways to make them better."
4. 4. If possible, use a document camera to project the peer review guide (PRG) on a screen.
5. Say, "We are going to use this peer review guide to give each other feedback." Point to the image.
6. Say, "I'm going to ask you to work in teams of three to do this. I will give each team a draft report to read. You three will then read the report together. Once you are done reading the report, I want you to answer each of the questions on the peer review guide." Point to the review questions on the image of the PRG.
7. Tell the students, "You can check No, Almost, or Yes after each question." Point to the checkboxes on the image of the PRG.
8. Say, "This will be your rating for this part of the report. Make sure you agree on the rating you give the author. If you mark No or Almost, then you need to tell the author what he or she needs to do to get a Yes." Point to the space for the reviewer feedback on the image of the peer review guide.
9. Say, "It is really important for you to give the authors feedback that is helpful. That means you need to tell them exactly what they need to do to make their report better."
10. Ask the students, "Do you have any questions about what you need to do?"
11. Answer any questions that come up.
12. Tell the students, "Please get into your teams of three."
13. Allow the students time to move into their triads.
14. Say, "Okay, I'm now going to give you one report to read." Give out one draft report to each triad. Make sure that the report you give a triad was not written by one of the students in that triad. Give each triad one PRG to fill out.

(continued)

What to Give Out

Peer Review Guide (one per group per review, so for this investigation you need the same number of peer review guides as the number of draft reports that will be reviewed)

What to Show

The "Peer Review Some Draft Reports" page in the ADI Learning Hub

What to Look For

Make sure students are reaching consensus about what rating to give each criterion. Also, be sure that students are giving one another feedback that is specific, detailed, and helpful when they think an aspect of a report needs to be improved.

15. Tell the students, "Okay, I'm going to give you 15 minutes to read the report I gave you and to fill out the peer review guide. Go ahead and get started."
16. Give the students 15 minutes to work. As they work, move around from team to team to check in and see how things are going, answer questions, and offer advice.
17. After 15 minutes pass, tell the students, "Okay, time is up. Please give me the report and the peer review guide that you filled out."
18. Collect the draft reports and the PRGs, or collect the workbooks if students are using them.
19. Say, "Okay, I am now going to give you a different report to read and a new peer review guide to fill out." Give out one more report to each triad. Make sure that the report you give a triad was not written by one of the students in that triad. Give each triad one new peer review guide to fill out as a team.
20. Say, "Okay, I'm going to give you 15 minutes to read this new report and to fill out the peer review guide. Go ahead and get started."
21. Give the students 15 minutes to work. As they work, move around from team to team to check in and see how things are going, answer questions, and offer advice.
22. After 15 minutes pass, tell the students, "Okay, time is up. Please give me the report and the peer review guide that you filled out."
23. Repeat this process for the remaining draft reports. Each triad should one have one more report to review (for a total of three reviews).
24. Collect the last draft reports and PRGs. Set all the draft reports on a table in separate stacks. Each stack should have a draft report on top (with the identification symbol facing up so students can find their reports) and the PRG on the bottom.

Teaching Tip

The peer review process is one of the best ways for students to learn how to write. When students read and review different reports, they have an opportunity to (1) see examples of texts written by others for the same purpose, (2) discuss what counts as quality and why some reports are stronger than others, (3) discuss ways to strengthen a report, and (4) pick up things that they can do to improve their reports. This helps the students learn new ways to organize and present information, which will help them write better on subsequent reports.

Attending to Equity

One helpful scaffold is to give students sentence stems for what to write in their feedback to the author. Ideal feedback is respectful, specific about what needs to be changed, and helpful in that it suggests how the author should go about making that change. You can review the writing of the emerging multilingual students in your class if they have limited English language proficiency, but make sure that all emerging multilingual students participate in the peer review process. You will see tremendous growth in their writing skills as they review their peers' writing and discuss what makes some reports stronger than others.

Activity C	Revise the Reports
Time Needed for the Activity	25 minutes
Grouping	Individuals

What to Do

1. Students should work on their own for this activity.
2. Show the "Reflect on the Feedback" page on a screen using the Presentation mode in the Learning Hub so everyone in the classroom can see it.
3. Say, "Please come pick up your reports from over here on this table." Point to the table.
4. Give the students a few minutes to pick up their reports and return to their seats.
5. Say, "Please take a few minutes to read over the peer review guide. You should use it to figure out what you need to change in your report and how you will change it."
6. Allow the students to read the PRG.
7. Show the "Create a Final Report" page on a screen using the Presentation mode in the Learning Hub so everyone in the classroom can see it.
8. Say, "Okay, let's revise our reports. I would like you to rewrite your report. You can use your draft report as a starting point but use the feedback on the peer review guide to make it better."
9. Ask the students, "Do you have any questions about what you need to do?"
10. Answer any questions that come up.
11. Tell the students, "Okay, let's write." Allow about 20 minutes for the students to revise their reports.
12. After about 20 minutes, give the following directions, "Okay, time's up. I will now come around and collect your first draft, final report, and peer review guide."
13. Collect all the first drafts, PRGs, and final reports.
14. Use the "Teacher Score" columns in the PRG to grade the final report. Be sure to give the students feedback about their writing in the "Teacher Comments" section.

What to Show First

The "Reflect on the Feedback" page in the ADI Learning Hub

What to Show Second

The "Create a Final Report" page in the ADI Learning Hub

Attending to Equity

Students don't always use the feedback they receive from others to improve the quality of their reports. Make sure you move from student to student and encourage them to make changes based on the feedback. If students are not sure how to revise their reports, model how to use feedback. For example, you might say, "Okay, based on the peer review guide, I need to do more to describe what I did to collect data. I bet if I add information about ________, readers will have a better idea of what I did."

Teaching Tip

The peer review guide (PRG) is designed to be used with any ADI investigation, thus allowing you to use the same scoring rubric throughout the entire year. This is beneficial for several reasons. First, the criteria for what counts as a high-quality report do not change from investigation to investigation. Students, therefore, quickly learn what is expected from them when they write a report, and you do not have to spend valuable class time explaining the various components of the PRG each time students write a report. Second, the PRG makes it clear which components of a report need to be improved next time because the grade is not based on a holistic evaluation of the report. The PRG format allows students to see which aspects of their writing are strong enough and which aspects need improvement. Finally, and perhaps most importantly, the PRG provides you with a standardized measure of student performance that can be compared over multiple reports across semesters, thus allowing you to track improvement over time.

Teacher Notes

Bark Beetles and Climate Change in Washington

✓ TASK

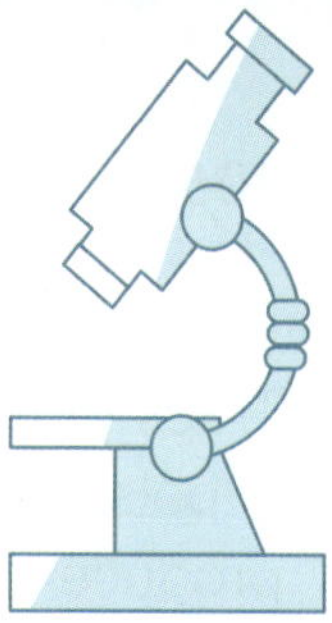

The Phenomenon

Bark beetles are tiny insects that live and feed underneath the bark of a tree. Take a few minutes to watch a video about bark beetle killing trees in Washington state. As you watch the video, write down some things you notice in the space below.

Things **I noticed...**

The Task

Adult bark beetles emerge from trees in August. They then look for a new tree to infest. Once they reach a new tree, they bore through its bark and create a vertical tunnel that is about three feet long as they feed. A pair of beetles will then mate and deposit about 100 eggs in the vertical tunnel. The adults then die. The eggs hatch in 10 to 14 days. The larvae start feeding on the inner bark right after they hatch, creating their own tunnels that split off the original. As they feed, the larvae destroy the tissues that transport the sugar and other carbohydrates that are produced in the leaves to the rest of the tree, which causes the cells in the trunk and the roots of the tree to starve. Bark beetles also carry a fungus that infects the trees. The fungus spreads through the water-transporting tissues in the tree and destroys it, which in turn, prevents water from moving from the roots to the leaves in the tree crown. The leaves then turn red and die from a lack of water. The larvae feed on the tree until May. They then create an oval cell at the end of their feeding tunnels and transform into pupae. The pupae mature into adults by July and chew an exit hole though the bark by August.

Trees defend against bark beetle attack by producing a sappy resin that fills a bore hole and expels the attacking beetle before it has a chance to mate. Healthy trees are usually able to produce enough of this resin to keep most bark beetles from chewing through its bark. However, the climate in Washington is changing. As a result, the summers are warmer, the winters are milder, and there is less rain. The lack of rain has caused severe and long-lasting droughts in the area. The droughts have results in numerous unhealthy trees and the warmer temperatures have resulted in more bark beetle outbreaks. The outbreaks, in turn, lead to many tree deaths, a rapid decline in the local tree population, and an increase in the number of wildfires. Your goal in this investigation is to figure out a way to reduce bark beetle outbreak occurrences and tree loss. The guiding question of this investigation is:

What can people do to reduce bark beetle outbreak occurrences and tree loss in Washington state?

Initial Ideas

Before you start this investigation, take a few minutes to think about all the different components of an ecosystem. Then draw a model in the space below that shows how you think these different components might cause the size of bark beetle population to increase suddenly. Be sure to include labels to help explain your thinking.

Some **ideas** that I have...

Share your model with the other members of your group. Once everyone has had a chance to share their idea, list some questions that your group has about this phenomenon.

Some questions my group has now...

Bark Beetles and Climate Change in Washington

IDEAS

Some Ideas You Can Use: Ecosystem Dynamics, Functioning, and Resilience

An **ecosystem** is a defined area that includes a community and the environment. A **community** is defined as two or more populations. A **population** is a group of organisms that belong to the same species and live in the same place at the same time. A **species** is a type of organism. The image below shows the difference between a species, a population, a community, and an ecosystem. In this image, there are three species: pine tree, deer, and the bark beetle. Since there is more than one individual in each species, there are also three populations. The three populations create a single community. The community along with the sunlight, water, and space is an ecosystem. Each ecosystem has a carrying capacity for each population due to such factors as the availability of living and nonliving resources and from such challenges such as predation, competition, and disease. The **carrying capacity** of an ecosystem is the maximum population size that a given ecosystem can support.

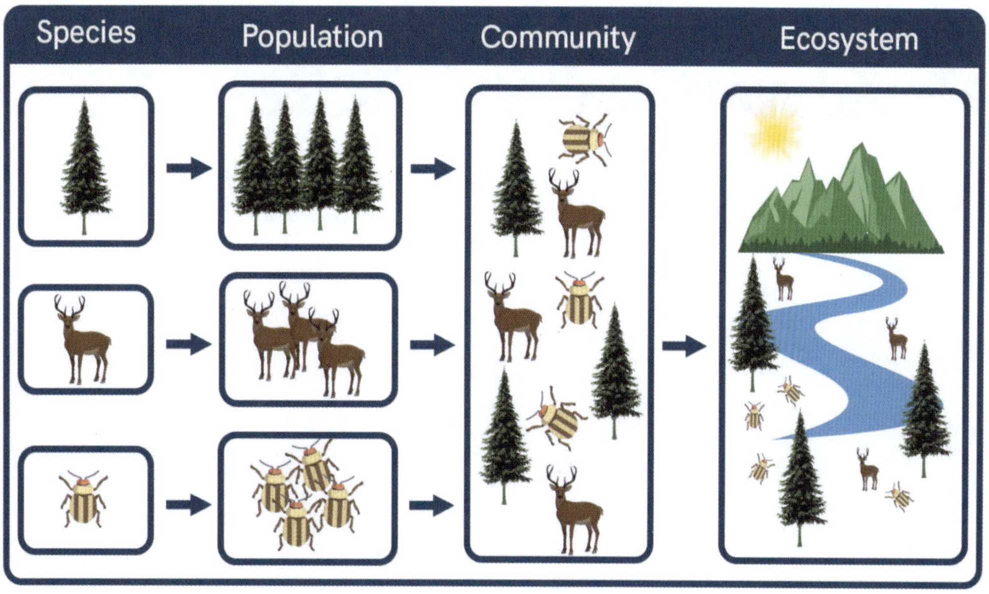

The number of species and the size of the different populations within an ecosystem stays relatively constant over long periods of time when conditions are stable. Even when there is a biological or physical disturbance to an ecosystem, the size of the populations within that ecosystem will only shift up or down for a brief period and then they will return to their original states. Examples of biological disturbances includes such things as population booms or a disease outbreak. Examples of physical disturbances include such things as a short drought or a harsh winter. Extreme changes in the environmental conditions or the size of any population over longer periods of time, however, can alter the functioning of an ecosystems in terms of resources and habitat availability. These extreme changes can cause an ecosystem to transform

ARGUMENT-DRIVEN INQUIRY

into a different one. In addition, changes in an ecosystem that are caused by human activity, which include such things as habitat destruction, pollution, introduction of invasive species, and climate change, will often alter the functioning of an ecosystem in a way that will lead to transformation in the ecosystem (see pictures below for an example). When an ecosystem transforms into a different one, there is often a corresponding loss of biodiversity in the area as different species leave or die off.

An example of ecosystem transformation in Sussex County, Delaware

Some **useful ideas** from what I read...

Some Ideas You Can Use: Developing Solutions to Problem of Decreasing Biodiversity

Humans depend on the living world for the resources and other benefits provided by biodiversity. Human activity, however, can have adverse impacts on biodiversity through overpopulation, overexploitation, habitat destruction, pollution, introduction of invasive species, and climate change. Thus, sustaining biodiversity in an ecosystem so that the ecosystem continues to function as it normally does is essential to supporting and enhancing life on Earth. Sustaining biodiversity also aids humanity by preserving landscapes of recreational or inspirational value. The climate-change driven bark beetle outbreak in Washington state is an example how human activity is having an adverse effect on the biodiversity of the forest ecosystems in that area. The photos below show two forests: one with isolated tree damage and one with widespread damage. As can be seen in the photos, bark beetle outbreaks can kill hundreds of trees.

People can slow the spread of bark beetles in several ways. Over the long term, people in Washington can adopt policies, such as limiting greenhouse gas emissions, that will reduce the amount the climate changes over time. In the short term, they can slow bark beetle reproduction. There are three different approaches or methods to slow bark beetle reproduction. The first method is to remove some trees from a forest before they are infested by bark beetles to help slow or reduce the anticipated spread by making it harder for beetles to find a new tree to infest. A second method is to completely remove any tree that is infested with bark beetles before they emerge in spring and spread to new trees. The third method is to remove parts of all the trees before they infested (such as low hanging branches or branches that touch the branches of other trees) or just the parts of a tree that show signs of infestation. Given the goal of this investigation, you will need to figure out which approach or combination of approaches is most likely to solve the problem of tree loss. When evaluating potential solutions to a problem (such as the potential loss of forest biodiversity or function), it is important to take into account a range of constraints including cost, safety, reliability, and aesthetics and to also consider the social and cultural impacts enacting the potential solution.

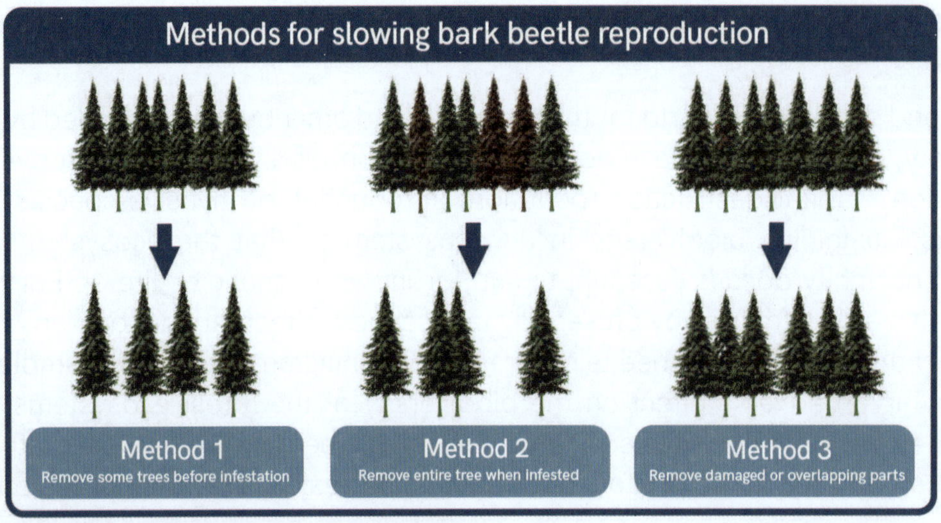

Some **useful ideas** from what I read...

Some Ideas You Can Use: System Models

Models can be used to represent a system and the way the components that make up that system interact with one another. A **model** can be physical, mathematical, visual, or a computer-based simulation. You will use a computer-based simulation as your model in this investigation (see image below). This simulation has been coded so it represents an ecosystem that contains a population of pine trees and a population of bark beetles. Each bark beetle can live up to two years but will die as soon as it reproduces. Some bark beetles will also die in the winter without reproducing. A bark beetle will produce 2 offspring beetles once it infests a tree. The temperature influences the death rate and reproduction rate of the beetles. The pine trees become reddish after being infested and die after they sustain too much damage from invading beetles. The severity of the drought determines how much beetle damage a tree can tolerate before it dies. The model also allows you to set different tree management policies for cutting and thinning. This model, as a result, will enable you to figure out what people in Washington state can do to reduce bark beetle outbreak occurrences and tree loss due to climate change.

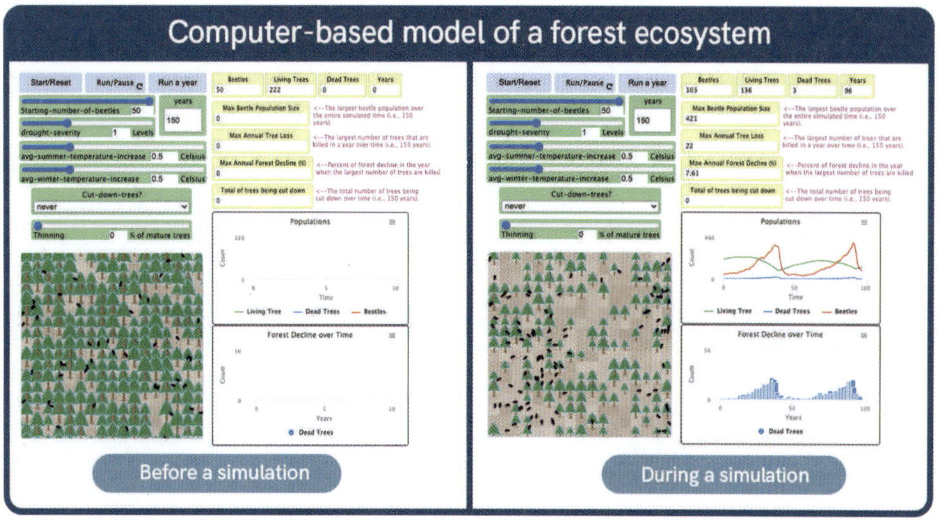

You can change an **input**, such as "cut-down-trees" and "% of mature trees to be thinned" in the ecosystem before each simulation. You can also change some of the interactions between the different parts of the ecosystem in the model by increasing the number of starting beetles, the drought severity (which affects tree health), the average summer temperature increase (which affects beetle reproduction rate), and the average winter temperature increase (which affects the beetle death rate). For this investigation, the bark beetle population size and the annual tree loss will be the **outputs of interest**. Models like this are useful for testing different solutions to a problem or to see which one is most efficient or economical; and in making a persuasive presentation to a client about how a given solution will meet his or her needs. However, it is important to always keep in mind that **these tests have limited precision and reliability** due to the assumptions and approximation inherent in any model.

Some **useful ideas** from what I read...

Bark Beetles and Climate Change in Washington

PLAN

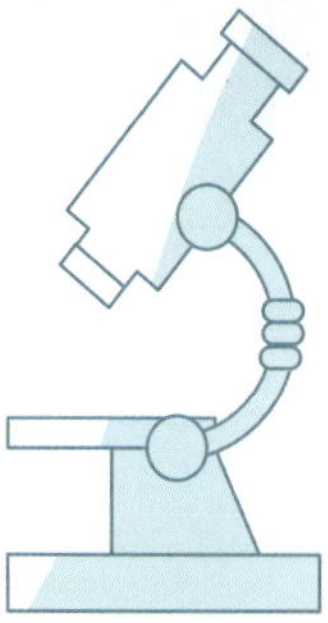

Plan Your Investigation

Prepare a plan for your investigation by filling out the chart below.

I am trying to answer the following **question**...

I will use the following **observations or measurements**...

I will **collect** these observations or measurements by...

I will **analyze** these observations or measurements by...

I approve of this investigation plan.

Teacher's signature

Date

Bark Beetles and Climate Change in Washington

DO

Collect Data

Keep a record of what you observe or measure during your investigation.

My **observations or measurements...**

Analyze Your Data

Use the space below to create any tables or graphs that will help you make sense of the data you collected so you can answer the guiding question.

Bark Beetles and Climate Change in Washington

SHARE

Draft Argument

Develop an argument on a whiteboard. It should include:

1. A *claim* that answers the guiding question,
2. *Evidence* in support of the claim that provides an analysis of the data and a description of what the analysis means,
3. A *justification of the evidence* that explains why your group thinks the evidence is important.

Our Guiding Question:	
Claim:	
Evidence:	Justification of the Evidence:

Argumentation Session

Share your argument with your classmates. In the space below, be sure to keep track of any ideas that you can use to revise your argument and make it better.

Some possible ways to **improve** our argument...

Bark Beetles and Climate Change in Washington

REFLECT

Reflective Discussion

In the space below, you can keep track of any ideas from the discussion that you think are important or will be useful in the future.

Some important **ideas**...

Bark Beetles and Climate Change in Washington

REPORT

Draft Report

Prepare a report to share what you figured out during your investigation.

Introduction

We have been studying ______________________________ in class. Before we started this investigation, we explored ____

____

____

We noticed ____

____

____

Our goal for this investigation was to figure out ______________________________

____

____

The guiding question was ____

____

Method

To answer this question, ____

____

____

____

____

____

____

____

Argument

We figured out ______________________________

____

____

____

The graph below includes information about ______________________________

____

____

____

This analysis suggests ______________________________

____

____

____

This evidence is based on several important scientific ideas. The first one is ______________________________

____

____

____

____

____

____

Chromosome Reduction in Reproductive Cells Lesson Plan

OVERVIEW

Description	Students figure out how organisms create reproductive cells.
Guiding Question	How do organisms create reproductive cells?
What Students Figure Out	A basic model for behavior of chromosomes during the process of meiosis.

STAGES

This investigation has seven stages. Each stage includes multiple activities. The flowchart below shows what happens during each stage and how much time is needed to complete each one.

Task (Day 1)
1 — **20 minutes**
Students are introduced to a phenomenon and a guiding question to answer. They also share their initial ideas.

Ideas (Day 1)
2 — **30 minutes**
Students read about DCI, a CC, and an SEP that they can use to help them figure out the phenomenon.

Plan (Day 2)
3 — **15 minutes**
Students formulate a plan collecting and analyzing the data they need to answer the guiding question.

Do (Day 2)
4 — **30 minutes**
Students carry out their plan, make sense of the data they collected, and answer the guiding question.

Share (Day 3)
5 — **45 minutes**
Students develop, share, and revise an evidence-based argument about what they figured out.

Reflect (Day 4)
6 — **20 minutes**
Students discuss a DCI and/or a CC and reflect on the SEPs they used during the investigation.

Report (Day 4)
7 — **25 minutes**
Students write investigation reports to share what they figured out during the investigation.

Report (Day 5)
7 — **45 minutes**
Students peer review the reports written by other students and offer feedback to improve.

Report (Day 6)
7 — **30 minutes**
Students revise their reports and then submit a final draft to their teacher for evaluation.

STANDARDS ALIGNMENT

NEXT GENERATION SCIENCE STANDARDS

Performance Expectation

Make and defend a claim based on evidence that inheritable genetic variations may result from new genetic combinations through meiosis, viable errors occurring during replication, and/or mutations caused by environmental factors (HS-LS3-2).

Core Ideas(s)	Crosscutting Concept(s)	Practices
Variation of Traits (LS3.B)	Systems and System Models (CC4)	Asking Questions (SEP1) Developing and Using Models (SEP2) Planning and Carrying Out Investigations (SEP3) Analyzing and Interpreting Data (SEP4) Using Mathematics and CT (SEP5) Constructing Explanations (SEP6) Engaging in Argument from Evidence (SEP7) Obtaining, Evaluating, and Communicating Information (SEP8)

WIDA ENGLISH LANGUAGE DEVELOPMENT STANDARDS FOR SCIENCE

Explain	Argue
Interpret scientific explanations (ELD-SC.9-12.Explain.Interpretive) Construct scientific explanations (ELD-SC.9-12.Explain.Expressive)	Interpret scientific arguments (ELD-SC.9-12.Argue.Interpretive) Construct scientific arguments (ELD-SC.9-12.Argue.Expressive)

COMMON CORE ENGLISH LANGUAGE ARTS

Reading	Writing	Speaking and Listening
Key ideas and details (RST.9-10.1 and RST.9-10.2) Craft and structure (RST.9-10.4) Integration of knowledge and ideas (RST.9-10.7, RST.9-10.8, and RST.9-10.9) Range of reading (RST.9-10.10)	Text types and purposes (WHST.9-10.1 and WHST.9-10.2) Production and distribution of writing (WHST.9-10.4, WHST.9-10.5, and WHST.9-10.6)	Comprehension and collaboration (SL.9-10.1 and SL.9-10.3) Presentation of knowledge and ideas (SL.9-10.4 and SL.9-10.6)

THINGS TO KNOW BEFORE STARTING THE LESSON

The Core Ideas

A reproductive cell contains genetic information that can be passed onto a new organism (or offspring). Sexual reproduction consists of two reproductive cells that fuse together and create a single cell during fertilization. While normal cells contain a full set of chromosomes, or the total genetic information of an organism, reproductive cells only contain half of a full set. These two halves then combine to create a full set during fertilization. This process is known as meiosis.

The Crosscutting Concepts

A system is a group of parts that interact with one another to carry out some function that the individual parts cannot achieve alone. A cell is a small but highly complex system. To better understand cells and their parts, we can construct a model. A model is a physical, visual, or computer-based representation of a system that we can test and use to make predictions about the system.

The Science and Engineering Practices

A practice of science is to **ask and refine questions** that lead to descriptions and explanations of how the natural world works. Another practice of science is to **use and construct models** as helpful tools for representing ideas and explanations. These tools include diagrams, drawings, physical replicas, mathematical representations, analogies, and computer simulations. Scientists **plan and carry out investigations** in the field or a laboratory, working collaboratively and individually. Their investigations are systematic and require clarifying what counts as data and identifying variables or parameters. Scientific investigations produce **data that must be analyzed and interpreted**. Because data patterns and trends are not always obvious, scientists use a range of tools and practices to identify the significant features and patterns in the data. Scientists also use **mathematics and computational thinking** for a range of tasks such as constructing simulations; statistically analyzing data; and recognizing, expressing, and applying quantitative relationships. The products of doing science are **explanations**. Scientific explanations often specify variables and are used by scientists to describe and predict different natural phenomena. **Scientific argumentation**, or the process of sharing, supporting, critiquing, and revising claims based on evidence is the process that scientists use to determine if an explanation is valid or acceptable. Finally, scientists must be able to **communicate clearly and persuasively** the ideas and methods they generate. Critiquing and communicating ideas individually and in groups is a critical professional activity.

PREPARATION

This lesson can be delivered paper-based, paperless, or remote. Use the paper-based delivery option of the investigation if students do not have 1:1 access to a computer and they are in the same classroom or physical space during the lesson. Use the paperless delivery option of the investigation if students have 1:1 access to a computer and they are in the same classroom or physical space during the lesson. Finally, if students are not in the same classroom or physical space during the lesson and have 1:1 access to a computer, use the remote delivery option.

Materials Needed

The materials needed by the teacher and each student in the class for each of the three modalities are listed in the table below.

Delivery Mode	Each Student Will Need	The Teacher Will Need
Paper-Based	• Copies of each student handout for the investigation (or a workbook) • A pencil	• A teacher account in the ADI Learning Hub • Access to a computer with high-speed internet and a browser (such as Chrome, Safari, Firefox, or Edge) • A projector or smart board (for displaying a computer screen)
Paperless	• A student account in the ADI Learning Hub • Access to a computer with high-speed internet and a browser (such as Chrome, Safari, Firefox, or Edge)	• A teacher account in the ADI Learning Hub • Access to a computer with high-speed internet and a browser (such as Chrome, Safari, Firefox, or Edge)
Remote	• A student account in the ADI Learning Hub • Access to a computer with high-speed internet and a browser (such as Chrome, Safari, Firefox, or Edge)	• A teacher account in the ADI Learning Hub • Access to a computer with high-speed internet and a browser (such as Chrome, Safari, Firefox, or Edge)

Additional Materials and Preparation Needed for In-Person Delivery

The materials needed for the in-person version of this investigation are listed in the table below. To purchase the **Chromosome Reduction in Reproductive Cells** kit from our store, please visit https://shop.argumentdriveninquiry.com/. The **Chromosomes Reduction in Reproductive Cells** kit comes with enough materials for eight groups.

Type of Material	Item	Quantity	
		For Entire Class	For Each Group
Equipment	Card 1: Meiosis Event A	8	1
	Card 2: Meiosis Event B	8	1
	Card 3: Meiosis Event C	8	1
	Card 4: Meiosis Event D	8	1
	Card 5: Meiosis Event E	8	1
	Card 6: Meiosis Event F	8	1
	Card 7: Meiosis Event G	8	1
	Card 8: Meiosis Event H	8	1
	Red pop beads	192	24
	Yellow pop beads	384	48
	Hook-and-loop cable ties, 4″	64	8
	Plastic cups	8	1

Preparation

Create a set of Meiosis Event cards for each group. Each set of cards should include one copy of cards 1–8 to make a total of eight cards in each set. Place 24 red pop beads, 48 yellow pop beads, and eight 4-inch hook-and-loop cable ties in each cup. Label each cup with the group's name or number.

During the Investigation

Students can use the cards to determine how the appearance of a cell changes as it divides. Examples of the cards are shown in the images below.

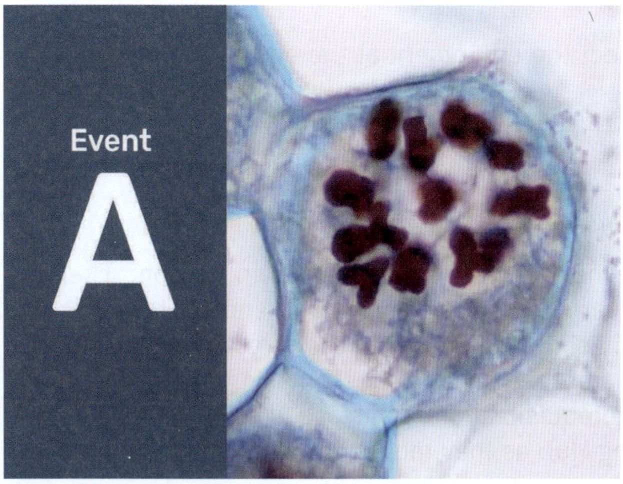

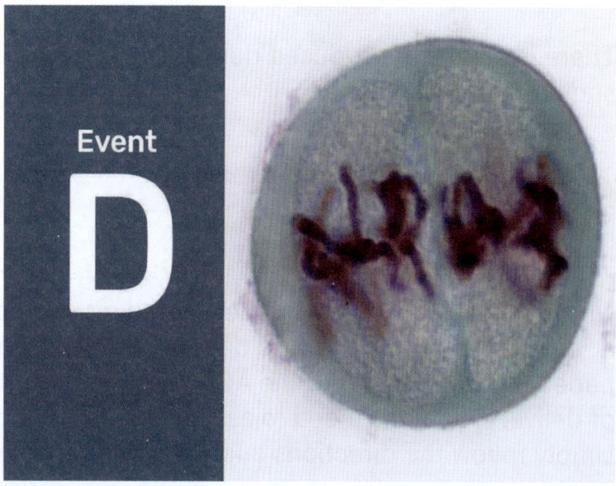

Students can then use pop beads and hook-and-loop cable ties to create a physical model of chromosomes. They can use the physical model to figure out what the chromosomes are doing during the different stages of meiosis. The image below is an example of a physical model. The names of each stage of meiosis are included in this image; however, students will likely use the same event letters as on the cards since they have not been formally introduced to each stage of meiosis at this time.

Example of a physical model of the process of meiosis

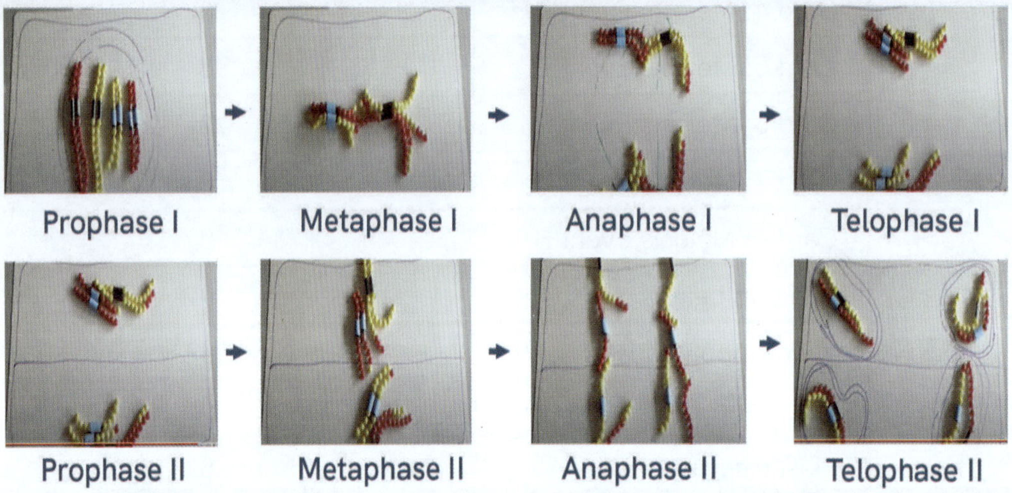

We recommend that you use a set routine for distributing and collecting the materials during the investigation. The equipment for each group, for example, can be set up at each group's lab station before class begins, or one member from each group can collect it from a table or a cart when needed during class.

Safety Precautions

Remind students to follow all normal lab safety rules.

Laboratory Waste Disposal

There is no waste produced during this activity. Collect the equipment at the end of the investigation and store for later use.

FACILITATION NOTES

The facilitation notes for this investigation describe how to deliver this lesson in person. They are written specifically for the paper-based mode of delivery. These same facilitation notes, however, can also be used for a paperless delivery of the lesson by simply substituting the directions that mention paper with the entry into the ADI Learning Hub. If you are teaching this lesson using the remote mode of delivery, you only need to assign due dates and prerequisites for each stage. The students can then simply follow the directions built into the ADI Learning Hub as they move through the investigation.

Chromosome Reduction in Reproductive Cells Lesson Plan

Stage 1 TASK

Total Time for the Stage	45 minutes
Grouping	Individual; Teams

Learning Goals

Students should make progress on their ability to do the following by the end of this stage of the lesson:

Science

1. Ask questions that arise from careful observation of phenomena to clarify and/or seek additional information (SEP1).
2. Develop and use a model to provide mechanistic accounts of phenomena (SEP2).
3. Evaluate the merits and limitations of different models of the same phenomenon (SEP2).
4. Ask questions that arise from examining models to clarify and/or seek additional information and relationships (SEP 1).
5. Ask questions to clarify and refine a model (SEP1).

English Language Development

None

English Language Arts

1. Initiate and participate effectively in small-group collaborative discussions with diverse partners by building on others' ideas and expressing their own clearly (SL.9-10.1).
2. Present information clearly, concisely, and logically such that listeners can follow the line of reasoning and the organization, substance, and style are appropriate to the purpose, audience, and task (SL.9-10.4).
3. Adapt speech to a variety of contexts and tasks, demonstrating command of formal English when indicated or appropriate (SL.9-10.6).

Activities within the Stage

Activity A	The Phenomenon
Time Needed for the Activity	5 minutes
Grouping	Teams

What to Do

1. Have students work in their assigned teams for this activity.
2. Begin the activity by giving out a copy of the Task handout to each student.
3. Show "The Phenomenon" page on a screen using the Presentation mode in the Learning Hub so everyone in the classroom can see it.
4. Read the introduction of the phenomenon out loud and then start the video. You can also have students read the introduction of the phenomenon and watch the video on their own devices.
5. Give the students a few moments to write down their observations on their handouts before moving on to the next activity.

What to Give Out

Task handout for **Chromosome Reduction in Reproductive Cells**

What to Show

"The Phenomenon" page in the ADI Learning Hub

Teaching Tip

Instead of assigning fixed roles to students by name at the beginning of the investigation, give each member of a group a different number so you can easily assign and change roles as needed during the various investigation stages. For example, you might give directions such as "1s will share out for your group" or "3s, will you please come back here and pick up these materials for your group?" or "2s and 3s, you will present the argument for your group while 1s and 4s travel to the other groups to offer feedback."

Attending to Equity

Be sure to intentionally group emerging multilingual students at times with peers who know the same languages as they do and at other times with peers whose English language development is slightly more advanced. Thoughtful grouping that varies throughout an investigation allows emerging multilingual students to benefit from the different linguistic resources of their peers.

Activity B	Share What You Noticed
Time Needed for the Activity	5 minutes
Grouping	Teams

What to Do

1. Show the "Share What You Noticed" page on a screen using the Presentation mode in the Learning Hub so everyone in the classroom can see it.
2. Say, "Okay, please share what you noticed with the others at your table."
3. Give the students about two minutes to share at their tables.
4. Say, "Okay, let's take a few minutes to share with the class. Team X (call on a team at random), what did you talk about?"
5. Allow Team X to share something they noticed. If it is related to how genetic information is passed from parent to offspring in reproduction, say, "That is really interesting. We are actually going to examine that in more detail during this investigation." If what is shared is not related to how genetic information is passed from parent to offspring in reproduction, say, "That is really interesting. Team Y (call on another team at random), what did you notice?" and allow that team to share. Follow this pattern until a team mentions something related to how the sperm and egg combine to form a new cell. If no team mentions this, say, "You know what I found interesting? I saw the sperm and egg combine to create a new cell. Did anyone notice that as well?"
6. Move on to the next activity.

What to Show

The "Share What You Noticed" page in the ADI Learning Hub

What to Look For

As soon as a group mentions something related to how the sperm and egg combine to form a new cell, move on to the next activity.

Attending to Equity

Be sure to provide students with opportunities to talk in small groups during this activity. Small-group structures offer all students, including emerging multilingual students, a chance to engage in sense-making with their peers and offers them space to use their verbal and nonverbal (e.g., writing, drawing pictures, and using gestures) resources to express their ideas. Small-group structures also give students a chance to learn from other students' uses of these resources.

Activity C	What You Need to Figure Out
Time Needed for the Activity	5 minutes
Grouping	Teams

What to Do

1. Show the "What You Need to Figure Out" page on a screen using the Presentation mode in the Learning Hub so everyone in the classroom can see it.
2. Read the task out loud and have students follow along on their handout, on their devices, or on the screen.
3. Allow students to ask clarifying questions about the task or the guiding question.
4. Once everyone understands what they will need to figure out during the investigation, move on to the next activity.

What to Show

The "What You Need to Figure Out" page in the ADI Learning Hub

Teaching Tip

Have students highlight, circle, or place a star next to the guiding question. Marking the guiding question helps students find it later during the Plan stage.

ADI Go Digital

The digital version of this lesson is available in the **ADI Learning Hub**. Be sure to check out the built-in teacher presentation mode.

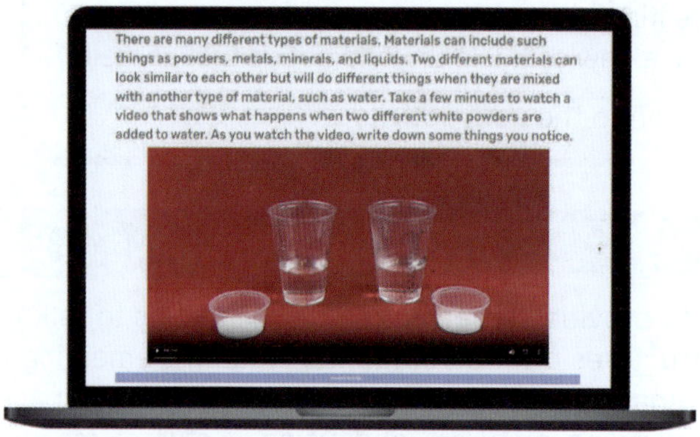

Activity D	Share Your Ideas
Time Needed for the Activity	15 minutes
Grouping	Teams

What to Do

1. Show the "Share Your ideas" page on a screen using the Presentation mode in the Learning Hub so everyone in the classroom can see it.
2. Read the directions on the handout out loud and then give students five minutes to draw their initial representations of the science.
3. Say, "Okay, will everyone share their representations with the rest of your team? Please make sure everyone has time to talk. Also, be sure to ask one another questions about the scientific ideas that are included in each representation."
4. Give the students about 8-9 minutes to talk and share. Be sure to walk around and listen to the discussions.
5. Say, "Okay, I heard some interesting ideas as I walked around the room."
6. Move on to the next stage.

What to Show

The "Share Your Ideas"page in the ADI Learning Hub

What to Look For

Each student's initial representation should include a combination of symbols, diagrams, graphs, or words. It should also clearly show how a scientific idea can be used to help complete the task and answer the guiding question.

Attending to Equity

When students are tasked with expressing their ideas, be sure to encourage them to do so through multiple modes (speaking, writing, drawings, and gestures) and to use both content-specific and everyday registers. This process helps students fully express themselves, allowing peers to evaluate, question, and build on ideas. Also, make sure that every student in every team has a chance to share their initial representation.

Activity E	More Questions
Time Needed for the Activity	15 minutes
Grouping	Teams

What to Do

1. Show the "More Questions" page on a screen using the Presentation mode in the Learning Hub so everyone in the classroom can see it.
2. Say, "Okay, I heard a bunch of interesting questions as I was walking around the room. Will you work with your team to identify two or three questions you think are important to learn more about that came up as you all discussed the different representations you made? You can write these questions on your handout."
3. Give students about five minutes to talk and write down questions. Be sure to walk around and listen to the discussions.
4. Say, "Okay, let's take a few minutes to share. Team X (call on a team at random), will you share one question you all have?"
5. Have each team share a question. Then say, "I want to share information about some useful scientific ideas with you. I bet this information will not only answer many of these questions but will also help you complete the task and answer the guiding question of the investigation."
6. Move on to the next stage.

What to Show

The "More Questions" page in the ADI Learning Hub

What to Look For

Each student should have 2-3 questions written on their handout.

Teaching Tip

When students are asked to create representations of the underlying scientific ideas, share these ideas with their peers, and then identify things they want to know more about, it sparks their curiosity and helps make the investigation meaningful. Also, identifying questions before reading (which students do during the Ideas stage) is an outstanding pre-reading activity that helps improve comprehension.

Chromosome Reduction in Reproductive Cells Lesson Plan

Stage 2 IDEAS

Total Time for the Stage	30 minutes
Grouping	Teams

Learning Goals

Students should make progress on their ability to do the following by the end of this stage of the lesson:

Science

1. Ask questions that arise from examining theory to clarify and/or seek additional information and relationships (SEP1).
2. Critically read scientific literature adapted for classroom use to determine the central ideas or conclusions and/or to obtain scientific and/or technical information to summarize complex concepts, processes, or information presented in a text by paraphrasing them in simpler but still accurate terms (SEP8).
3. Integrate sources of information presented in different media or formats (e.g., visually and quantitatively) as well as in words to address a scientific question (SEP8).
4. Gather, read, and evaluate scientific and/or technical information from multiple authoritative sources, assessing the evidence and usefulness of each source (SEP8).
5. Describe how individual cells in multicellular organisms grow and then divide in a process called mitosis (LS1.B).
6. Use models to simulate systems and interactions within and between systems at different scales (CC4).

English Language Development

1. Paraphrase central ideas in complex evidence, concepts, processes, and information to help explain how or why a phenomenon occurs (ELD-SLC.9-12.Explain.Interpretive).

English Language Arts

1. Determine the central ideas or conclusions of a text; trace the text's explanation or depiction of a complex process, phenomenon, or concept; provide an accurate summary of the text (RST.9-10.2).
2. Determine the meaning of symbols, key terms, and other domain-specific words and phrases as they are used in a specific scientific or technical context (RST.9-10.4).
3. Read and comprehend scientific/technical texts (RST.9-10.10).
4. Initiate and participate effectively in small-group collaborative discussions with diverse partners by building on others' ideas and expressing their own clearly (SL.9-10.1).

Activities

Activity A	Read About a Core Idea
Time Needed for the Activity	10 minutes
Grouping	Teams

What to Do

1. Have students work in their assigned teams during this stage. Give out a copy of the Ideas handout to each student.
2. The first activity provides students with information about fertilization.
3. Show the "Read About a Core Idea You Can Use" page on a screen using the Presentation mode in the Learning Hub so everyone in the classroom can see it. You can also click on the images to make them full screen and easier for students to see.
4. Have students read the text on their own, either on their handouts or on the screen.
5. Give the students about three minutes to discuss the text with the others in their teams and develop a shared list of ideas that they think are important or useful.
6. Ask a team to share an idea they thought was useful. Have the remaining teams share a different idea until there are no more ideas to share.
7. Move on to the next activity.

Teaching Tip

Students do not need to do a close read of this text or completely understand all the ideas in it before moving on to the next activity. They will be encouraged to revisit this text later in the investigation when creating their arguments and writing their reports. All students need to be able to do during this activity is to work with the other members of their group to identify one or two ideas that they think are important to keep in mind or are potentially helpful.

What to Give Out

Ideas handout for **Chromosome Reduction in Reproduction Cells**

What to Show

The "Read About a Core Idea You Can Use" page in the ADI Learning Hub

What to Look For

Students should write down some of the following ideas in the box labeled, "Some things I know from what I read" on their handouts:

- A chromosome is a piece of genetic information.
- Human cells have 46 chromosomes, or 23 pairs.
- Fertilization is a process where two reproductive cells fuse together to create a single cell that has the same number of chromosomes as its parents.
- Reproductive cells have only half of a full set of chromosomes, so when the cells combine, the offspring receives a full set.
- Meiosis is the process in which chromosomes are reduced when creating reproductive cells.

Activity B	Read About a Core Idea
Time Needed for the Activity	10 minutes
Grouping	Teams

What to Do

1. The second activity provides students with information about meiosis.
2. Show the "Read About Another Core Idea You Can Use" page on a screen using the Presentation mode in the Learning Hub so everyone in the classroom can see it. You can also click on the images to make them full screen and easier for students to see.
3. Have students read the text on their own, either on their handouts or on the screen.
4. Give the students about three minutes to discuss the text with the others in their teams and develop a shared list of ideas that they think are important or useful.
5. Ask a team to share an idea they thought was useful. Have the remaining teams share a different idea until there are no more ideas to share.
6. Move on to the next activity.

What to Show

The "Read About Another Core Idea You Can Use" page in the ADI Learning Hub

What to Look For

Students should write down some of the following ideas in the box labeled, "Some things I know from what I read" on their handouts:

- Meiosis consists of eight major events.
- In some stages of meiosis, the chromosomes are grouped together.
- In other stages of meiosis, the chromosomes are separated and at the ends of the cell.

Attending to Equity

There are many supports for helping students comprehend what they read already embedded into this activity (e.g., activating prior knowledge, providing a shared experience, making connections, synthesizing, and talking with peers). So, you might not need to provide much extra support. If you are concerned about some students not understanding what they are reading because of their scores on past reading comprehension tests, you can read it out loud as they follow along. As you read the text out loud, be sure to stop at each important idea and ask the students to put a star (or other annotation) next to it in the margin of their handouts. They can then discuss these ideas in their small groups.

Activity C	Read About a Final Core Idea
Time Needed for the Activity	10 minutes
Grouping	Teams

What to Do

1. The third activity provides students with information about using models to represent a process.
2. Show the "Read About a Final Core Idea You Can Use" page on a screen using the Presentation mode in the Learning Hub so everyone in the classroom can see it. You can also click on the images to make them full screen and easier for students to see.
3. Have students read the text on their own, either on their handouts or on the screen.
4. Give the students about three minutes to discuss the text with the others in their teams and develop a shared list of ideas that they think are important or useful.
5. Ask a team to share an idea they thought was useful. Have the remaining teams share a different idea until there are no more ideas to share.
6. Move on to the next activity

What to Show

The "Read About a Final Core Idea You Can Use" page in the ADI Learning Hub

What to Look For

Students should write down some of the following ideas in the box labeled, "Some things I know from what I read" on their handouts:

- Models are used to represent an object, event, or process.
- Models can be physical, visual, or a computer-based simulation.
- A visual model includes images, symbols, words, and numbers to help explain a process.
- A visual model provides a way to make predictions about what will happen in the future.

Attending to Equity

To support emerging multilingual students during this activity, be sure to create a word bank on the wall of high-frequency, high-need concrete vocabulary that is needed to communicate scientific ideas. The word bank highlights terms needed for basic communication in an academic context. It also helps with the spelling. You can also encourage your emerging multilingual students to:

1. Use strategic learning techniques such as concept mapping, drawing, memorizing, comparing, contrasting, and reviewing.
2. Decode (sound out) words using skills such as recognizing sound-letter relationships and identifying cognates, affixes, roots, and base words.
3. Use visual and contextual support and support from peers and teachers to enhance and confirm understanding of the texts.

Chromosome Reduction in Reproductive Cells Lesson Plan

Stage 3 PLAN

Total Time for the Stage	15 minutes
Grouping	Teams

Learning Goals

Students should make progress on their ability to do the following by the end of this stage of the lesson:

Science

1. Evaluate a question to determine if it is testable and relevant (SEP1).
2. Plan an investigation individually and collaboratively to produce data to serve as the basis for evidence as part of building and revising models (SEP3).
3. Consider possible variables or effects and evaluate the confounding investigation's design to ensure variables are controlled (SEP3).
4. Ask questions to determine relationships, including quantitative relationships, between independent and dependent variables (SEP1).
5. Decide on types, quantity, and accuracy of data needed and consider limitations on the precision of the data (e.g., number of trials, cost, risk, and time), and refine the design accordingly (SEP3).
6. Select appropriate tools to collect, record, and analyze data (SEP3).

English Language Development

1. Define investigable questions on observations, information, and/or data about a phenomenon (ELD-SLC.9-12.Explain.Interpretive).
2. Develop reasoning to illustrate and/or predict the relationships between variables (ELD-SLC.9-12.Explain.Expressive).

English Language Arts

1. Initiate and participate effectively in small-group collaborative discussions with diverse partners by building on others' ideas and expressing their own clearly (SL.9-10.1).
2. Present information clearly, concisely, and logically such that listeners can follow the line of reasoning and the organization, substance, and style are appropriate to the purpose, audience, and task (SL.9-10.4).
3. Adapt speech to a variety of contexts and tasks, demonstrating command of formal English when indicated or appropriate (SL.9-10.6).

Activities

Activity A	Some Materials You Can Use
Time Needed for the Activity	5 minutes
Grouping	Teams

What to Do

1. Show the "Some Materials You Can Use" page on a screen using the Presentation mode in the Learning Hub so everyone in the classroom can see it. You can also click on the images to make them full screen and easier for students to see.
2. Say, "Okay, it sounds like we now have enough information to plan an investigation that will help us figure out how organisms create reproductive cells."
3. Say, "Before we get started, I want to show you some materials that we are going to use during our investigation. Will you all please come over to this table so I can show you what we have?"
4. Direct students to stand around the table where all the materials are set up.
5. Hold up the Meiosis Event card set.
6. Ask, "What do you think we could use these for?"
7. Allow students to share ideas. Then say, "We can use these cards to look at cells undergoing different stages of meiosis."
8. Hold up the red pop beads, yellow pop beads, and hook-and-loop cable ties.
9. Ask, "What do you think we could use these for?"
10. Allow students to share ideas. Then say, "We can use these to create a model of the chromosomes."
11. Hold up the plastic cups.
12. Ask, "What do you think we could use these for?"
13. Allow students to share ideas. Then say, "We can use these to hold any extra beads."
14. Say, "Okay, that is all the stuff we can use. Will you all head back to your teams so we can plan our investigation?"
15. Move on to the next activity.

What to Set Up

Place the following materials on a table so students can see them:

1. Meiosis Event card set
2. Red pop beads
3. Yellow pop beads
4. Hook-and-loop cable ties
5. Plastic cups

What to Show

The "Create a Final Report" page in the ADI Learning Hub

Teaching Tip

This activity helps students see for themselves all the materials they will be able to use during the hands-on part of the investigation instead of just reading about them or looking at pictures. It is important to show each material to the students and then point out how they can use it during the investigation so they can participate in developing a plan for the investigation during the next activity.

Activity B	Make a Plan
Time Needed for the Activity	10 minutes
Grouping	Teams

What to Do

1. Have students work in their assigned teams during this activity. Give a copy of the Plan handout to each student.
2. Show the "Make a Plan" page on a screen using the Presentation mode in the Learning Hub so everyone in the classroom can see it.
3. Say, "Work as a team to create a plan for your investigation by filling out the chart on your handout. As part of your plan, be sure to include which tools and techniques you will use to solve problems."
4. Give the students at least 10 minutes to complete this task. As students work in small teams to create their investigation proposals, be sure to move from team to team to answer questions and offer advice.
5. When a team thinks that they have a good plan, read it over. If you agree, sign off at the bottom of their handouts to let them know that their plan has been approved. If you do not agree, tell them what they will need to add or change before you will approve it. Once they have made the changes you can sign their handouts.
6. Once all the teams are done, you can move on to the next activity. You can also use an approved investigation plan as an "exit ticket" out of class.

What to Give Out

Plan handout for **Chromosome Reduction in Reproductive Cells**

What to Show

The "Make a Plan" page in the ADI Learning Hub

What to Look For

The investigation plan should, at a minimum, include:

- The question they are trying to answer.
- How they will make a model using the materials.
- How they will match their model to the cards.
- How they will analyze the data they collect from the model and cards.
- How they will use their analysis to answer the guiding question.

Attending to Equity

To support emerging multilingual students during this stage, be sure to create a word bank on the wall of high-frequency, high-need concrete vocabulary that is needed to communicate scientific processes and ideas. The word bank highlights terms needed for basic communication in an academic context. It also helps with the spelling. Also, you can encourage your emerging multilingual students to use modes (speaking, writing, drawings, and gestures) and to use both content-specific and everyday registers to discuss and create their plans. This process helps students fully express themselves, allowing peers to evaluate, question, and build on ideas.

Teaching Tip

One way to do a quick check-in with a group during this activity is to use the "two questions, a nudge, and a goal" technique. This technique includes the following steps:

1. Sit down with a group of students.
2. Ask, "Would one of you tell me a little about what you all have been doing or thinking about?" (Question 1)
3. A student answers on behalf of the group.
4. Ask, "Would anyone else like to add anything to that?" (Question 2)
5. A different student provides additional information on behalf of the group.
6. Ask, "Have you thought about [a hint or something you want to highlight]?" (A nudge)
7. Say, "Okay, I'm going to check on the other groups. When I get back, I want to see [something that you want the group to accomplish]." (A goal)

When you have completed the check-in with one group, move on to the next group.

Teacher Notes

Chromosome Reduction in Reproductive Cells Lesson Plan

Stage 4 DO

Total Time for the Stage	45 minutes
Grouping	Teams

Learning Goals

Students should make progress on their ability to do the following by the end of this stage of the lesson:

Science

1. Conduct an investigation individually and collaboratively to produce data to serve as the basis for evidence (SEP3).
2. Analyze data using tools and/or technologies to make valid and reliable scientific claims (SEP4).
3. Consider limitations of data analysis (e.g., measurement error and sample selection) when analyzing and interpreting data (SEP4).
4. Evaluate the impact of new data on a working explanation and/or model of a proposed process or system (SEP4).
5. Apply ratios, rates, percentages, and unit conversions in the context of complicated measurement problems involving quantities with derived or compound units (SEP5).
6. Make a quantitative and/or qualitative claim regarding the relationship between dependent and independent variables (SEP6).
7. Construct and revise an explanation based on valid and reliable evidence obtained from a variety of sources and the assumption that theories and laws that describe the natural world operate today as they did in the past and will continue to do so in the future (SEP6).
8. Apply scientific ideas, principles, and/or evidence to provide an explanation of phenomena taking into account possible unanticipated effects (SEP6).

English Language Development

1. Paraphrase central ideas in complex evidence, concepts, processes, and information to help explain how or why a phenomenon occurs (ELD-SLC.9-12.Explain.Interpretive).
2. Evaluate the extent to which reasoning, theory, and/or models link evidence to claims and support conclusions (ELD-SLC.9-12.Explain.Interpretive).
3. Develop reasoning to illustrate the relationships between variables (ELD-SLC.9-12.Explain.Expressive).

English Language Arts

1. Initiate and participate effectively in small-group collaborative discussions with diverse partners by building on others' ideas and expressing their own clearly (SL.9-10.1).
2. Present information clearly, concisely, and logically such that listeners can follow the line of reasoning and the organization, substance, and style are appropriate to the purpose, audience, and task (SL.9-10.4).
3. Adapt speech to a variety of contexts and tasks, demonstrating command of formal English when indicated or appropriate (SL.9-10.6).

Activities

Activity A	Carry Out Your Plan
Time Needed for the Activity	25 minutes
Grouping	Teams

What to Do

1. Students should work in their assigned teams for this activity. Begin this activity by giving out a copy of the Do handout to each student.
2. Show the "Carry Out Your Plan" page on a screen using the Presentation mode in the Learning Hub so everyone in the classroom can see it.
3. Say, "Okay, it sounds like we are now ready to carry out our plans and collect the data or observations we need to answer the guiding question. You can get the materials and tools you need to use from this table." Point to the table where you have placed them.
4. Ask, "Does anyone have any questions about these materials or tools?"
5. Answer any questions that come up.
6. Say, "As you work, be sure to record your observations and measurements on your handout so you can keep track of everything you do. Does anyone have any final questions?"
7. Answer any questions that come up.
8. Say, "Okay, go ahead and get started. You have the next 25 minutes to work. I'll be available to help if any team needs me."
9. Give the students at least 25 minutes to carry out their plans and solve the problem. As students work in small teams, be sure to move from team to team to answer questions and offer advice.

What to Set Up

Place the following materials on a table for each group to use:

1. Meiosis Event card set
2. Red pop beads
3. Yellow pop beads
4. Hook-and-loop cable ties
5. Plastic cups

What to Give Out

Do handout for **Chromosome Reduction in Reproductive Cells**

What to Show

The "Carry Out Your Plan" page in the ADI Learning Hub

Teaching Tip

One way to do a quick check-in with a group during this activity is to use the "two questions, a nudge, and a goal" technique. This technique includes the following steps:

1. Sit down with a group of students.
2. Ask, "Would one of you tell me a little about what you all have been doing or thinking about?" (Question 1)
3. A student answers on behalf of the group.
4. Ask, "Would anyone else like to add anything to that?" (Question 2)
5. A different student provides additional information on behalf of the group.
6. Ask, "Have you thought about [a hint or something you want to highlight]?" (A nudge)
7. Say, "Okay, I'm going to check on the other groups. When I get back, I want to see [something that you want the group to accomplish]." (A goal)

When you have completed the check-in with one group, move on to the next group.

Attending to Equity

Be sure to intentionally group emerging multilingual students at times with peers who know the same languages as they do and at other times with peers whose English language development is slightly more advanced. Thoughtful grouping that varies during different investigation (or activities within an investigation) allows emerging multilingual students to benefit from the different linguistic resources of their peers. We also recommend assessing emerging multilingual students based on how much they contribute to the group during this activity, rather than solely on what they write or draw on their handouts.

Teacher Notes

Activity B	Make Sense of Your Data
Time Needed for the Activity	20 minutes
Grouping	Teams

What to Do

1. Students should work in their assigned teams for this activity. Begin this activity by showing the "Make Sense of Your Data" page on a screen using the Presentation mode in the Learning Hub so everyone in the classroom can see it.
2. Say, "Okay, it sounds like we are now ready to make sense of the data we collected. One way to do this is to create a table, graph, or other visualization that shows a trend or change over time, a comparison between teams, or a relationship between things you tested. You can create a table, graph, or other visualization here on your handout."
3. Point to the box on the handout.
4. Say, "Please work together in your teams to figure out the best way to make sense of your data. Remember, you are looking for any trends, changes, differences, or relationships in the data that will help you answer the guiding question. As you work, be sure to use the space on your handout to perform any calculations that you need to do and create any representations that you need. Does anyone have any last questions?"
5. Answer any questions that come up.
6. Say, "Okay, go ahead and get started. You have the rest of the period to work. I'll be available to help if any team needs me."
7. Give the students at least 15 minutes to make sense of their data. As students work together in small teams, be sure to move from team to team to answer questions and offer advice.
8. Make sure students return the materials and tools to the table where they found them before dismissing class. Cleanup should take no longer than five minutes.

What to Show

The "Make Sense of Your Data" page in the ADI Learning Hub

What to Look For

One of the most useful ways to make sense of the data during this activity is to make a diagram or model that shows the process of meiosis.

Teaching Tip

Collecting and returning materials or tools to a central table works best if you assign the job to a specific number within each team. For example, you can say, "2s, please come up and grab the materials and tools that your team needs" at the start of the activity. Then you can say, "3s, please return the materials and tools your team used back to this table," at the end of the activity. Assigning jobs like this not only cuts down on how many students need to be up and moving around, but also allows you to hold specific students accountable for the materials and tools.

Attending to Equity

Encourage your emerging multilingual students to speak using learning strategies such as requesting assistance, employing nonverbal cues, and using synonyms and circumlocution (conveying ideas by defining or describing when exact English words are not known) during this activity. It is also helpful to point out and/or draw emerging multilingual learners' attention to elements of the English sound system in newly acquired vocabulary such as long and short vowels, silent letters, and consonant clusters as they come up naturally during this activity.

Teacher Notes

Chromosome Reduction in Reproductive Cells Lesson Plan

Stage 5 SHARE

Total Time Needed for the Stage	45 minutes
Grouping	Teams

Learning Goals

Students should make progress on their ability to do the following by the end of this stage of the lesson:

Science

1. Make and defend a claim based on evidence about the natural world that reflects scientific knowledge and student-generated evidence (SEP7).
2. Present an oral argument based on evidence (SEP7).
3. Evaluate the claims, evidence, and/or reasoning to determine the merits of arguments (SEP7).
4. Ask questions to clarify relationships, including quantitative relationships, between independent and dependent variables (SEP1).
5. Ask questions that challenge the interpretation of a dataset or the suitability of a method (SEP1).
6. Respectfully give critiques on scientific arguments by probing reasoning and evidence and challenging ideas and conclusions (SEP7).
7. Respectfully receive critiques on scientific arguments by responding thoughtfully to diverse perspectives and determining what additional information is required to resolve contradictions. (SEP7).
8. Communicate scientific information or ideas in multiple formats including orally, graphically, textually, and mathematically (SEP8).

English Language Development

1. Identify appropriate and sufficient evidence from data, models, and/or information from investigations of a phenomenon (ELD-SLC.9-12.Argue.Interpretive).
2. Compare reasoning and claims based on evidence from competing arguments (ELD-SLC.9-12.Argue.Interpretive).
3. Defend or refute a claim based on data and evidence (ELD-SLC.9-12.Argue.Expressive).
4. Establish and maintain an appropriate tone and stance (ELD-SLC.9-12.Argue.Expressive).
5. Signal logical relationships among reasoning, evidence, data, and/or models when making and defending a claim, counterclaim, and/or rebuttal (ELD-SLC.9-12.Argue.Expressive).

English Language Arts

1. Initiate and participate effectively in small-group collaborative discussions with diverse partners by building on others' ideas and expressing their own clearly (SL.9-10.1).
2. Integrate multiple sources of information presented in diverse media or formats (e.g., visually, quantitatively, and orally) evaluating the credibility and accuracy of each source (SL.9-10.2).
3. Evaluate a speaker's reasoning and use of evidence and rhetoric, identifying any fallacious reasoning or exaggerated or distorted evidence (SL.9-10.3).

(continued)

4. Present information clearly, concisely, and logically such that listeners can follow the line of reasoning and the organization, substance, and style are appropriate to the purpose, audience, and task (SL.9-10.4).
5. Make strategic use of media (e.g., textual, graphical, and visual) in presentations to enhance understanding of findings, reasoning, and evidence and to add interest (SL.9.10.5).
6. Adapt speech to a variety of contexts and tasks, demonstrating command of formal English when indicated or appropriate (SL.9-10.6).

Activities

Activity A	Create a Draft Argument
Time Needed for the Activity	15 minutes
Grouping	Teams

What to Do

1. Students should work in their assigned teams for this activity. Begin this activity by giving out a copy of the Share handout to each student.
2. Show the "Create a Draft Argument" page on a screen using the Presentation mode in the Learning Hub so everyone in the classroom can see it.
3. Say, "Okay, it sounds like we have all solved the task. I would now like you to create a draft argument to share what you figured out with the other teams. Your argument should include everything in this image here." Point to the image of a science-based argument projected on the screen. Be sure to go over what should be included in each component of the argument and where students can find the information they need.
4. Say, "I would like you to make your draft argument on a whiteboard. You can use the materials on this table to make your argument." Point to the table where you have placed the materials they need.
5. Ask, "Does anyone have any questions?"
6. Answer any questions that come up.
7. Say, "I'm going to give you exactly 15 minutes to get this done. Your draft argument doesn't need to be perfect. We just need something down on the whiteboards so we can share our ideas."
8. Give each team of students 15 minutes to create a draft argument. As students work in small teams, be sure to move from team to team to answer questions and offer advice.

What to Set Up

Place the following items on a table so students can use them during this stage:

- 24″ x 36″ whiteboard
- 3-4 dry-erase pens in different colors
- Eraser or cloth towel

What to Give Out

Share handout for **Chromosome Reduction in Reproductive Cells**

What to Show

The "Create a Draft Argument" page in the ADI Learning Hub

What to Look for

Students should include core ideas that they read about during the Ideas stage in their justification.

Teaching Tip

Be sure to plant some different ideas with each group about how to make a strong argument as you move from group to group to check in, answer questions, and offer advice. This strategy helps to ensure that there will be some really strong examples of the evidence and justification components of the argument for students to see and discuss during the next activity. For example, you could help two groups have a really strong evidence component by telling one group, "I think a graph would be really helpful here" and telling another group, "I think your evidence would be stronger if you explained, in words, what this graph shows." You could help a third group have a really strong justification by saying, "Let's brainstorm some ideas that will help you defend your choice of evidence," or "Let's go back to our Ideas handout and see if there are any concepts we can include."

It is important you make sure that all the arguments are different at this point in the investigation by planting different ideas with different groups, because the students will have nothing to critique or discuss during the next activity if they all have the same argument. The goal during this activity is to set the stage for the next activity by ensuring that the various components of a strong argument can be found on different whiteboards within the class. These strong components will then be taken up by all the groups because they reflect good ideas. Be sure to give students a voice and choice during this activity, so they can make mistakes and learn from them.

Attending to Equity

It is often helpful to assign and change roles at least once during this activity. For example, you might assign the 1s to be "in charge of writing" at the beginning of the activity. You can then easily switch roles about five minutes into the activity by saying, "1s, please hand the pens to the 2s. 2s, you are now in charge of writing for the group." By changing roles, you can ensure that all students have a chance to participate and contribute their ideas to the creation of a draft argument.

ADI Go Digital

The **ADI Learning Hub** has tons of features that allow students to share their thinking with each other when they are doing the investigation paperless or remote.

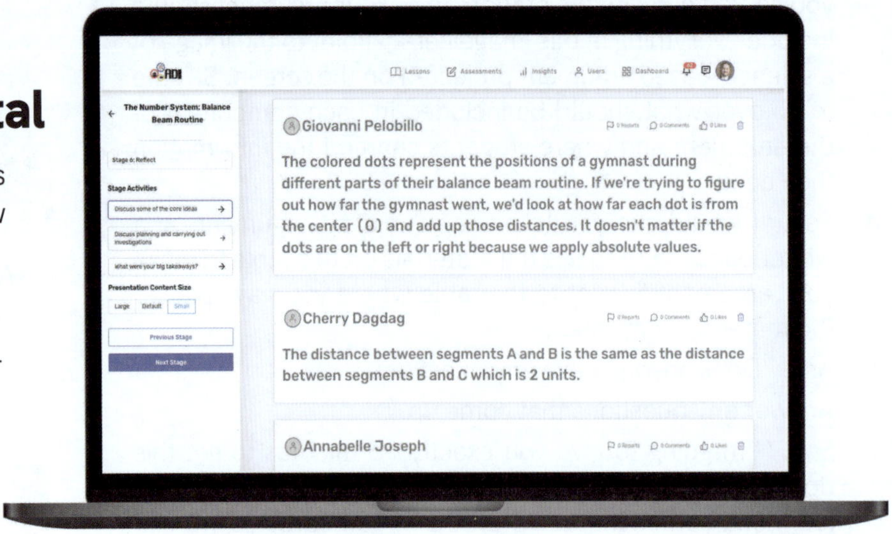

Activity B	Present Your Draft Argument
Time Needed for the Activity	15 minutes
Grouping	Teams

What to Do

1. Students should work in their assigned teams for this activity.
2. Show the "Present Your Draft Argument" page on a screen using the Presentation mode in the Learning Hub so everyone in the classroom can see it.
3. Say, "We are now going to share our arguments. Please set up your whiteboards so everyone can see them." Allow students time to set up their whiteboards. They should be set up on stands or chairs, so they are facing toward the outside of the room.
4. Say, "Okay, before we get started, I want to explain what we are going to do next. I'm going to ask some of you to present your arguments to your classmates. If you are presenting your argument, your job is to share your team's claim, evidence, and justification of the evidence. The rest of you will be travelers. If you are a traveler, your job is to listen to the presenters, ask them questions if you do not understand something, and then offer them some suggestions about ways to make their argument better. Do you have any questions about these instructions?"
5. Answer any questions that come up.
6. Say, "As you talk to one another, make sure you write down any good ideas or suggestions that are important in the space on your handout, so you don't forget about them." Point to the box labeled, "Some possible ways to improve our argument" on the handout.
7. Say, "After we have a chance to learn from one another, I'm going to give you some time to revise your arguments and make them better. Do you have any questions about what we are doing?"
8. Answer any questions that come up.
9. Say, "Okay. Let's get started. Travelers, move one team to the left. Don't move to the next team until I tell you to move. Presenters, go ahead and share your argument with the travelers when they get there."
10. Give each team of presenters and travelers about 3-4 minutes to talk.
11. Tell the students, "Okay. Let's move on to the next argument. Travelers, move one team to the left."
12. Again, give each team of presenters and travelers about 3-4 minutes to talk.

What to Show

The "Present Your Draft Argument" page in the ADI Learning Hub

What to Look for

The critiques and feedback that students give one another during this activity should:

- Be shared in a way that is kind (respectful),
- Provide information about what, exactly, needs to be changed or improved (specific), and
- Offer a suggestion about how to make something better rather than just saying something is wrong (helpful).

In addition, students should be focusing their critiques on the science included in the arguments, not the appearance of the arguments.

(continued)

13. Repeat this process until each team has had their argument read and critiqued three times. Then tell the students to return to their workstations.
14. Be sure to move between workstations during this activity to keep everyone involved and model what it means to communicate and critique scientific ideas and reasoning.

Attending to Equity

Be sure to encourage students to provide feedback that is respectful, specific, and helpful during this activity. It is important to encourage student-to-student talk because it helps students develop speaking and listening skills and enables them to make sense of what and how they know. It also provides a context for emerging multilingual students to use both productive (speaking and writing) and receptive (listening and reading) language and learn science through meaningful interactions and scientific sense-making.

Teaching Tip

There are two types of formats for sharing and critiquing arguments. The first format is called a *gallery walk.* In this format, each group sets up their argument at their workstation so others can see it. The entire group then moves to a different workstation. While at a different workstation, the group is expected to read the argument at that workstation (which was created by a different group) and offer critiques and suggestions for improvement. The students should move to a different workstation after a few minutes to read and critique an argument written by a different group. Students should also be encouraged to keep a record of good ideas or potential ways to improve their own arguments as they move from workstation to workstation to offer critiques and feedback. After repeating this process in the gallery walk 3-4 times, every group will have had an opportunity to read and critique three or four different arguments and receive feedback from three or four different groups.

The second format is called a *modified gallery walk.* In the modified gallery walk format, one or two members of the group stay at their workstation to share their group's ideas (these students are called presenters) while the other members of their group go to different groups, one group at a time, to listen to and critique the arguments developed by their classmates (these students are called travelers). The travelers should visit at least three different workstations during the argumentation session. We also recommend that the presenters keep a record of the critiques made by their classmates and any suggestions for improvement. The travelers should also be encouraged to keep a record of good ideas or potential ways to improve their own arguments as they travel from group to group.

Attending to Equity

Another good way to have students, especially emerging multilingual learners, give one another feedback is to have them use four different color sticky notes to make comments. Give each group a single pink, blue, yellow, and green sticky note. Then direct the students to use the pink sticky note to suggest "something to add" to the argument, the blue sticky note to suggest "something to change" in the argument, the yellow sticky note to highlight "something they like" in the argument, and the green sticky note to ask a question about "something they don't understand". This strategy is useful because everyone in the classroom will be actively engaged during this activity, and everyone will have a chance to see different arguments. This format, however, does not give students a chance to support, critique, and challenge the ideas, solutions, scientific reasoning of other groups through talk because two different groups are never at the same workstation at the same time.

Activity C	Revise Your Argument
Time Needed for the Activity	15 minutes
Grouping	Teams

What to Do

1. Students should work in their assigned teams for this activity.
2. Show the "Revise Your Argument" page on a screen using the Presentation mode in the Learning Hub so everyone in the classroom can see it.
3. Say, "I'm now going to give you all about 10 minutes to revise your argument. Please return to your original teams."
4. Wait for the students to move back into their original teams.
5. Say, "Okay, take a few minutes to talk in your teams and determine what you want to change to make your argument better. Once you have decided what to change, go ahead and make the changes to your whiteboard."
6. Ask, "Do you have any questions about what you need to do?"
7. Answer any questions that come up.
8. Tell the students, "Okay. Let's get started."
9. Give the students 10 minutes to work in their teams on their arguments. As they work, move from team to team to check in, ask probing questions, and offer a suggestion if a team gets stuck.
10. When students are finished with their final arguments, have them take a picture of it or have them put the boards in a safe spot so they can refer to the picture or the boards during the Report stage.

What to Show

The "Revise Your Argument" page in the ADI Learning Hub

What to Look For

A strong argument includes the following information:

- A claim that states the stages of meiosis.
- A diagram or model that shows the process of meiosis.
- An explanation of the diagram or model.
- An explanation of the chosen evidence that uses core ideas from the investigation.

Teaching Tip

It is important to give students time to meet with their original groups so they can discuss what they learned by interacting with individuals from the other groups. At this point in the process, students often realize that their initial problem-solving model was flawed in some way. It is therefore important to encourage students to "redo the science" as needed. At the end of this activity, each group should have a final argument that is much better than their original one.

Chromosome Reduction in Reproductive Cells Lesson Plan

Stage 6 REFLECT

Total Time for the Stage	25 minutes
Grouping	Whole class, Teams

Learning Goals

Students should make progress on their ability to do the following by the end of this stage of the lesson:

Science

1. Ask questions that arise from careful observation of phenomena, or unexpected results, to clarify and/or seek additional information as needed (SEP1).
2. Describe how individual cells in multicellular organisms grow and then divide in a process called mitosis (LS1.B).
3. Use models to simulate systems and interactions within and between systems at different scales (CC4).

English Language Development

1. Paraphrase central ideas in complex evidence, concepts, processes, and information to help explain how or why a phenomenon occurs (ELD-SLC.9-12.Explain.Interpretive).
2. Use scientific explanations about a phenomenon in a different context or situation (ELD-SLC.9-12.Explain.Expressive).

English Language Arts

1. Participate effectively in small-group collaborative discussions with diverse partners on texts by building on others' ideas and expressing their own clearly and persuasively (SL.9-10.1).

Teacher Notes

Activities

Activity A	Discuss a Core Idea
Time Needed for the Activity	10 minutes
Grouping	Whole class/Teams

What to Do

1. Students should be in small teams for this activity. However, they do not need to be in the same teams as they were during previous stages.
2. Begin this activity by giving out a copy of the Reflect handout to each student.
3. Show the "Discuss a Core Idea" page on a screen using the Presentation mode in the Learning Hub so everyone in the classroom can see it.
4. Say, "Okay. Now that we have all revised our arguments and made them better, let's take some time to discuss some of the core ideas you used during this investigation and to justify the evidence you used in your arguments, so we are all on the same page. Look at the picture on the screen."
5. Pause for about 30 seconds.
6. Say, "Take a few minutes to talk in your teams and see if you can use what you know about fertilization and the meiosis process to explain what is happening in the picture."
7. Give the students at least three minutes to talk in their small teams.
8. Say, "Okay. Let's share with the whole class. Team A (or whichever team you want to share first), what did you talk about?"
9. Allow the first team to share.
10. Say, "Interesting. Would anyone like to add to, challenge, or refine the ideas shared by Team A?"
11. Allow students to respond.
12. Say, "Okay. Team B (or whichever team you want to share second), what did you talk about?"
13. Allow the second team to share.

(continued)

What to Give Out

Reflect handout for **Chromosome Reduction in Reproduction Cells**

What to Show

The "Discuss a Core Idea" page in the ADI Learning Hub

Explanation of the Image

The image that the students discuss during this activity shows four reproductive cells.

What to Look For During the Discussion

Students who understand the content will likely mention some, if not all, of the following ideas when providing an explanation for the image:

- A chromosome is a piece of genetic information.
- Human cells have 46 chromosomes, or 23 pairs.
- Fertilization is a process where two reproductive cells fuse together to create a single cell that has the same number of chromosomes as its parents.

(continued)

14. Say, "Interesting. Would anyone like to add to, challenge, or refine the ideas shared by Team B?"
15. Allow students to respond.
16. Keep up this pattern of questioning until all teams have shared. Be sure to keep encouraging students to talk about the content until they reach a correct and consensual explanation for what is shown in the image. If they are unable to reach a consensus or reach consensus using an inaccurate idea, be sure to provide an accurate explanation for the image before moving on to the next activity.

- Reproductive cells have only half of a full set of chromosomes, so when the cells combine the offspring receives a full set.
- Meiosis is the process in which chromosomes are reduced when creating reproductive cells.

Teaching Tip

This activity is a discussion, not a presentation about what the students "should have seen" or "should have learned" during the investigation. The more students talk during this activity, the more meaningful the experience will be for them and the more you can learn about their thinking. Be sure to use talk goals and moves to help facilitate the discussion. Below are some examples of talk goals that you can use to help facilitate the discussion and some questions that you can ask the students to help accomplish each goal. These talk goals and questions are only suggestions.

Goal 1: Encourage students to share, expand on, or clarify an idea.

- What do you see going on here?
- What do you think is happening?
- How might we explain that?
- Can you tell me more about that?
- What do you mean by that?
- Can you give me an example of that?
- Can you say that in another way?

Goal 2: Encourage students to support an idea.

- Why do you think that?
- What is your evidence for that?
- How did you arrive at that conclusion?
- What makes you say that?
- How do you know that?

Goal 3: Encourage students to think with others.

- Would anyone like to add to that idea?
- What should we keep or change about that idea?
- Can anyone else give me some examples of that?
- Can anyone take that idea and push it a little further?
- Does anyone want to respond to that idea?
- Do you agree or disagree with that idea, and why?
- What do we think about that idea?
- Does that fit with what we have figured out so far?
- Is there another way to do it?

Goal 4: Encourage students to think about what they know and don't know yet.

- I'm confused now. What have we figured out so far?
- What do we need to figure out now?
- So, what do we agree on right now, and what are we still trying to figure out?
- What are we sure and not so sure about?
- What questions do you have?

Attending to Equity

Be sure to intentionally group emerging multilingual students at times with peers who know the same languages as they do and at other times with peers whose English language development is slightly more advanced. Also, be sure to ask students how they have seen the core idea in the everyday lives; this will make the topic of discussion more engaging and relatable.

Activity B	Discuss the Practices
Time Needed for the Activity	10 minutes
Grouping	Whole class/Teams

What to Do	What to Show
1. Show the first page of "Discuss the Practices" on a screen using the Presentation mode in the Learning Hub so everyone in the classroom can see it. 2. Say, "Okay. Now that we have talked about some of the core ideas that we used during our investigation, let's take a few minutes to talk about the science practices that we used during the investigation. An example of a scientific practice used was analyzing and interpreting data. On the screen there is a question that I want you all to think about: What other things did you do during your investigation that you feel were good examples of a scientific practice or attempting to figure out something in a scientific way?" 3. Pause for about 30 seconds. 4. Say, "Take a few minutes to talk about this question in your teams." 5. Give the students at least three minutes to talk in their small teams.	**What to Show First** The first page of "Discuss the Practices" in the ADI Learning Hub
6. Say, "Okay. Let's share with the whole class. Team A (or whichever team you want to share first), what did you talk about?" 7. Allow the first team to share. 8. Say, "Okay. Team B (or whichever team you want to share second), what did you talk about?" 9. Allow the second team to share. 10. Continue asking teams until every team has a chance to share.	**What to Show Second** The second page of "Discuss the Practices" in the ADI Learning Hub
11. Show the second page of "Discuss the Practices" on a screen using the Presentation mode in the Learning Hub so everyone in the classroom can see it. 12. Say, "Here is another question that I want you all to think about: What things did you do during your investigation that were not good examples of a scientific practice or not very scientific?" 13. Pause for about 30 seconds.	**What to Show Third** The third page of "Discuss the Practices" in the ADI Learning Hub

(*continued*)

14. Say, "Take a few minutes to talk about this question in your team."
15. Give the students at least three minutes to talk in their small teams.
16. Say, "Okay. Let's share with the whole class. Team A (or whichever team you want to share first), what did you talk about?"
17. Allow the first team to share.
18. Say, "Okay. Team B (or whichever team you want to share second), what did you talk about?"
19. Allow the second team to share.
20. Continue asking teams until every team has a chance to share.
21. Show the third page of "Discuss the Practices" on a screen using the Presentation mode in the Learning Hub so everyone in the classroom can see it.
22. Say, "Here is a final question that I want you all to think about: What should we all do during our next investigation to make sure we are using scientific practices and being scientific in the way we try to figure something out?"
23. Pause for about 30 seconds.
24. Say, "Take a few minutes to talk about this question in your team."
25. Give the students at least three minutes to talk in their small teams.
26. Say, "Okay. Let's share with the whole class. Team A (or whichever team you want to share first), what did you talk about?"
27. Allow the first team to share.
28. Say, "Okay. Team B (or whichever team you want to share second), what did you talk about?"
29. Allow the second team to share.
30. Continue asking teams until every team has a chance to share.
31. Once everyone has shared, offer any suggestions or advice about what they can do to improve their next investigation and then move on to the next stage.

Teaching Tip

A great way to wrap up this activity is for the students to contribute to a list of class norms or "gotta-haves" to keep in mind when planning and carrying out an investigation in science. To create this list, each group can suggest for a potential norm or "gotta-have" to add to the list at the end of this activity, then the class votes on the one that they think is most important. The most important one gets added to the list. The list then serves as reminder for students during the Plan and Do stages of subsequent investigations. Over time, the list will grow and each addition to the list will be more nuanced than the last as students develop expertise with scientific processes.

Attending to Equity

Be sure to give all students, especially emerging multilingual students, enough time to talk in small groups before asking anyone to share with the entire class during this activity. Small-group structures offer all students, including emerging multilingual students, a chance to engage in sense-making with their peers as well as space to use their verbal and nonverbal (e.g., writing, drawing pictures, and using gestures) resources to express their ideas. Small-group structures also give students a chance to learn from other students' uses of these resources. When students are tasked with expressing their ideas, be sure that they know it is okay for them to do so in a mode or combination of modes (speaking, writing, drawings, and gestures).

Chromosome Reduction in Reproductive Cells Lesson Plan

Stage 7 REPORT

Total Time Needed for the Stage	100 minutes (over three days)
Grouping	Individual, Teams

Learning Goals

Students should make progress on their ability to do the following by the end of this stage of the lesson:

Science

1. Make and defend a claim based on evidence that inheritable genetic variations may result from new genetic combinations through meiosis, viable errors occurring during replication, and/or mutations caused by environmental factors (HS-LS3-2).
2. Make and defend a claim based on evidence about the natural world that reflects scientific knowledge and student-generated evidence (SEP7).
3. Construct a written argument based on evidence (SEP7).
4. Evaluate the claims, evidence, and/or reasoning to determine the merits of arguments (SEP7).
5. Respectfully give critiques on scientific arguments by and challenging ideas and conclusions (SEP7).
6. Communicate scientific information or ideas in multiple formats including graphically and textually (SEP8).

English Language Development

1. Identify appropriate and sufficient evidence from data, models, and/or information from investigations of a phenomenon (ELD-SLC.9-12.Argue.Interpretive).
2. Compare reasoning and claims based on evidence from competing arguments (ELD-SLC.9-12.Argue.Interpretive).
3. Defend or refute a claim based on data and evidence (ELD-SLC.9-12.Argue.Expressive).
4. Establish and maintain an appropriate tone and stance (ELD-SLC.9-12.Argue.Expressive).
5. Signal logical relationships among reasoning, evidence, data, and/or models when making and defending a claim (ELD-SLC.9-12.Argue.Expressive).

English Language Arts

1. Write routinely over extended time frames (time for reflection and revision) and shorter time frames for a range of discipline-specific tasks, purposes, and audiences (WHST.9-10.10).
2. Write arguments focused on discipline-specific content (WHST.9-10.1).
3. Write informative/explanatory texts, including the narration of scientific procedures/experiments or technical processes (WHST.9-10.2).
4. Produce clear and coherent writing in which the development, organization, and style are appropriate to task, purpose, and audience (WHST.9-10.4).
5. Strengthen writing as needed by revising, editing, rewriting, and focusing on addressing what is most significant for a specific purpose and audience (WHST.9-10.5).

(continued)

6. Determine the central ideas or conclusions of a text; trace the text's explanation or depiction of a complex process, phenomenon, or concept; provide an accurate summary of the text (RST.9-10.2).
7. Determine the meaning of symbols, key terms, and other domain-specific words and phrases as they are used in a specific scientific or technical context (RST.9-10.4).
8. Analyze the structure of the relationships among concepts in a text, including relationships among key terms (RST.9-10.5).
9. Assess the extent to which the reasoning and evidence in a text support the author's claim (RST.9-10.8).
10. Compare and contrast findings presented in a text to those from other sources (including their own experiments), noting when the findings support or contradict other accounts (RST.9-10.9).

Activities

Activity A	Draft Report
Time Needed for the Activity	20 minutes
Grouping	Individual

What to Do

1. Have students work on their own during this activity. Give a copy of the Report handout to each student.
2. Show the "Create a Draft Report" page on a screen using the Presentation mode in the Learning Hub so everyone in the classroom can see it.
3. Say, "You are now going to write an investigation report to share what you figured out. The first part of the report is called the 'Introduction'. In this section, you should explain to the reader what you were investigating, why you were investigating it, and what question you were trying to answer. All this information can be found on your Task handout."
4. Hold up the Task handout.
5. Ask the students, "Do you have any questions about what you need to do?"
6. Answer any questions that come up.
7. Say, "Here are some sentence starters to help you begin writing."
8. Hold up the Report handout and point to the sentence starters.
9. Tell the students, "Okay, let's write."
10. Give the students five minutes to write the "Introduction" section of the report. As they work, move from student to student to check in, ask probing questions, and offer a suggestion if a student gets stuck.

What to Give Out

Report handout for **Chromosome Reduction in Reproduction Cells**

What to Show

The "Create a Draft Report" page in the ADI Learning Hub

(continued)

11. Say, "Okay, let's move on to the second part of the report. The second part of the report is called the 'Method'. In this section, you should explain to the reader what you did during the investigation to complete the task and answer the guiding question. All this information can be found in the Plan handout. Remember that you all planned and carried out different investigations, so do not assume that the reader will know what you did."
12. Ask the students, "Do you have any questions about what you need to do?"
13. Answer any questions that come up.
14. Say, "Here are some sentence starters to help you begin writing."
15. Hold up the Report handout and point to the sentence starters.
16. Tell the students, "Okay, let's write."
17. Give the students five minutes to write the "Method" section of the report. As they work, move from student to student to check in, ask probing questions, and offer a suggestion if a student gets stuck.
18. Say, "Okay, let's move on to the last part of the report. The third part of the report is called the 'Argument'. In this section, you should explain to the reader what you figured out and how you know you are correct. All this information can be found on your Do handout and on your team's whiteboard."
19. Hold up the Do handout and point to the boards (or remind them that they have a picture of their board).
20. Ask the students, "Do you have any questions about what you need to do?"
21. Answer any questions that come up.
22. Say, "Here are some sentence starters to help you begin writing."
23. Hold up the Report handout and point to the sentence starters.
24. Tell the students, "Okay, let's write."
25. Give the students 10 minutes to write the "Argument" section of the report. As they work, move from student to student to check in, ask probing questions, and offer a suggestion if a student gets stuck.
26. Collect all the draft reports before moving on to the next activity. Tell students to mark their draft reports with a symbol that they can use to identify it later. They should NOT write their names on their draft reports.

Teaching Tip

Writing is an important component of this model. When a student writes the report on their own, it not only helps the student learn to write in the context of science (an important literacy skill) but also allows that student to write to learn (develop a better understanding of scientific concepts by writing about them). There are many writing supports (e.g., sentence starters) already embedded into this activity, so you might not need to provide much extra support for your students.

This activity can be done completely in class or started in class and then finished as homework. Do not assign the entire activity as homework unless you know that every student in your class has dedicated time, space, and resources to write outside of class and everyone understands and is comfortable with the report writing process.

If you are concerned about students feeling overwhelmed, you can develop a class "gotta-have-it" list of information that needs to be included in each section of the report. Students can then refer to this list as they write each section.

Don't worry if students do not produce a perfect report at this point. All the reports will be reviewed and revised before they are turned in to you. The reports should therefore be viewed as a starting point. Be sure to encourage students to "just get something down on paper" so "we can work on it together."

Attending to Equity

Emerging multilingual students' access to instructional tasks that require complex thinking, such as writing a report, is enhanced when linguistic complexity and instructional support match their current level of language proficiency. Therefore, it is important to scaffold the writing of the reports for the emerging multilingual students based on their current English language proficiency level. The use of sentence starters to model written language production skills can help emerging multilingual students develop English language skills because the development of academic language and academic content knowledge are interrelated processes. However, it is important that such scaffolds be used purposefully and removed when no longer needed.

ADI Go Digital

Use the **ADI Learning Hub** to go paperless. When you create a class and add students to it, students can complete all the stages right in the system.

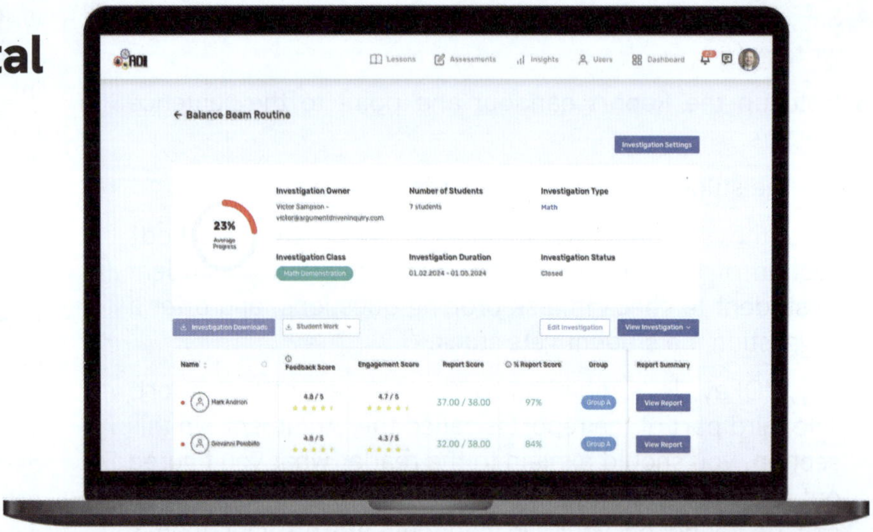

Activity B	Peer Review the Reports
Time Needed for the Activity	45 minutes
Grouping	Teams

What to Do

1. Students should work in teams of three for this activity. These can be completely different teams.
2. Show the "Peer Review Some Draft Reports" page on a screen using the Presentation mode in the Learning Hub so everyone in the classroom can see it.
3. Say, "We are now going to review our reports to find ways to make them better."
4. If possible, use a document camera to project the peer review guide (PRG) on a screen.
5. Say, "We are going to use this peer review guide to give each other feedback." Point to the image.
6. Say, "I'm going to ask you to work in teams of three to do this. I will give each team a draft report to read. You three will then read the report together. Once you are done reading the report, I want you to answer each of the questions on the peer review guide." Point to the review questions on the image of the PRG.
7. Tell the students, "You can check No, Almost, or Yes after each question." Point to the checkboxes on the image of the PRG.
8. Say, "This will be your rating for this part of the report. Make sure you agree on the rating you give the author. If you mark No or Almost, then you need to tell the author what he or she needs to do to get a Yes." Point to the space for the reviewer feedback on the image of the peer review guide.
9. Say, "It is really important for you to give the authors feedback that is helpful. That means you need to tell them exactly what they need to do to make their report better."
10. Ask the students, "Do you have any questions about what you need to do?"
11. Answer any questions that come up.
12. Tell the students, "Please get into your teams of three."
13. Allow the students time to move into their triads.
14. Say, "Okay, I'm now going to give you one report to read." Give out one draft report to each triad. Make sure that the report you give a triad was not written by one of the students in that triad. Give each triad one PRG to fill out.

(continued)

What to Give Out

Peer Review Guide (one per group per review, so for this investigation you need the same number of peer review guides as the number of draft reports that will be reviewed)

What to Show

The "Peer Review Some Draft Reports" page in the ADI Learning Hub

What to Look For

Make sure students are reaching consensus about what rating to give each criterion. Also, be sure that students are giving one another feedback that is specific, detailed, and helpful when they think an aspect of a report needs to be improved.

15. Tell the students, "Okay, I'm going to give you 15 minutes to read the report I gave you and to fill out the peer review guide. Go ahead and get started."
16. Give the students 15 minutes to work. As they work, move around from team to team to check in and see how things are going, answer questions, and offer advice.
17. After 15 minutes pass, tell the students, "Okay, time is up. Please give me the report and the peer review guide that you filled out."
18. Collect the draft reports and the PRGs, or collect the workbooks if students are using them.
19. Say, "Okay, I am now going to give you a different report to read and a new peer review guide to fill out." Give out one more report to each triad. Make sure that the report you give a triad was not written by one of the students in that triad. Give each triad one new peer review guide to fill out as a team.
20. Say, "Okay, I'm going to give you 15 minutes to read this new report and to fill out the peer review guide. Go ahead and get started."
21. Give the students 15 minutes to work. As they work, move around from team to team to check in and see how things are going, answer questions, and offer advice.
22. After 15 minutes pass, tell the students, "Okay, time is up. Please give me the report and the peer review guide that you filled out."
23. Repeat this process for the remaining draft reports. Each triad should one have one more report to review (for a total of three reviews).
24. Collect the last draft reports and PRGs. Set all the draft reports on a table in separate stacks. Each stack should have a draft report on top (with the identification symbol facing up so students can find their reports) and the PRG on the bottom.

Teaching Tip

The peer review process is one of the best ways for students to learn how to write. When students read and review different reports, they have an opportunity to (1) see examples of texts written by others for the same purpose, (2) discuss what counts as quality and why some reports are stronger than others, (3) discuss ways to strengthen a report, and (4) pick up things that they can do to improve their reports. This helps the students learn new ways to organize and present information, which will help them write better on subsequent reports.

Attending to Equity

One helpful scaffold is to give students sentence stems for what to write in their feedback to the author. Ideal feedback is respectful, specific about what needs to be changed, and helpful in that it suggests how the author should go about making that change. You can review the writing of the emerging multilingual students in your class if they have limited English language proficiency, but make sure that all emerging multilingual students participate in the peer review process. You will see tremendous growth in their writing skills as they review their peers' writing and discuss what makes some reports stronger than others.

Activity C	Revise the Reports
Time Needed for the Activity	25 minutes
Grouping	Individuals

What to Do

1. Students should work on their own for this activity.
2. Show the "Reflect on the Feedback" page on a screen using the Presentation mode in the Learning Hub so everyone in the classroom can see it.
3. Say, "Please come pick up your reports from over here on this table." Point to the table.
4. Give the students a few minutes to pick up their reports and return to their seats.
5. Say, "Please take a few minutes to read over the peer review guide. You should use it to figure out what you need to change in your report and how you will change it."
6. Allow the students to read the PRG.
7. Show the "Create a Final Report" page on a screen using the Presentation mode in the Learning Hub so everyone in the classroom can see it.
8. Say, "Okay, let's revise our reports. I would like you to rewrite your report. You can use your draft report as a starting point but use the feedback on the peer review guide to make it better."
9. Ask the students, "Do you have any questions about what you need to do?"
10. Answer any questions that come up.
11. Tell the students, "Okay, let's write." Allow about 20 minutes for the students to revise their reports.
12. After about 20 minutes, give the following directions, "Okay, time's up. I will now come around and collect your first draft, final report, and peer review guide."
13. Collect all the first drafts, PRGs, and final reports.
14. Use the "Teacher Score" columns in the PRG to grade the final report. Be sure to give the students feedback about their writing in the "Teacher Comments" section.

What to Show First

The "Reflect on the Feedback" page in the ADI Learning Hub

What to Show Second

The "Create a Final Report" page in the ADI Learning Hub

Attending to Equity

Students don't always use the feedback they receive from others to improve the quality of their reports. Make sure you move from student to student and encourage them to make changes based on the feedback. If students are not sure how to revise their reports, model how to use feedback. For example, you might say, "Okay, based on the peer review guide, I need to do more to describe what I did to collect data. I bet if I add information about ________, readers will have a better idea of what I did."

Teaching Tip

The peer review guide (PRG) is designed to be used with any ADI investigation, thus allowing you to use the same scoring rubric throughout the entire year. This is beneficial for several reasons. First, the criteria for what counts as a high-quality report do not change from investigation to investigation. Students, therefore, quickly learn what is expected from them when they write a report, and you do not have to spend valuable class time explaining the various components of the PRG each time students write a report. Second, the PRG makes it clear which components of a report need to be improved next time because the grade is not based on a holistic evaluation of the report. The PRG format allows students to see which aspects of their writing are strong enough and which aspects need improvement. Finally, and perhaps most importantly, the PRG provides you with a standardized measure of student performance that can be compared over multiple reports across semesters, thus allowing you to track improvement over time.

Teacher Notes

Chromosome Reduction in Reproductive Cells

TASK

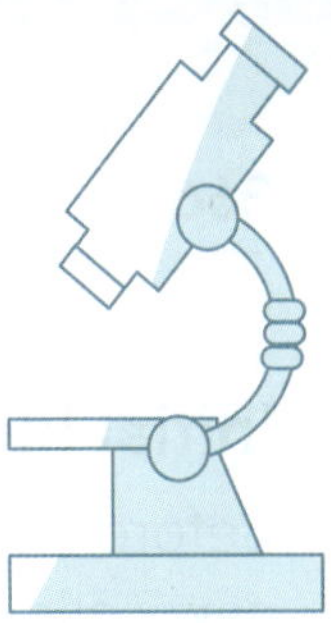

The Phenomenon

Sexual reproduction is a process that creates a new organism by fusing reproductive cells produced by two existing organisms. Take a few minutes to watch a video that shows what happens when two reproductive cells come into physical contact with each other. As you watch the video, write down some things you notice in the space below.

Things I **noticed**...

The Task

The video you watched showed how two different reproductive cells, one called an egg and one called a sperm, fuse together and form a zygote. The organisms that created the different types of reproductive cells are called the "parents", and the zygote that is produced by fusing the two reproductive cells together is called an "offspring". When an offspring is created through sexual reproduction, it will have the same amount of genetic information, in terms of molecules of DNA, as both of its parents. An offspring, however, never receives all, or even most, of its genetics information from only one of its parents; it receives exactly half of its genetic information from each one. Organisms that reproduce sexually must therefore have a way to produce reproductive cells that (a) contain half their genetic information and (b) will complete the genetic information that is missing from a second reproductive cell. Your goal in this investigation is to develop a model that you can use to explain how reproductive cells are created by organisms that reproduce sexually. The guiding question of this investigation is:

How do organisms create reproductive cells?

Initial Ideas

Before you start this investigation, take a few minutes to think about how an organism makes new cells so that all the new cells have an exact copy of all its genetic information found in its existing cells. Then, draw a model in the space below that shows how you think an organism might use a similar, but different, process to create a reproductive cell from one of its existing cells. Be sure to include labels to help explain your thinking.

Some **ideas** that I have...

Share your model with the rest of your group. Once everyone has had a chance to share their ideas, list some questions that your group has about this phenomenon.

Some **questions** we have now...

Chromosome Reduction in Reproductive Cells

IDEAS

Some Ideas You Can Use: Fertilization

Fertilization is a process where two reproductive cells fuse together to create a single cell that has the same number of chromosomes as the parent organisms. The number of chromosomes donated from each parent organism is also equal. After fertilization is complete, chromosomes are found in homologous pairs. Each chromosome in a homologous pair looks the same because they are the same length and have a centromere in the same location. Each chromosome in a homologous pair also has the same genes in the same location. These genes, however, can have variations. The image below shows the 23 pairs of homologous chromosomes that are found in the cells of humans. Different species have different numbers of homologous chromosomes.

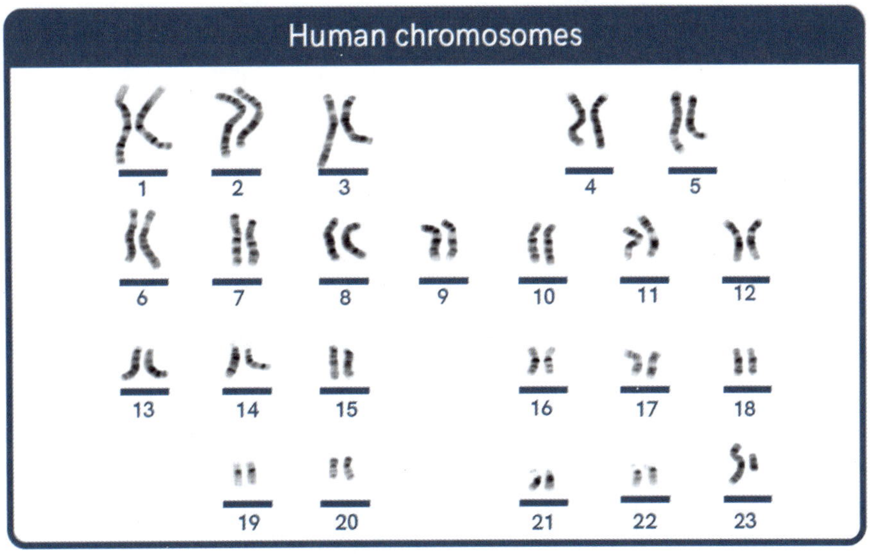

If a reproductive cell from one parent and the reproductive cell from the other parent each carried the same number of chromosomes that are found in the existing cells of the parents, then the chromosome number of the species would double with each generation. For example, if the reproductive cells produced by humans each contained 46 chromosomes rather than 23, the offspring would have 92 chromosomes after the process of fertilization. Yet the doubling of chromosomes doesn't happen with each new generation; instead, the chromosome number within a species stays constant from one generation to the next. Therefore, a mechanism that reduces the number of chromosomes found in reproductive cells is needed to prevent the chromosome number from doubling as the result of fertilization. This mechanism is called meiosis. **Meiosis** is the process in which chromosomes are duplicated and then separated into four reproductive cells that have exactly half the number of chromosomes of the original organism. This process ensures that there are no chromosomes pairs found in the reproductive cells. In other words, there is only one copy of each chromosome in reproductive cells instead of two.

Some **useful ideas** from what I read...

Some Ideas You Can Use: Images of the Meiosis Process

The image below shows cells in various stages of meiosis. This image was taken using a microscope. To create this image, a very thin slice of tissue was placed onto a microscope slide. The tissue was then stained with a dye that made the chromosomes visible. A picture was then taken of the cells that make up the tissue under 1600x magnification. The image, as a result, provides us with a clear view of the various events that happen within a cell as it undergoes meiosis. Meiosis consists of eight major events, or stages, and some of these stages are marked with a letter (C, E, and G) in the image. You will be given images of cells during each stage of meiosis during this investigation so you can use this information as you attempt to figure out how the process of meiosis reduces the number of chromosomes in reproductive cells.

Different cells during the process of meiosis

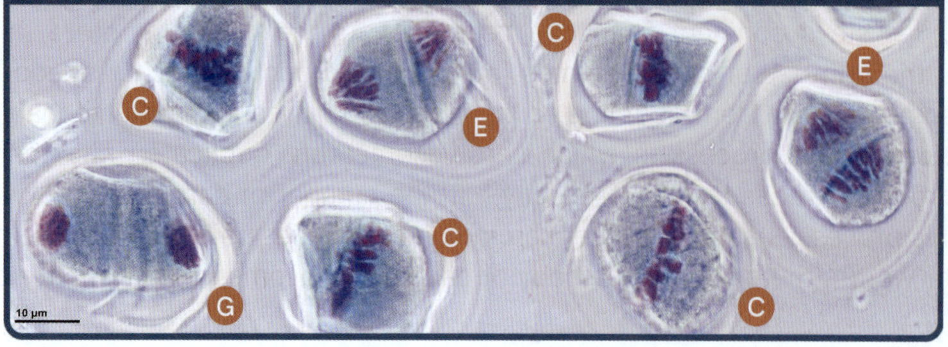

Some **useful ideas** from what I read...

Some Ideas You Can Use: Using Models to Represent a Process

Models can be used to represent an object, event, or process. A **model** can be physical, visual, or a computer-based simulation. In this investigation you will need to use both a physical model and a visual model to answer the guiding question. First, you will need to make a physical model of the chromosomes of a cell to figure out how the process of meiosis is able to reduce the number of chromosomes in reproductive cells. The image below shows how you can make a physical model. It is important to note that this example of a physical model is just a starting point. You can add other things to your physical model to help you make sense of what is happening to the chromosomes of a cell during each stage of meiosis.

A physical model of the chromosomes in a cell

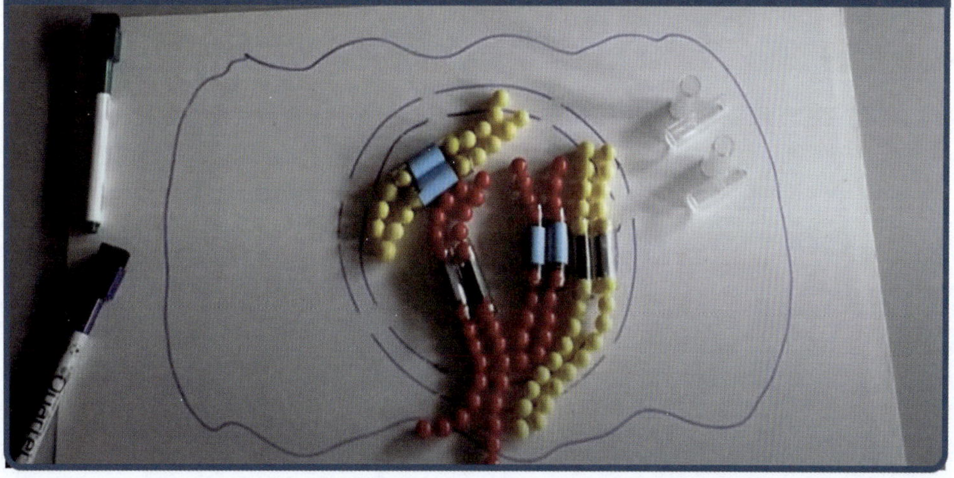

Once you are finished testing your ideas with your physical model, you can use what you figured out to create a visual model. A visual model includes images, symbols, words, and numbers to help explain a process. A visual model does not just replicate or duplicate something you saw like a physical model does; instead, a visual model provides a way to make predictions about what will happen in the future. You should be able to use your visual model to predict what happens when the existing cells of other organisms, which have different numbers of chromosomes, make new cells.

Some **useful ideas** from what I read...

Chromosome Reduction in Reproductive Cells

PLAN

Plan Your Investigation

Prepare a plan for your investigation by filling out the chart below.

I am trying to answer the following **question**...

I will use the following **observations or measurements**...

I will **analyze** the observations or measurements that I collect by...

I approve of this investigation plan.

Teacher's signature

Date

Chromosome Reduction in Reproductive Cells

DO

Collect Data

Keep a record of what you observe or measure during your investigation.

My **observations or measurements...**

Make Sense of Your Data

Use the space below to create your model so you can answer the guiding question.

My **model...**

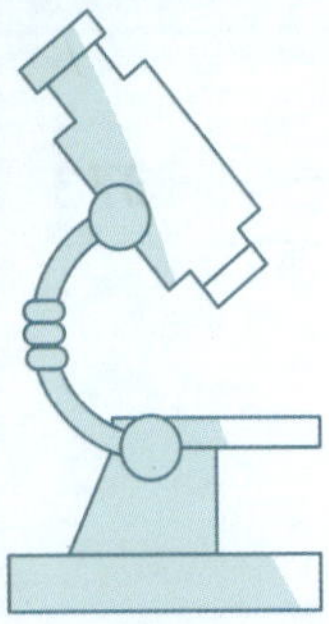

Chromosome Reduction in Reproductive Cells

SHARE

Draft Argument

Develop an argument on a whiteboard. It should include:

1. A *claim* that answers the guiding question,
2. *Evidence* in support of the claim that provides an analysis of the data and a description of what the analysis means,
3. A *justification of the evidence* that explains why your group thinks the evidence is important.

Our Guiding Question:

Claim:

Evidence:	Justification of the Evidence:

Argumentation Session

Share your argument with your classmates. In the space below, be sure to keep track of any ideas that you can use to revise your argument and make it better.

Some possible ways to **improve** our argument...

Chromosome Reduction in Reproductive Cells

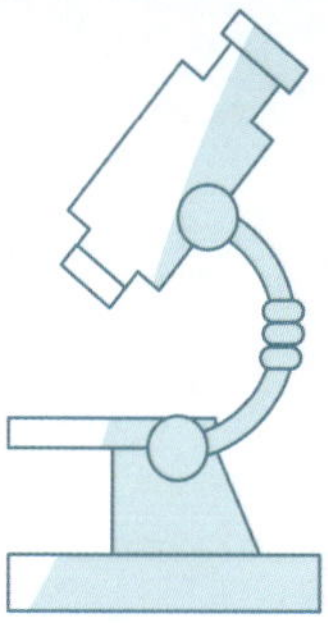

REFLECT

Reflective Discussion

In the space below, you can keep track of any ideas from the discussion that you think are important or will be useful in the future.

Some important **ideas**...

Chromosome Reduction in Reproductive Cells

REPORT

Draft Report

Prepare a report to share what you figured out during your investigation.

Introduction

We have been studying ______________________ in class. Before we started this investigation, we explored ______________________

____

____

We noticed ______________________

____

____

Our goal for this investigation was to figure out ______________________

____

____

The guiding question was ______________________

____

Method

To answer this question, ______________________

____

____

____

____

____

____

____

____

____

____

Argument

The model below shows ________________________________

My evidence for this model is ________________________________

This evidence is based on several important scientific ideas. The first one is ________________

Fruit Fly Body Color and Wing Shape Lesson Plan

OVERVIEW

Description	Students figure out how to model the inheritance of traits between flies.
Guiding Question	What model or models of inheritance best explain how the yellow body trait and the curly wing trait in fruit flies are passed down from parent to offspring?
What Students Figure Out	The yellow body color trait follows the sex-linked inheritance patterns, and it is determined by the recessive allele. The curly wings trait follows the dominant-recessive inheritance pattern, and it is determined by the dominant allele.

STAGES

This investigation has seven stages. Each stage includes multiple activities. The flowchart below shows what happens during each stage and how much time is needed to complete each one.

Task (Day 1)
1 **20 minutes**
Students are introduced to a phenomenon and a guiding question to answer. They also share their initial ideas.

Ideas (Day 1)
2 **30 minutes**
Students read about DCI, a CC, and an SEP that they can use to help them figure out the phenomenon.

Plan (Day 2)
3 **15 minutes**
Students formulate a plan collecting and analyzing the data they need to answer the guiding question.

Do (Day 2)
4 **30 minutes**
Students carry out their plan, make sense of the data they collected, and answer the guiding question.

Share (Day 3)
5 **45 minutes**
Students develop, share, and revise an evidence-based argument about what they figured out.

Reflect (Day 4)
6 **20 minutes**
Students discuss a DCI and/or a CC and reflect on the SEPs they used during the investigation.

Report (Day 4)
7 **25 minutes**
Students write investigation reports to share what they figured out during the investigation.

Report (Day 5)
7 **45 minutes**
Students peer review the reports written by other students and offer feedback to improve.

Report (Day 6)
7 **30 minutes**
Students revise their reports and then submit a final draft to their teacher for evaluation.

STANDARDS ALIGNMENT

NEXT GENERATION SCIENCE STANDARDS

Performance Expectation
Make and defend a claim based on evidence that inheritable genetic variations may result from new genetic combinations through meiosis, viable errors occurring during replication, and/or mutations caused by environmental factors (HS-LS3-2).

Core Ideas(s)	Crosscutting Concept(s)	Practices
Variation of Traits (LS3.B)	Patterns (CC1) Cause and Effect (CC2)	Asking Questions (SEP1) Developing and Using Models (SEP2) Planning and Carrying Out Investigations (SEP3) Analyzing and Interpreting Data (SEP4) Using Mathematics and CT (SEP5) Constructing Explanations (SEP6) Engaging in Argument from Evidence (SEP7) Obtaining, Evaluating, and Communicating Information (SEP8)

WIDA ENGLISH LANGUAGE DEVELOPMENT STANDARDS FOR SCIENCE

Explain	Argue
Interpret scientific explanations (ELD-SC.9-12.Explain.Interpretive) Construct scientific explanations (ELD-SC.9-12.Explain.Expressive)	Interpret scientific arguments (ELD-SC.9-12.Argue.Interpretive) Construct scientific arguments (ELD-SC.9-12.Argue.Expressive)

COMMON CORE ENGLISH LANGUAGE ARTS

Reading	Writing	Speaking and Listening
Key ideas and details (RST.9-10.1 and RST.9-10.2) Craft and structure (RST.9-10.4) Integration of knowledge and ideas (RST.9-10.7, RST.9-10.8, and RST.9-10.9) Range of reading (RST.9-10.10)	Text types and purposes (WHST.9-10.1 and WHST.9-10.2) Production and distribution of writing (WHST.9-10.4, WHST.9-10.5, and WHST.9-10.6)	Comprehension and collaboration (SL.9-10.1 and SL.9-10.3) Presentation of knowledge and ideas (SL.9-10.4 and SL.9-10.6)

THINGS TO KNOW BEFORE STARTING THE LESSON

The Core Ideas

Organisms are made up of one or more cells. A cell contains chromosomes, which in turn contain the cells' genetic material, or DNA. Organisms inherit chromosomes from one or both parents. A gene describes a segment of DNA that encodes for the production of a certain protein. Between organisms, there are alternate forms of genes known as alleles. When chromosomes are passed down from parents to offspring in sexual reproduction, the organism inherits its parents' genes. How the organism inherits these genes depends on the model of inheritance. Five different models of inheritance are dominant-recessive, incomplete dominance, codominance, multiple alleles, and sex-linked.

In the dominant-recessive model, an organism inherits two alleles. If an allele is dominant, it is the one that manifests as a trait in the organism. The recessive allele will only manifest as a trait if the organism inherits two recessive alleles. On the other hand, in the incomplete dominance model, an organism still inherits two alleles. Even if one allele is dominant and the other is recessive, one allele does not completely cancel out the other. Instead, the alleles combine and create a hybrid trait. The codominance model is similar to incomplete dominance, but the alleles do not combine. Instead, they can both be seen in the offspring.

In the multiple allele model, there are more than two versions of a gene. Therefore, the alleles can be dominant, recessive, incompletely dominant, or codominant. Lastly, in the sex-linked model, inheritance of traits only depends on the organism's sex chromosomes. Depending on the sex of the organism, the offspring will either inherit only one copy or two copies of the gene.

The Crosscutting Concepts

A system is a group of components that interact with one another to achieve some function. Because some systems are too complex to work with directly, scientists create models of the system so that they can test these models and use them to identify cause-and-effect relationships. However, because these models simplify complex systems, they inherently have a limited reliability and precision.

Teacher Notes

ARGUMENT-DRIVEN INQUIRY

The Science and Engineering Practices

A practice of science is to **ask and refine questions** that lead to descriptions and explanations of how the natural world works. Another practice of science is to **use and construct models** as helpful tools for representing ideas and explanations. These tools include diagrams, drawings, physical replicas, mathematical representations, analogies, and computer simulations. Scientists **plan and carry out investigations** in the field or a laboratory, working collaboratively and individually. Their investigations are systematic and require clarifying what counts as data and identifying variables or parameters. Scientific investigations produce **data that must be analyzed and interpreted**. Because data patterns and trends are not always obvious, scientists use a range of tools and practices to identify the significant features and patterns in the data. Scientists also use **mathematics and computational thinking** for a range of tasks such as constructing simulations; statistically analyzing data; and recognizing, expressing, and applying quantitative relationships. The products of doing science are **explanations**. Scientific explanations often specify variables and are used by scientists to describe and predict different natural phenomena. **Scientific argumentation**, or the process of sharing, supporting, critiquing, and revising claims based on evidence is the process that scientists use to determine if an explanation is valid or acceptable. Finally, scientists must be able to **communicate clearly and persuasively** the ideas and methods they generate. Critiquing and communicating ideas individually and in groups is a critical professional activity.

PREPARATION

This lesson can be delivered paper-based, paperless, or remote. Use the paper-based delivery option of the investigation if students do not have 1:1 access to a computer and they are in the same classroom or physical space during the lesson. Use the paperless delivery option of the investigation if students have 1:1 access to a computer and they are in the same classroom or physical space during the lesson. Finally, if students are not in the same classroom or physical space during the lesson and have 1:1 access to a computer, use the remote delivery option.

Materials Needed

The materials needed by the teacher and each student in the class for each of the three modalities are listed in the table below.

Teacher Notes

Delivery Mode	Each Student Will Need	The Teacher Will Need
Paper-Based	• Copies of each student handout for the investigation (or a workbook) • A pencil	• A teacher account in the ADI Learning Hub • Access to a computer with high-speed internet and a browser (such as Chrome, Safari, Firefox, or Edge) • A projector or smart board (for displaying a computer screen)
Paperless	• A student account in the ADI Learning Hub • Access to a computer with high-speed internet and a browser (such as Chrome, Safari, Firefox, or Edge)	• A teacher account in the ADI Learning Hub • Access to a computer with high-speed internet and a browser (such as Chrome, Safari, Firefox, or Edge)
Remote	• A student account in the ADI Learning Hub • Access to a computer with high-speed internet and a browser (such as Chrome, Safari, Firefox, or Edge)	• A teacher account in the ADI Learning Hub • Access to a computer with high-speed internet and a browser (such as Chrome, Safari, Firefox, or Edge)

Additional Materials and Preparation Needed for In-Person Delivery

This in-person investigation will require students to have access to a Chromebook, desktop, laptop, iPad, or other electronic device in order to use the simulation. The simulation is embedded in the Learning Hub, so students do not need access to external sites. If students do not have access to the Learning Hub, they can use the simulation by visiting *FlyLab JS.* The teacher can determine if each student will use an electronic device or if each group will share.

How the Simulation Works

This simulation has been coded to model how different traits are inherited in fruit flies. Students can design crosses between female and male fruit flies carrying one or more genetic mutations. They can make hypotheses for the mode of inheritance of genetic traits and test their hypotheses by selecting fruit flies with different visible mutations, mating them, and analyzing the phenotypic ratios of the offspring.

How to Use the Simulation

The simulation interface is shown in the image below. To use the simulation, simply follow these steps:

1. If it is not already selected, click on the **Design** tab. It should default to a female fly.
2. Click on the button for **a category of traits** on the left side of the view. The available traits within each category will be displayed.

3. Select the **radio** button to select an option. The appearance of the designed fly will change to match the selected trait.
4. Click the **Select Fly for Mating** button in the lower right corner of the Design view. The view will change to the Mate view. You have now selected a female fly with a specific set of traits for mating.
5. Click the **Design Fly** button below the gray image of a male fly. The view will change back to the Design view.
6. Click on the button for **a category of traits** on the left side of the view. The available traits within each category will be displayed.
7. Select the **radio** button to select an option. The appearance of the designed fly will change to match the selected trait.
8. Click the **Select Fly for Mating** button in the lower right corner of the Design view. The view will change to the Mate view. You have now selected a male fly with a specific set of traits for mating.
9. Click the **Mate Flies** button in the center of the view. The two flies will be mated, and a cross will appear in the Crosses view. The parents and offspring are shown.
10. You can then select a female and male fly from the offspring for mating. Click the **Select to Mate** button below the female offspring. A green check mark will appear on the fly image.
11. Click the **Select to Mate** button below the male offspring. The view will change to the Mate view showing the female and male flies that were selected.
12. Click the **Mate Flies** button in the center of the view. The two flies will be mated, and a second cross will appear in the Crosses view. The parents and offspring are shown. You will need to scroll down to see all the offspring.
13. Click on the **Analyze** tab. A table will be displayed with the numbers and proportions of each phenotype by sex among the offspring. Click the check box option to **Ignore sex of flies.** The data are now organized only by phenotype.

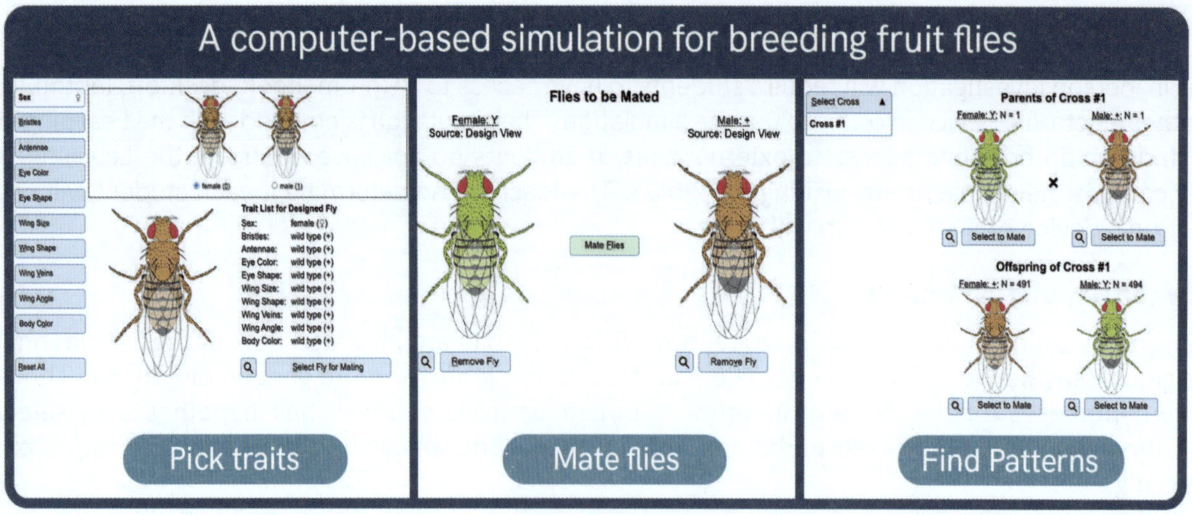

Preparation

Embedded in the Learning Hub are both the simulation as well as a short instructional video explaining how to use the simulation. We recommend showing the video first to the whole class and then giving students time to explore the simulation before completing the Plan stage, because viewing these variables will aid students in determining what data they will collect and how they will analyze their data.

FACILITATION NOTES

The facilitation notes for this investigation describe how to deliver this lesson in person. They are written specifically for the paper-based mode of delivery. These same facilitation notes, however, can also be used for a paperless delivery of the lesson by simply substituting the directions that mention paper with the entry into the ADI Learning Hub. If you are teaching this lesson using the remote mode of delivery, you only need to assign due dates and prerequisites for each stage. The students can then simply follow the directions built into the ADI Learning Hub as they move through the investigation.

ADI Go Digital

The digital version of this lesson is available in the **ADI Learning Hub**. Be sure to check out the built-in teacher presentation mode.

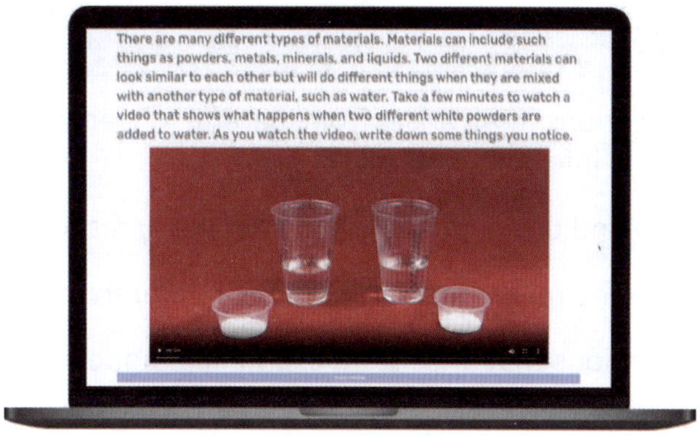

Teacher Notes

Fruit Fly Body Color and Wing Shape Lesson Plan

Stage 1 TASK

Total Time for the Stage	45 minutes
Grouping	Individual; Teams

Learning Goals

Students should make progress on their ability to do the following by the end of this stage of the lesson:

Science

1. Ask questions that arise from careful observation of phenomena to clarify and/or seek additional information (SEP1).
2. Develop and use a model to provide mechanistic accounts of phenomena (SEP2).
3. Evaluate the merits and limitations of different models of the same phenomenon (SEP2).
4. Ask questions that arise from examining models to clarify and/or seek additional information and relationships (SEP 1).
5. Ask questions to clarify and refine a model (SEP1).

English Language Development

None

English Language Arts

1. Initiate and participate effectively in small-group collaborative discussions with diverse partners by building on others' ideas and expressing their own clearly (SL.9-10.1).
2. Present information clearly, concisely, and logically such that listeners can follow the line of reasoning and the organization, substance, and style are appropriate to the purpose, audience, and task (SL.9-10.4).
3. Adapt speech to a variety of contexts and tasks, demonstrating command of formal English when indicated or appropriate (SL.9-10.6).

Activities within the Stage

Activity A	The Phenomenon
Time Needed for the Activity	5 minutes
Grouping	Teams

What to Do

1. Have students work in their assigned teams for this activity.
2. Begin the activity by giving out a copy of the Task handout to each student.
3. Show "The Phenomenon" page on a screen using the Presentation mode in the Learning Hub so everyone in the classroom can see it.
4. Read the introduction of the phenomenon out loud and then start the video. You can also have students read the introduction of the phenomenon and watch the video on their own devices.
5. Give the students a few moments to write down their observations on their handouts before moving on to the next activity.

What to Give Out

Task handout for **Fruit Fly Body Color and Wing Shape**

What to Show

"The Phenomenon" page in the ADI Learning Hub

Teaching Tip

Instead of assigning fixed roles to students by name at the beginning of the investigation, give each member of a group a different number so you can easily assign and change roles as needed during the various investigation stages. For example, you might give directions such as "1s will share out for your group" or "3s, will you please come back here and pick up these materials for your group?" or "2s and 3s, you will present the argument for your group while 1s and 4s travel to the other groups to offer feedback."

Attending to Equity

Be sure to intentionally group emerging multilingual students at times with peers who know the same languages as they do and at other times with peers whose English language development is slightly more advanced. Thoughtful grouping that varies throughout an investigation allows emerging multilingual students to benefit from the different linguistic resources of their peers.

Activity B	Share What You Noticed
Time Needed for the Activity	5 minutes
Grouping	Teams

What to Do

1. Show the "Share What You Noticed" page on a screen using the Presentation mode in the Learning Hub so everyone in the classroom can see it.
2. Say, "Okay, please share what you noticed with the others at your table."
3. Give the students about two minutes to share at their tables.
4. Say, "Okay, let's take a few minutes to share with the class. Team X (call on a team at random), what did you talk about?"
5. Allow Team X to share something they noticed. If it is related to the appearance of the fruit flies or inheritance of traits, say, "That is really interesting. We are actually going to examine that in more detail during this investigation." If what is shared is not related to the appearance of the fruit flies or inheritance of traits, say, "That is really interesting. Team Y (call on another team at random), what did you notice?" and allow that team to share. Follow this pattern until a team mentions something related to how not all fruit flies look the same. If no team mentions this, then say "You know what I found interesting? I saw that not all fruit flies look the same. Did anyone notice that as well?"
6. Move on to the next activity.

What to Show

The "Share What You Noticed" page in the ADI Learning Hub

What to Look For

As soon as a group mentions that not all fruit flies look the same, move on to the next activity.

Attending to Equity

Be sure to provide students with opportunities to talk in small groups during this activity. Small-group structures offer all students, including emerging multilingual students, a chance to engage in sense-making with their peers and offers them space to use their verbal and nonverbal (e.g., writing, drawing pictures, and using gestures) resources to express their ideas. Small-group structures also give students a chance to learn from other students' uses of these resources.

Activity C	What You Need to Figure Out
Time Needed for the Activity	5 minutes
Grouping	Teams

What to Do

1. Show the "What You Need to Figure Out" page on a screen using the Presentation mode in the Learning Hub so everyone in the classroom can see it.
2. Read the task out loud and have students follow along on their handout, on their devices, or on the screen.
3. Allow students to ask clarifying questions about the task or the guiding question.
4. Once everyone understands what they will need to figure out during the investigation, move on to the next activity.

What to Show

The "What You Need to Figure Out" page in the ADI Learning Hub

Teaching Tip

Have students highlight, circle, or place a star next to the guiding question. Marking the guiding question helps students find it later during the Plan stage.

Teacher Notes

Activity D	Share Your Ideas
Time Needed for the Activity	15 minutes
Grouping	Teams

What to Do

1. Show the "Share Your ideas" page on a screen using the Presentation mode in the Learning Hub so everyone in the classroom can see it.
2. Read the directions on the handout out loud and then give students five minutes to draw their initial representations of the science.
3. Say, "Okay, will everyone share their representations with the rest of your team? Please make sure everyone has time to talk. Also, be sure to ask one another questions about the scientific ideas that are included in each representation."
4. Give the students about 8-9 minutes to talk and share. Be sure to walk around and listen to the discussions.
5. Say, "Okay, I heard some interesting ideas as I walked around the room."
6. Move on to the next stage.

What to Show

The "Share Your Ideas"page in the ADI Learning Hub

What to Look For

Each student's initial representation should include a combination of symbols, diagrams, graphs, or words. It should also clearly show how a scientific idea can be used to help complete the task and answer the guiding question.

Attending to Equity

When students are tasked with expressing their ideas, be sure to encourage them to do so through multiple modes (speaking, writing, drawings, and gestures) and to use both content-specific and everyday registers. This process helps students fully express themselves, allowing peers to evaluate, question, and build on ideas. Also, make sure that every student in every team has a chance to share their initial representation.

Activity E	More Questions
Time Needed for the Activity	15 minutes
Grouping	Teams

What to Do

1. Show the "More Questions" page on a screen using the Presentation mode in the Learning Hub so everyone in the classroom can see it.
2. Say, "Okay, I heard a bunch of interesting questions as I was walking around the room. Will you work with your team to identify two or three questions that you think are important to learn more about that came up as you all discussed the different representations you made? You can write these questions on your handout."
3. Give students about five minutes to talk and write down questions. Be sure to walk around and listen to the discussions.
4. Say, "Okay, let's take a few minutes to share. Team X (call on a team at random), will you share one question you all have?"
5. Have each team share a question. Then say, "I want to share information about some useful scientific ideas with you. I bet this information will not only answer many of these questions but will also help you complete the task and answer the guiding question of the investigation."
6. Move on to the next stage.

What to Show

The "More Questions" page in the ADI Learning Hub

What to Look For

Each student should have 2-3 questions written on their handout.

Teaching Tip

When students are asked to create representations of the underlying scientific ideas, share these ideas with their peers, and then identify things they want to know more about, it sparks their curiosity and helps make the investigation meaningful. Also, identifying questions before reading (which students do during the Ideas stage) is an outstanding pre-reading activity that helps improve comprehension.

Fruit Fly Body Color and Wing Shape Lesson Plan

Stage 2 IDEAS

Total Time for the Stage	30 minutes
Grouping	Teams

Learning Goals

Students should make progress on their ability to do the following by the end of this stage of the lesson:

Science

1. Ask questions that arise from examining theory to clarify and/or seek additional information and relationships (SEP1).
2. Critically read scientific literature adapted for classroom use to determine the central ideas or conclusions and/or to obtain scientific and/or technical information to summarize complex concepts, processes, or information presented in a text by paraphrasing them in simpler but still accurate terms (SEP8).
3. Integrate sources of information presented in different media or formats (e.g., visually and quantitatively) as well as in words to address a scientific question (SEP8).
4. Gather, read, and evaluate scientific and/or technical information from multiple authoritative sources, assessing the evidence and usefulness of each source (SEP8).
5. Describe how in sexual reproduction, chromosomes can swap sections during meiosis and create new genetic combinations and thus more genetic variation (LS3.B).
6. Explain how DNA replication is a remarkably accurate process that occasionally allows for mutations that provide genetic variation (LS3.B).
7. Identify how environmental factors can cause genetic mutations and affect the expression of traits that can then be inherited (LS3.B).
8. Describe how different patterns may be observed at each of the scales at which a system is studied and can provide evidence for causality in explanations of phenomena (CC1).
9. Use empirical evidence to differentiate between cause and correlation and make claims about specific causes and effects (CC2).

English Language Development

1. Paraphrase central ideas in complex evidence, concepts, processes, and information to help explain how or why a phenomenon occurs (ELD-SLC.9-12.Explain.Interpretive).

English Language Arts

1. Determine the central ideas or conclusions of a text; trace the text's explanation or depiction of a complex process, phenomenon, or concept; provide an accurate summary of the text (RST.9-10.2).
2. Determine the meaning of symbols, key terms, and other domain-specific words and phrases as they are used in a specific scientific or technical context (RST.9-10.4).
3. Read and comprehend scientific/technical texts (RST.9-10.10).
4. Initiate and participate effectively in small-group collaborative discussions with diverse partners by building on others' ideas and expressing their own clearly (SL.9-10.1).

Teacher Notes

Activities

Activity A	Read About a Core Idea
Time Needed for the Activity	10 minutes
Grouping	Teams

What to Do

1. Have students work in their assigned teams during this stage. Give out a copy of the Ideas handout to each student.
2. The first activity provides students with information about chromosomes and genes.
3. Show the "Read About a Core Idea You Can Use" page on a screen using the Presentation mode in the Learning Hub so everyone in the classroom can see it. You can also click on the images to make them full screen and easier for students to see.
4. Have students read the text on their own, either on their handouts or on the screen.
5. Give the students about three minutes to discuss the text with the others in their teams and develop a shared list of ideas that they think are important or useful.
6. Ask a team to share an idea they thought was useful. Have the remaining teams share a different idea until there are no more ideas to share.
7. Move on to the next activity.

What to Give Out

Ideas handout for **Fruit Fly Body Color and Wing Shape**

What to Show

The "Read About a Core Idea You Can Use" page in the ADI Learning Hub

What to Look For

Students should write down some of the following ideas in the box labeled, "Some things I know from what I read" on their handouts:

- Chromosomes are structures that contain the genetic material, or DNA, of a cell.
- In sexual reproduction, chromosomes are passed down from two parents to their offspring.

Teaching Tip

Students do not need to do a close read of this text or completely understand all the ideas in it before moving on to the next activity. They will be encouraged to revisit this text later in the investigation when creating their arguments and writing their reports. All students need to be able to do during this activity is to work with the other members of their group to identify one or two ideas that they think are important to keep in mind or are potentially helpful.

Activity B	Read About a Core Idea
Time Needed for the Activity	10 minutes
Grouping	Teams

What to Do

1. The second activity provides students with information about models of inheritance.
2. Show the "Read About Another Core Idea You Can Use" page on a screen using the Presentation mode in the Learning Hub so everyone in the classroom can see it. You can also click on the images to make them full screen and easier for students to see.
3. Have students read the text on their own, either on their handouts or on the screen.
4. Give the students about three minutes to discuss the text with the others in their teams and develop a shared list of ideas that they think are important or useful.
5. Ask a team to share an idea they thought was useful. Have the remaining teams share a different idea until there are no more ideas to share.
6. Move on to the next activity.

What to Show

The "Read About Another Core Idea You Can Use" page in the ADI Learning Hub

What to Look For

Students should write down some of the following ideas in the box labeled, "Some things I know from what I read" on their handouts:

- A gene is a region in a DNA molecule that codes for the production of a certain protein.
- Alternative versions of a gene are called alleles.
- How alleles interact with other alleles determines how a trait will manifest in an organism.
- Five common models of inheritance are dominant-recessive, incomplete dominance, codominance, multiple-allele, and sex-linked.

Attending to Equity

There are many supports for helping students comprehend what they read already embedded into this activity (e.g., activating prior knowledge, providing a shared experience, making connections, synthesizing, and talking with peers). So, you might not need to provide much extra support. If you are concerned about some students not understanding what they are reading because of their scores on past reading comprehension tests, you can read it out loud as they follow along. As you read the text out loud, be sure to stop at each important idea and ask the students to put a star (or other annotation) next to it in the margin of their handouts. They can then discuss these ideas in their small groups.

Activity C	Read About a Final Core Idea
Time Needed for the Activity	10 minutes
Grouping	Teams

What to Do

1. The third activity provides students with information about using patterns to develop cause-and-effect relationships.
2. Show the "Read About a Final Core Idea You Can Use" page on a screen using the Presentation mode in the Learning Hub so everyone in the classroom can see it. You can also click on the images to make them full screen and easier for students to see.
3. Have students read the text on their own, either on their handouts or on the screen.
4. Give the students about three minutes to discuss the text with the others in their teams and develop a shared list of ideas that they think are important or useful.
5. Ask a team to share an idea they thought was useful. Have the remaining teams share a different idea until there are no more ideas to share.
6. Move on to the next activity

What to Show

The "Read About a Final Core Idea You Can Use" page in the ADI Learning Hub

What to Look For

Students should write down some of the following ideas in the box labeled, "Some things I know from what I read" on their handouts:

- Biologists use patterns to identify cause-and-effect relationships.
- Biologists can use cause-and-effect relationships to make predictions about inheritance.
- Biologists developed Punnett squares to make predictions about inheritance.

Attending to Equity

To support emerging multilingual students during this activity, be sure to create a word bank on the wall of high-frequency, high-need concrete vocabulary that is needed to communicate scientific ideas. The word bank highlights terms needed for basic communication in an academic context. It also helps with the spelling. You can also encourage your emerging multilingual students to:

1. Use strategic learning techniques such as concept mapping, drawing, memorizing, comparing, contrasting, and reviewing.
2. Decode (sound out) words using skills such as recognizing sound-letter relationships and identifying cognates, affixes, roots, and base words.
3. Use visual and contextual support and support from peers and teachers to enhance and confirm understanding of the texts.

Fruit Fly Body Color and Wing Shape Lesson Plan

Stage 3 PLAN

Total Time for the Stage	15 minutes
Grouping	Teams

Learning Goals

Students should make progress on their ability to do the following by the end of this stage of the lesson:

Science

1. Evaluate a question to determine if it is testable and relevant (SEP1).
2. Plan an investigation individually and collaboratively to produce data to serve as the basis for evidence as part of building and revising models (SEP3).
3. Consider possible variables or effects and evaluate the confounding investigation's design to ensure variables are controlled (SEP3).
4. Ask questions to determine relationships, including quantitative relationships, between independent and dependent variables (SEP1).
5. Decide on types, quantity, and accuracy of data needed and consider limitations on the precision of the data (e.g., number of trials, cost, risk and time), and refine the design accordingly (SEP3).
6. Select appropriate tools to collect, record, and analyze data (SEP3).

English Language Development

1. Define investigable questions on observations, information, and/or data about a phenomenon (ELD-SLC.9-12.Explain.Interpretive).
2. Develop reasoning to illustrate and/or predict the relationships between variables (ELD-SLC.9-12.Explain.Expressive).

English Language Arts

1. Initiate and participate effectively in small-group collaborative discussions with diverse partners by building on others' ideas and expressing their own clearly (SL.9-10.1).
2. Present information clearly, concisely, and logically such that listeners can follow the line of reasoning and the organization, substance, and style are appropriate to the purpose, audience, and task (SL.9-10.4).
3. Adapt speech to a variety of contexts and tasks, demonstrating command of formal English when indicated or appropriate (SL.9-10.6).

Activities

Activity A	Some Materials You Can Use
Time Needed for the Activity	5 minutes
Grouping	Teams

What to Do

1. Show the "Some Materials You Can Use" page on a screen using the Presentation mode in the Learning Hub so everyone in the classroom can see it. You can also click on the images to make them full screen and easier for students to see.
2. Say, "Okay, it sounds like we now have enough information to plan out an investigation that will help us figure out which model of inheritance best describes how fruit flies inherit their body color and wing patterns."
3. Say, "Before we get started, I want to show you some materials that we are going to be able to use during our investigation. Will you all please look at the screen so I can show you the simulation we will be using?"
4. Direct students to look at the screen where the simulation is displayed.
5. Ask, "How do you think we can use this simulation?"
6. Allow students to share ideas. Then say, "We can use this simulation to test different models of inheritance on fruit flies."
7. Say, "Okay, that is all the stuff we can use. Will you all head back to your teams so we can plan out our investigation?"
8. Move on to the next activity.

What to Set Up

Pull up the simulation in the Learning Hub on a screen so that all students can see it. If you do not have a Learning Hub subscription, you can access the simulation by visiting *FlyLab JS.*

What to Show

The "Create a Final Report" page in the ADI Learning Hub

Teaching Tip

This activity helps students see for themselves all the materials they will be able to use during the hands-on part of the investigation instead of just reading about them or looking at pictures. It is important to show each material to the students and then point out how they can use it during the investigation so they can participate in developing a plan for the investigation during the next activity.

Activity B	Make a Plan
Time Needed for the Activity	10 minutes
Grouping	Teams

What to Do

1. Have students work in their assigned teams during this activity. Give a copy of the Plan handout to each student.
2. Show the "Make a Plan" page on a screen using the Presentation mode in the Learning Hub so everyone in the classroom can see it.
3. Say, "Work as a team to create a plan for your investigation by filling out the chart on your handout. As part of your plan, be sure to include which tools and techniques you will use to solve problems."
4. Give the students at least 10 minutes to complete this task. As students work in small teams to create their investigation proposals, be sure to move from team to team to answer questions and offer advice.
5. When a team thinks that they have a good plan, read it over. If you agree, sign off at the bottom of their handouts to let them know that their plan has been approved. If you do not agree, tell them what they will need to add or change before you will approve it. Once they have made the changes you can sign their handouts.
6. Once all the teams are done, you can move on to the next activity. You can also use an approved investigation plan as an "exit ticket" out of class.

What to Give Out

Plan handout for **Fruit Fly Body Color and Wings**

What to Show

The "Make a Plan" page in the ADI Learning Hub

What to Look For

The investigation plan should, at a minimum, include:

- The question they are trying to answer.
- What data students will collect from the simulation.
- How students will collect their data.
- How they will use this data to answer the guiding question.

Attending to Equity

To support emerging multilingual students during this stage, be sure to create a word bank on the wall of high-frequency, high-need concrete vocabulary that is needed to communicate scientific processes and ideas. The word bank highlights terms needed for basic communication in an academic context. It also helps with the spelling. Also, you can encourage your emerging multilingual students to use modes (speaking, writing, drawings, and gestures) and to use both content-specific and everyday registers to discuss and create their plans. This process helps students fully express themselves, allowing peers to evaluate, question, and build on ideas.

Teaching Tip

One way to do a quick check-in with a group during this activity is to use the "two questions, a nudge, and a goal" technique. This technique includes the following steps:

1. Sit down with a group of students.
2. Ask, "Would one of you tell me a little about what you all have been doing or thinking about?" (Question 1)
3. A student answers on behalf of the group.
4. Ask, "Would anyone else like to add anything to that?" (Question 2)
5. A different student provides additional information on behalf of the group.
6. Ask, "Have you thought about [a hint or something you want to highlight]?" (A nudge)
7. Say, "Okay, I'm going to check on the other groups. When I get back, I want to see [something that you want the group to accomplish]." (A goal)

When you have completed the check-in with one group, move on to the next group.

Teacher Notes

Fruit Fly Body Color and Wing Shape Lesson Plan

Stage 4 DO

Total Time for the Stage	45 minutes
Grouping	Teams

Learning Goals

Students should make progress on their ability to do the following by the end of this stage of the lesson:

Science

1. Conduct an investigation individually and collaboratively to produce data to serve as the basis for evidence (SEP3).
2. Analyze data using tools and/or technologies to make valid and reliable scientific claims (SEP4).
3. Consider limitations of data analysis (e.g., measurement error and sample selection) when analyzing and interpreting data (SEP4).
4. Evaluate the impact of new data on a working explanation and/or a model of a proposed process or system (SEP4).
5. Make a quantitative and/or qualitative claim regarding the relationship between dependent and independent variables (SEP6).
6. Construct and revise an explanation based on valid and reliable evidence obtained from a variety of sources and the assumption that theories and laws that describe the natural world operate today as they did in the past and will continue to do so in the future (SEP6).
7. Apply scientific ideas, principles, and/or evidence to provide an explanation of phenomena taking into account possible unanticipated effects (SEP6).

English Language Development

1. Paraphrase central ideas in complex evidence, concepts, processes, and information to help explain how or why a phenomenon occurs (ELD-SLC.9-12.Explain.Interpretive).
2. Evaluate the extent to which reasoning, theory, and/or models link evidence to claims and support conclusions (ELD-SLC.9-12.Explain.Interpretive).
3. Develop reasoning to illustrate the relationships between variables (ELD-SLC.9-12.Explain.Expressive).

English Language Arts

1. Initiate and participate effectively in small-group collaborative discussions with diverse partners by building on others' ideas and expressing their own clearly (SL.9-10.1).
2. Present information clearly, concisely, and logically such that listeners can follow the line of reasoning and the organization, substance, and style are appropriate to the purpose, audience, and task (SL.9-10.4).
3. Adapt speech to a variety of contexts and tasks, demonstrating command of formal English when indicated or appropriate (SL.9-10.6).

Activities

Activity A	Carry Out Your Plan
Time Needed for the Activity	25 minutes
Grouping	Teams

What to Do

1. Students should work in their assigned teams for this activity. Begin this activity by giving out a copy of the Do handout to each student.
2. Show the "Carry Out Your Plan" page on a screen using the Presentation mode in the Learning Hub so everyone in the classroom can see it.
3. Say, "Okay, it sounds like we are now ready to carry out our plans and collect the data or observations we need to answer the guiding question. You can get the materials and tools you need to use from this table." Point to the table where you have placed them.
4. Ask, "Does anyone have any questions about these materials or tools?"
5. Answer any questions that come up.
6. Say, "As you work, be sure to record your observations and measurements on your handout so you can keep track of everything you do. Does anyone have any final questions?"
7. Answer any questions that come up.
8. Say, "Okay, go ahead and get started. You have the next 25 minutes to work. I'll be available to help if any team needs me."
9. Give the students at least 25 minutes to carry out their plans and solve the problem. As students work in small teams, be sure to move from team to team to answer questions and offer advice.

What to Set Up

Provide each student, or group of students, with access to a computer so they can use the computer-based model. Students can access the model in the Learning Hub using their accounts. If students do not have a Learning Hub account, they can access the simulation by going to the *FlyLab JS* simulation.

What to Give Out

Do handout for **Fruit Fly Body Color and Wing Shape**

What to Show

The "Carry Out Your Plan" page in the ADI Learning Hub

Teaching Tip

One way to do a quick check-in with a group during this activity is to use the "two questions, a nudge, and a goal" technique. This technique includes the following steps:

1. Sit down with a group of students.
2. Ask, "Would one of you tell me a little about what you all have been doing or thinking about?" (Question 1)
3. A student answers on behalf of the group.
4. Ask, "Would anyone else like to add anything to that?" (Question 2)
5. A different student provides additional information on behalf of the group.
6. Ask, "Have you thought about [a hint or something you want to highlight]?" (A nudge)
7. Say, "Okay, I'm going to check on the other groups. When I get back, I want to see [something that you want the group to accomplish]." (A goal)

When you have completed the check-in with one group, move on to the next group.

Attending to Equity

Be sure to intentionally group emerging multilingual students at times with peers who know the same languages as they do and at other times with peers whose English language development is slightly more advanced. Thoughtful grouping that varies during different investigation (or activities within an investigation) allows emerging multilingual students to benefit from the different linguistic resources of their peers. We also recommend assessing emerging multilingual students based on how much they contribute to the group during this activity, rather than solely on what they write or draw on their handouts.

Teacher Notes

Activity B	Make Sense of Your Data
Time Needed for the Activity	20 minutes
Grouping	Teams

What to Do

1. Students should work in their assigned teams for this activity. Begin this activity by showing the "Make Sense of Your Data" page on a screen using the Presentation mode in the Learning Hub so everyone in the classroom can see it.
2. Say, "Okay, it sounds like we are now ready to make sense of the data we collected. One way to do this is to create a table, graph, or other visualization that shows a trend or change over time, a comparison between teams, or a relationship between things you tested. You can create a table, graph, or other visualization here on your handout."
3. Point to the box on the handout.
4. Say, "Please work together in your teams to figure out the best way to make sense of your data. Remember, you are looking for any trends, changes, differences, or relationships in the data that will help you answer the guiding question. As you work, be sure to use the space on your handout to perform any calculations that you need to do and create any representations that you need. Does anyone have any last questions?"
5. Answer any questions that come up.
6. Say, "Okay, go ahead and get started. You have the rest of the period to work. I'll be available to help if any team needs me."
7. Give the students at least 15 minutes to make sense of their data. As students work together in small teams, be sure to move from team to team to answer questions and offer advice.
8. Make sure students return the materials and tools to the table where they found them before dismissing class. Cleanup should take no longer than five minutes.

What to Show

The "Make Sense of Your Data" page in the ADI Learning Hub

What to Look For

One of the most useful ways to make sense of the data during this activity is a table that compares the parents' traits to that of their offspring.

Teaching Tip

Collecting and returning materials or tools to a central table works best if you assign the job to a specific number within each team. For example, you can say, "2s, please come up and grab the materials and tools that your team needs" at the start of the activity. Then you can say, "3s, please return the materials and tools your team used back to this table," at the end of the activity. Assigning jobs like this not only cuts down on how many students need to be up and moving around, but also allows you to hold specific students accountable for the materials and tools.

Attending to Equity

Encourage your emerging multilingual students to speak using learning strategies such as requesting assistance, employing nonverbal cues, and using synonyms and circumlocution (conveying ideas by defining or describing when exact English words are not known) during this activity. It is also helpful to point out and/or draw emerging multilingual learners' attention to elements of the English sound system in newly acquired vocabulary such as long and short vowels, silent letters, and consonant clusters as they come up naturally during this activity.

Teacher Notes

Fruit Fly Body Color and Wing Shape Lesson Plan

Stage 5 SHARE

Total Time Needed for the Stage	45 minutes
Grouping	Teams

Learning Goals

Students should make progress on their ability to do the following by the end of this stage of the lesson:

Science

1. Conduct an investigation individually and collaboratively to produce data to serve as the basis for evidence (SEP3).
2. Analyze data using tools and/or technologies to make valid and reliable scientific claims (SEP4).
3. Consider limitations of data analysis (e.g., measurement error and sample selection) when analyzing and interpreting data (SEP4).
4. Evaluate the impact of new data on a working explanation and/or model of a proposed process or system (SEP4).
5. Apply ratios, rates, percentages, and unit conversions in the context of complicated measurement problems involving quantities with derived or compound units (SEP5).
6. Make a quantitative and/or qualitative claim regarding the relationship between dependent and independent variables (SEP6).
7. Construct and revise an explanation based on valid and reliable evidence obtained from a variety of sources and the assumption that theories and laws that describe the natural world operate today as they did in the past and will continue to do so in the future (SEP6).
8. Apply scientific ideas, principles, and/or evidence to provide an explanation of phenomena taking into account possible unanticipated effects (SEP6).

English Language Development

1. Identify appropriate and sufficient evidence from data, models, and/or information from investigations of a phenomenon (ELD-SLC.9-12.Argue.Interpretive).
2. Compare reasoning and claims based on evidence from competing arguments (ELD-SLC.9-12.Argue.Interpretive).
3. Defend or refute a claim based on data and evidence (ELD-SLC.9-12.Argue.Expressive).
4. Establish and maintain an appropriate tone and stance (ELD-SLC.9-12.Argue.Expressive).
5. Signal logical relationships among reasoning, evidence, data, and/or models when making and defending a claim, counterclaim, and/or rebuttal (ELD-SLC.9-12.Argue.Expressive).

English Language Arts

1. Initiate and participate effectively in small-group collaborative discussions with diverse partners by building on others' ideas and expressing their own clearly (SL.9-10.1).
2. Integrate multiple sources of information presented in diverse media or formats (e.g., visually, quantitatively, and orally) evaluating the credibility and accuracy of each source (SL.9-10.2).
3. Evaluate a speaker's reasoning and use of evidence and rhetoric, identifying any fallacious reasoning or exaggerated or distorted evidence (SL.9-10.3).

(continued)

4. Present information clearly, concisely, and logically such that listeners can follow the line of reasoning, and the organization, substance, and style are appropriate to the purpose, audience, and task (SL.9-10.4).
5. Make strategic use of media (e.g., textual, graphical, and visual) in presentations to enhance understanding of findings, reasoning, and evidence and to add interest (SL.9.10.5).
6. Adapt speech to a variety of contexts and tasks, demonstrating command of formal English when indicated or appropriate (SL.9-10.6).

Activities

Activity A	Create a Draft Argument
Time Needed for the Activity	15 minutes
Grouping	Teams

What to Do

1. Students should work in their assigned teams for this activity. Begin this activity by giving out a copy of the Share handout to each student.
2. Show the "Create a Draft Argument" page on a screen using the Presentation mode in the Learning Hub so everyone in the classroom can see it.
3. Say, "Okay, it sounds like we have all solved the task. I would now like you to create a draft argument to share what you figured out with the other teams. Your argument should include everything in this image here." Point to the image of a science-based argument projected on the screen. Be sure to go over what should be included in each component of the argument and where students can find the information they need.
4. Say, "I would like you to make your draft argument on a whiteboard. You can use the materials on this table to make your argument." Point to the table where you have placed the materials they need.
5. Ask, "Does anyone have any questions?"
6. Answer any questions that come up.
7. Say, "I'm going to give you exactly 15 minutes to get this done. Your draft argument doesn't need to be perfect. We just need something down on the whiteboards so we can share our ideas."
8. Give each team of students 15 minutes to create a draft argument. As students work in small teams, be sure to move from team to team to answer questions and offer advice.

What to Set Up

Place the following items on a table so students can use them during this stage:

- 24″ x 36″ whiteboard
- 3-4 dry-erase pens in different colors
- Eraser or cloth towel

What to Give Out

Share handout for **Fruit Fly Body Color and Wing Shape**

What to Show

The "Create a Draft Argument" page in the ADI Learning Hub

What to Look for

Students should include core ideas that they read about during the Ideas stage in their justification.

ARGUMENT-DRIVEN INQUIRY

Teaching Tip

Be sure to plant some different ideas with each group about how to make a strong argument as you move from group to group to check in, answer questions, and offer advice. This strategy helps to ensure that there will be some really strong examples of the evidence and justification components of the argument for students to see and discuss during the next activity. For example, you could help two groups have a really strong evidence component by telling one group, "I think a graph would be really helpful here" and telling another group, "I think your evidence would be stronger if you explained, in words, what this graph shows." You could help a third group have a really strong justification by saying, "Let's brainstorm some ideas that will help you defend your choice of evidence," or "Let's go back to our Ideas handout and see if there are any concepts we can include."

It is important you make sure that all the arguments are different at this point in the investigation by planting different ideas with different groups, because the students will have nothing to critique or discuss during the next activity if they all have the same argument. The goal during this activity is to set the stage for the next activity by ensuring that the various components of a strong argument can be found on different whiteboards within the class. These strong components will then be taken up by all the groups because they reflect good ideas. Be sure to give students a voice and choice during this activity, so they can make mistakes and learn from them.

Attending to Equity

It is often helpful to assign and change roles at least once during this activity. For example, you might assign the 1s to be "in charge of writing" at the beginning of the activity. You can then easily switch roles about five minutes into the activity by saying, "1s, please hand the pens to the 2s. 2s, you are now in charge of writing for the group." By changing roles, you can ensure that all students have a chance to participate and contribute their ideas to the creation of a draft argument.

ADI Go Digital

The **ADI Learning Hub** has tons of features that allow students to share their thinking with each other when they are doing the investigation paperless or remote.

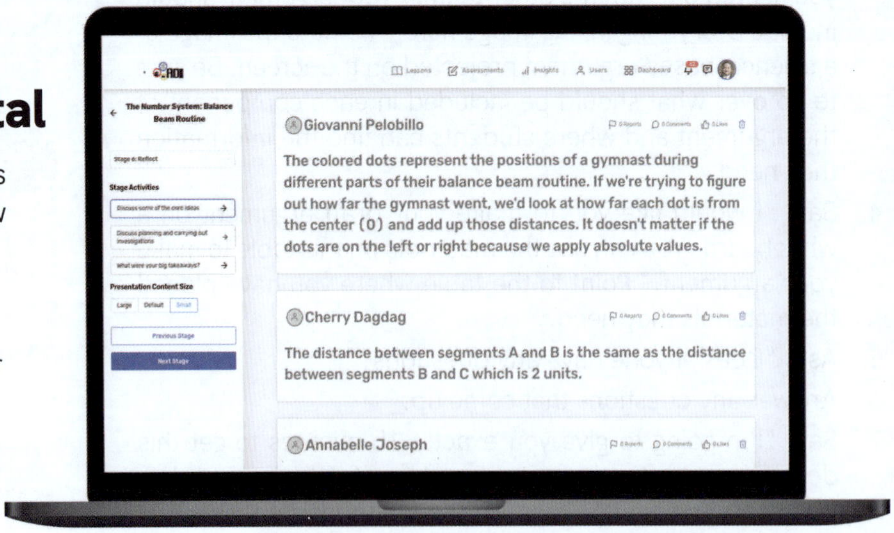

Activity B	Present Your Draft Argument
Time Needed for the Activity	15 minutes
Grouping	Teams

What to Do

1. Students should work in their assigned teams for this activity.
2. Show the "Present Your Draft Argument" page on a screen using the Presentation mode in the Learning Hub so everyone in the classroom can see it.
3. Say, "We are now going to share our arguments. Please set up your whiteboards so everyone can see them." Allow students time to set up their whiteboards. They should be set up on stands or chairs, so they are facing toward the outside of the room.
4. Say, "Okay, before we get started, I want to explain what we are going to do next. I'm going to ask some of you to present your arguments to your classmates. If you are presenting your argument, your job is to share your team's claim, evidence, and justification of the evidence. The rest of you will be travelers. If you are a traveler, your job is to listen to the presenters, ask them questions if you do not understand something, and then offer them some suggestions about ways to make their argument better. Do you have any questions about these instructions?"
5. Answer any questions that come up.
6. Say, "As you talk to one another, make sure you write down any good ideas or suggestions that are important in the space on your handout, so you don't forget about them." Point to the box labeled, "Some possible ways to improve our argument" on the handout.
7. Say, "After we have a chance to learn from one another, I'm going to give you some time to revise your arguments and make them better. Do you have any questions about what we are doing?"
8. Answer any questions that come up.
9. Say, "Okay. Let's get started. Travelers, move one team to the left. Don't move to the next team until I tell you to move. Presenters, go ahead and share your argument with the travelers when they get there."
10. Give each team of presenters and travelers about 3-4 minutes to talk.
11. Tell the students, "Okay. Let's move on to the next argument. Travelers, move one team to the left."
12. Again, give each team of presenters and travelers about 3-4 minutes to talk.

What to Show

The "Present Your Draft Argument" page in the ADI Learning Hub

What to Look for

The critiques and feedback that students give one another during this activity should:

- Be shared in a way that is kind (respectful),
- Provide information about what, exactly, needs to be changed or improved (specific), and
- Offer a suggestion about how to make something better rather than just saying something is wrong (helpful).

In addition, students should be focusing their critiques on the science included in the arguments, not the appearance of the arguments.

(continued)

13. Repeat this process until each team has had their argument read and critiqued three times. Then tell the students to return to their workstations.
14. Be sure to move between workstations during this activity to keep everyone involved and model what it means to communicate and critique scientific ideas and reasoning.

Attending to Equity

Be sure to encourage students to provide feedback that is respectful, specific, and helpful during this activity. It is important to encourage student-to-student talk because it helps students develop speaking and listening skills and enables them to make sense of what and how they know. It also provides a context for emerging multilingual students to use both productive (speaking and writing) and receptive (listening and reading) language and learn science through meaningful interactions and scientific sense-making.

Teaching Tip

There are two types of formats for sharing and critiquing arguments. The first format is called a *gallery walk*. In this format, each group sets up their argument at their workstation so others can see it. The entire group then moves to a different workstation. While at a different workstation, the group is expected to read the argument at that workstation (which was created by a different group) and offer critiques and suggestions for improvement. The students should move to a different workstation after a few minutes to read and critique an argument written by a different group. Students should also be encouraged to keep a record of good ideas or potential ways to improve their own arguments as they move from workstation to workstation to offer critiques and feedback. After repeating this process in the gallery walk 3-4 times, every group will have had an opportunity to read and critique three or four different arguments and receive feedback from three or four different groups.

The second format is called a *modified gallery walk*. In the modified gallery walk format, one or two members of the group stay at their workstation to share their group's ideas (these students are called presenters) while the other members of their group go to different groups, one group at a time, to listen to and critique the arguments developed by their classmates (these students are called travelers). The travelers should visit at least three different workstations during the argumentation session. We also recommend that the presenters keep a record of the critiques made by their classmates and any suggestions for improvement. The travelers should also be encouraged to keep a record of good ideas or potential ways to improve their own arguments as they travel from group to group.

Attending to Equity

Another good way to have students, especially emerging multilingual learners, give one another feedback is to have them use four different color sticky notes to make comments. Give each group a single pink, blue, yellow, and green sticky note. Then direct the students to use the pink sticky note to suggest "something to add" to the argument, the blue sticky note to suggest "something to change" in the argument, the yellow sticky note to highlight "something they like" in the argument, and the green sticky note to ask a question about "something they don't understand". This strategy is useful because everyone in the classroom will be actively engaged during this activity, and everyone will have a chance to see different arguments. This format, however, does not give students a chance to support, critique, and challenge the ideas, solutions, scientific reasoning of other groups through talk because two different groups are never at the same workstation at the same time.

Activity C	Revise Your Argument
Time Needed for the Activity	15 minutes
Grouping	Teams

What to Do

1. Students should work in their assigned teams for this activity.
2. Show the "Revise Your Argument" page on a screen using the Presentation mode in the Learning Hub so everyone in the classroom can see it.
3. Say, "I'm now going to give you all about 10 minutes to revise your argument. Please return to your original teams."
4. Wait for the students to move back into their original teams.
5. Say, "Okay, take a few minutes to talk in your teams and determine what you want to change to make your argument better. Once you have decided what to change, go ahead and make the changes to your whiteboard."
6. Ask, "Do you have any questions about what you need to do?"
7. Answer any questions that come up.
8. Tell the students, "Okay. Let's get started."
9. Give the students 10 minutes to work in their teams on their arguments. As they work, move from team to team to check in, ask probing questions, and offer a suggestion if a team gets stuck.
10. When students are finished with their final arguments, have them take a picture of it or have them put the boards in a safe spot so they can refer to the picture or the boards during the Report stage.

What to Show

The "Revise Your Argument" page in the ADI Learning Hub

What to Look For

A strong argument includes the following information:

- A claim that states, "The yellow body color trait follows the sex-linked inheritance patterns, and it is determined by the recessive allele. The curly wings trait follows the dominant-recessive inheritance pattern, and it is determined by the dominant allele."
- A table that compares the parents' traits to that of their offspring.
- An explanation of the table.
- An explanation of the chosen evidence that uses core ideas from the investigation.

Teaching Tip

It is important to give students time to meet with their original groups so they can discuss what they learned by interacting with individuals from the other groups. At this point in the process, students often realize that their initial problem-solving model was flawed in some way. It is therefore important to encourage students to "redo the science" as needed. At the end of this activity, each group should have a final argument that is much better than their original one.

Fruit Fly Body Color and Wing Shape Lesson Plan

Stage 6 REFLECT

Total Time for the Stage	25 minutes
Grouping	Whole class, Teams

Learning Goals

Students should make progress on their ability to do the following by the end of this stage of the lesson:

Science

1. Ask questions that arise from careful observation of phenomena, or unexpected results, to clarify and/or seek additional information as needed (SEP1).
2. Describe how in sexual reproduction, chromosomes can swap sections during meiosis and create new genetic combinations and thus more genetic variation (LS3.B).
3. Explain how DNA replication is a remarkably accurate process that occasionally allows for mutations that provide genetic variation (LS3.B).
4. Identify how environmental factors can cause genetic mutations and affect the expression of traits that can then be inherited (LS3.B).
5. Describe how different patterns may be observed at each of the scales at which a system is studied and can provide evidence for causality in explanations of phenomena (CC1).
6. Use empirical evidence to differentiate between cause and correlation and make claims about specific causes and effects (CC2).

English Language Development

1. Paraphrase central ideas in complex evidence, concepts, processes, and information to help explain how or why a phenomenon occurs (ELD-SLC.9-12.Explain.Interpretive).
2. Use scientific explanations about a phenomenon in a different context or situation (ELD-SLC.9-12.Explain.Expressive).

English Language Arts

1. Participate effectively in small-group collaborative discussions with diverse partners on texts by building on others' ideas and expressing their own clearly and persuasively (SL.9-10.1).

Activities

Activity A	Discuss a Core Idea
Time Needed for the Activity	10 minutes
Grouping	Whole class/Teams

What to Do

1. Students should be in small teams for this activity. However, they do not need to be in the same teams as they were during previous stages.
2. Begin this activity by giving out a copy of the Reflect handout to each student.
3. Show the "Discuss a Core Idea" page on a screen using the Presentation mode in the Learning Hub so everyone in the classroom can see it.
4. Say, "Okay. Now that we have all revised our arguments and made them better, let's take some time to discuss some of the core ideas you used during this investigation and to justify the evidence in your arguments so we are all on the same page. Look at the picture on the screen."
5. Pause for about 30 seconds.
6. Say, "Take a few minutes to talk in team and see if you can use what you know about chromosomes, genes, proteins, and traits and models of inheritance to explain what is happening in the picture."
7. Give the students at least three minutes to talk in their small teams.
8. Say, "Okay. Let's share with the whole class. Team A (or whichever team you want to share first), what did you talk about?"
9. Allow the first team to share.
10. Say, "Interesting. Would anyone like to add to, challenge, or refine the ideas shared by Team A?"
11. Allow students to respond.
12. Say, "Okay. Team B (or whichever team you want to share second), what did you talk about?"
13. Allow the second team to share.

What to Give Out

Reflect handout for **Fruit Fly Body Color and Wing Shape**

What to Show

The "Discuss a Core Idea" page in the ADI Learning Hub

Explanation of the Image

The image that the students discuss during this activity shows multiple allele inheritance.

What to Look For During the Discussion

Students who understand the content will likely mention some, if not all, of the following ideas when providing an explanation for the image:

- Chromosomes contain an organism's genetic material or DNA.
- A gene is a segment of DNA that encodes for a certain protein, and variations in a gene are called alleles.
- The way that a trait manifests in an organism depends on how alleles interact with one another.
- Five common models of inheritance are dominant-recessive, incomplete dominance, codominance, multiple-allele, and sex-linked.

(continued)

14. Say, "Interesting. Would anyone like to add to, challenge, or refine the ideas shared by Team B?"
15. Allow students to respond.
16. Keep up this pattern of questioning until all teams have shared. Be sure to keep encouraging students to talk about the content until they reach a correct and consensual explanation for what is shown in the image. If they are unable to reach a consensus or reach consensus using an inaccurate idea, be sure to provide an accurate explanation for the image before moving on to the next activity.

Teaching Tip

This activity is a discussion, not a presentation about what the students "should have seen" or "should have learned" during the investigation. The more students talk during this activity, the more meaningful the experience will be for them and the more you can learn about their thinking. Be sure to use talk goals and moves to help facilitate the discussion. Below are some examples of talk goals that you can use to help facilitate the discussion and some questions that you can ask the students to help accomplish each goal. These talk goals and questions are only suggestions.

Goal 1: Encourage students to share, expand on, or clarify an idea.

- What do you see going on here?
- What do you think is happening?
- How might we explain that?
- Can you tell me more about that?
- What do you mean by that?
- Can you give me an example of that?
- Can you say that in another way?

Goal 2: Encourage students to support an idea.

- Why do you think that?
- What is your evidence for that?
- How did you arrive at that conclusion?
- What makes you say that?
- How do you know that?

Goal 3: Encourage students to think with others.

- Would anyone like to add to that idea?
- What should we keep or change about that idea?
- Can anyone else give me some examples of that?
- Can anyone take that idea and push it a little further?
- Does anyone want to respond to that idea?
- Do you agree or disagree with that idea, and why?
- What do we think about that idea?
- Does that fit with what we have figured out so far?
- Is there another way to do it?

Goal 4: Encourage students to think about what they know and don't know yet.

- I'm confused now. What have we figured out so far?
- What do we need to figure out now?
- So, what do we agree on right now, and what are we still trying to figure out?
- What are we sure and not so sure about?
- What questions do you have?

Activity B	Discuss the Practices
Time Needed for the Activity	10 minutes
Grouping	Whole class/Teams

What to Do	What to Show
1. Show the first page of "Discuss the Practices" on a screen using the Presentation mode in the Learning Hub so everyone in the classroom can see it. 2. Say, "Okay. Now that we have talked about some of the core ideas that we used during our investigation, let's take a few minutes to talk about the science practices that we used during the investigation. An example of a scientific practice used was analyzing and interpreting data. On the screen there is a question that I want you all to think about: What other things did you do during your investigation that you feel were good examples of a scientific practice or attempting to figure out something in a scientific way?" 3. Pause for about 30 seconds. 4. Say, "Take a few minutes to talk about this question in your teams." 5. Give the students at least three minutes to talk in their small teams. 6. Say, "Okay. Let's share with the whole class. Team A (or whichever team you want to share first), what did you talk about?" 7. Allow the first team to share. 8. Say, "Okay. Team B (or whichever team you want to share second), what did you talk about?" 9. Allow the second team to share. 10. Continue asking teams until every team has a chance to share. 11. Show the second page of "Discuss the Practices" on a screen using the Presentation mode in the Learning Hub so everyone in the classroom can see it. 12. Say, "Here is another question that I want you all to think about: What things did you do during your investigation that were not good examples of a scientific practice or not very scientific?" 13. Pause for about 30 seconds.	**What to Show First** The first page of "Discuss the Practices" in the ADI Learning Hub **What to Show Second** The second page of "Discuss the Practices" in the ADI Learning Hub **What to Show Third** The third page of "Discuss the Practices" in the ADI Learning Hub

(*continued*)

14. Say, "Take a few minutes to talk about this question in your team."
15. Give the students at least three minutes to talk in their small teams.
16. Say, "Okay. Let's share with the whole class. Team A (or whichever team you want to share first), what did you talk about?"
17. Allow the first team to share.
18. Say, "Okay. Team B (or whichever team you want to share second), what did you talk about?"
19. Allow the second team to share.
20. Continue asking teams until every team has a chance to share.
21. Show the third page of "Discuss the Practices" on a screen using the Presentation mode in the Learning Hub so everyone in the classroom can see it.
22. Say, "Here is a final question that I want you all to think about: What should we all do during our next investigation to make sure we are using scientific practices and being scientific in the way we try to figure something out?"
23. Pause for about 30 seconds.
24. Say, "Take a few minutes to talk about this question in your team."
25. Give the students at least three minutes to talk in their small teams.
26. Say, "Okay. Let's share with the whole class. Team A (or whichever team you want to share first), what did you talk about?"
27. Allow the first team to share.
28. Say, "Okay. Team B (or whichever team you want to share second), what did you talk about?"
29. Allow the second team to share.
30. Continue asking teams until every team has a chance to share.
31. Once everyone has shared, offer any suggestions or advice about what they can do to improve their next investigation and then move on to the next stage.

Teaching Tip

A great way to wrap up this activity is for the students to contribute to a list of class norms or "gotta-haves" to keep in mind when planning and carrying out an investigation in science. To create this list, each group can suggest for a potential norm or "gotta-have" to add to the list at the end of this activity, then the class votes on the one that they think is most important. The most important one gets added to the list. The list then serves as reminder for students during the Plan and Do stages of subsequent investigations. Over time, the list will grow and each addition to the list will be more nuanced than the last as students develop expertise with scientific processes.

Attending to Equity

Be sure to give all students, especially emerging multilingual students, enough time to talk in small groups before asking anyone to share with the entire class during this activity. Small-group structures offer all students, including emerging multilingual students, a chance to engage in sense-making with their peers as well as space to use their verbal and nonverbal (e.g., writing, drawing pictures, and using gestures) resources to express their ideas. Small-group structures also give students a chance to learn from other students' uses of these resources. When students are tasked with expressing their ideas, be sure that they know it is okay for them to do so in a mode or combination of modes (speaking, writing, drawings, and gestures).

Fruit Fly Body Color and Wing Shape Lesson Plan

Stage 7 REPORT

Total Time Needed for the Stage	100 minutes (over three days)
Grouping	Individual, Teams

Learning Goals

Students should make progress on their ability to do the following by the end of this stage of the lesson:

Science

1. Make and defend a claim based on evidence that inheritable genetic variations may result from new genetic combinations through meiosis, viable errors occurring during replication, and/or mutations caused by environmental factors (HS-LS3-2).
2. Make and defend a claim based on evidence about the natural world that reflects scientific knowledge and student-generated evidence (SEP7).
3. Construct a written argument based on evidence (SEP7).
4. Evaluate the claims, evidence, and/or reasoning to determine the merits of arguments (SEP7).
5. Respectfully give critiques on scientific arguments by and challenging ideas and conclusions (SEP7).
6. Communicate scientific information or ideas in multiple formats including graphically and textually (SEP8).

English Language Development

1. Identify appropriate and sufficient evidence from data, models, and/or information from investigations of a phenomenon (ELD-SLC.9-12.Argue.Interpretive).
2. Compare reasoning and claims based on evidence from competing arguments (ELD-SLC.9-12.Argue.Interpretive).
3. Defend or refute a claim based on data and evidence (ELD-SLC.9-12.Argue.Expressive).
4. Establish and maintain an appropriate tone and stance (ELD-SLC.9-12.Argue.Expressive).
5. Signal logical relationships among reasoning, evidence, data, and/or models when making and defending a claim (ELD-SLC.9-12.Argue.Expressive).

English Language Arts

1. Write routinely over extended time frames (time for reflection and revision) and shorter time frames for a range of discipline-specific tasks, purposes, and audiences (WHST.9-10.10).
2. Write arguments focused on discipline-specific content (WHST.9-10.1).
3. Write informative/explanatory texts, including the narration of scientific procedures/experiments or technical processes (WHST.9-10.2).
4. Produce clear and coherent writing in which the development, organization, and style are appropriate to task, purpose, and audience (WHST.9-10.4).

(continued)

5. Strengthen writing as needed by revising, editing, rewriting, and focusing on addressing what is most significant for a specific purpose and audience (WHST.9-10.5).
6. Determine the central ideas or conclusions of a text; trace the text's explanation or depiction of a complex process, phenomenon, or concept; provide an accurate summary of the text (RST.9-10.2).
7. Determine the meaning of symbols, key terms, and other domain-specific words and phrases as they are used in a specific scientific or technical context (RST.9-10.4).
8. Analyze the structure of the relationships among concepts in a text, including relationships among key terms (RST.9-10.5).
9. Assess the extent to which the reasoning and evidence in a text support the author's claim (RST.9-10.8).
10. Compare and contrast findings presented in a text to those from other sources (including their own experiments), noting when the findings support or contradict other accounts (RST.9-10.9).

Activities

Activity A	Draft Report
Time Needed for the Activity	20 minutes
Grouping	Individual

What to Do

1. Have students work on their own during this activity. Give a copy of the Report handout to each student.
2. Show the "Create a Draft Report" page on a screen using the Presentation mode in the Learning Hub so everyone in the classroom can see it.
3. Say, "You are now going to write an investigation report to share what you figured out. The first part of the report is called the 'Introduction'. In this section, you should explain to the reader what you were investigating, why you were investigating it, and what question you were trying to answer. All this information can be found on your Task handout."
4. Hold up the Task handout.
5. Ask the students, "Do you have any questions about what you need to do?"
6. Answer any questions that come up.
7. Say, "Here are some sentence starters to help you begin writing."
8. Hold up the Report handout and point to the sentence starters.
9. Tell the students, "Okay, let's write."
10. Give the students five minutes to write the "Introduction" section of the report. As they work, move from student to student to check in, ask probing questions, and offer a suggestion if a student gets stuck.

What to Give Out

Report handout for **Fruit Fly Body Color and Wing Shape**

What to Show

The "Create a Draft Report" page in the ADI Learning Hub

(continued)

11. Say, "Okay, let's move on to the second part of the report. The second part of the report is called the 'Method'. In this section, you should explain to the reader what you did during the investigation to complete the task and answer the guiding question. All this information can be found in the Plan handout. Remember that you all planned and carried out different investigations, so do not assume that the reader will know what you did."
12. Ask the students, "Do you have any questions about what you need to do?"
13. Answer any questions that come up.
14. Say, "Here are some sentence starters to help you begin writing."
15. Hold up the Report handout and point to the sentence starters.
16. Tell the students, "Okay, let's write."
17. Give the students five minutes to write the "Method" section of the report. As they work, move from student to student to check in, ask probing questions, and offer a suggestion if a student gets stuck.
18. Say, "Okay, let's move on to the last part of the report. The third part of the report is called the 'Argument'. In this section, you should explain to the reader what you figured out and how you know you are correct. All this information can be found on your Do handout and on your team's whiteboard."
19. Hold up the Do handout and point to the boards (or remind them that they have a picture of their board).
20. Ask the students, "Do you have any questions about what you need to do?"
21. Answer any questions that come up.
22. Say, "Here are some sentence starters to help you begin writing."
23. Hold up the Report handout and point to the sentence starters.
24. Tell the students, "Okay, let's write."
25. Give the students 10 minutes to write the "Argument" section of the report. As they work, move from student to student to check in, ask probing questions, and offer a suggestion if a student gets stuck.
26. Collect all the draft reports before moving on to the next activity. Tell students to mark their draft reports with a symbol that they can use to identify it later. They should NOT write their names on their draft reports.

Teaching Tip

Writing is an important component of this model. When a student writes the report on their own, it not only helps the student learn to write in the context of science (an important literacy skill) but also allows that student to write to learn (develop a better understanding of scientific concepts by writing about them). There are many writing supports (e.g., sentence starters) already embedded into this activity, so you might not need to provide much extra support for your students.

This activity can be done completely in class or started in class and then finished as homework. Do not assign the entire activity as homework unless you know that every student in your class has dedicated time, space, and resources to write outside of class and everyone understands and is comfortable with the report writing process.

If you are concerned about students feeling overwhelmed, you can develop a class "gotta-have-it" list of information that needs to be included in each section of the report. Students can then refer to this list as they write each section.

Don't worry if students do not produce a perfect report at this point. All the reports will be reviewed and revised before they are turned in to you. The reports should therefore be viewed as a starting point. Be sure to encourage students to "just get something down on paper" so "we can work on it together."

Attending to Equity

Emerging multilingual students' access to instructional tasks that require complex thinking, such as writing a report, is enhanced when linguistic complexity and instructional support match their current level of language proficiency. Therefore, it is important to scaffold the writing of the reports for the emerging multilingual students based on their current English language proficiency level. The use of sentence starters to model written language production skills can help emerging multilingual students develop English language skills because the development of academic language and academic content knowledge are interrelated processes. However, it is important that such scaffolds be used purposefully and removed when no longer needed.

ADI Go Digital

Use the **ADI Learning Hub** to go paperless. When you create a class and add students to it, students can complete all the stages right in the system.

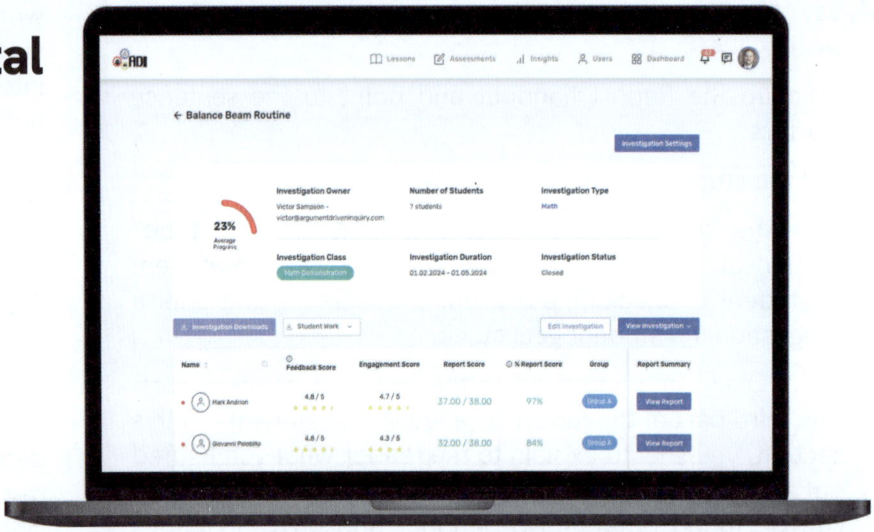

Activity B	Peer Review the Reports
Time Needed for the Activity	45 minutes
Grouping	Teams

What to Do

1. Students should work in teams of three for this activity. These can be completely different teams.
2. Show the "Peer Review Some Draft Reports" page on a screen using the Presentation mode in the Learning Hub so everyone in the classroom can see it.
3. Say, "We are now going to review our reports to find ways to make them better."
4. 4. If possible, use a document camera to project the peer review guide (PRG) on a screen.
5. Say, "We are going to use this peer review guide to give each other feedback." Point to the image.
6. Say, "I'm going to ask you to work in teams of three to do this. I will give each team a draft report to read. You three will then read the report together. Once you are done reading the report, I want you to answer each of the questions on the peer review guide." Point to the review questions on the image of the PRG.
7. Tell the students, "You can check No, Almost, or Yes after each question." Point to the checkboxes on the image of the PRG.
8. Say, "This will be your rating for this part of the report. Make sure you agree on the rating you give the author. If you mark No or Almost, then you need to tell the author what he or she needs to do to get a Yes." Point to the space for the reviewer feedback on the image of the peer review guide.
9. Say, "It is really important for you to give the authors feedback that is helpful. That means you need to tell them exactly what they need to do to make their report better."
10. Ask the students, "Do you have any questions about what you need to do?"
11. Answer any questions that come up.
12. Tell the students, "Please get into your teams of three."
13. Allow the students time to move into their triads.
14. Say, "Okay, I'm now going to give you one report to read." Give out one draft report to each triad. Make sure that the report you give a triad was not written by one of the students in that triad. Give each triad one PRG to fill out.

(*continued*)

What to Give Out

Peer Review Guide (one per group per review, so for this investigation you need the same number of peer review guides as the number of draft reports that will be reviewed)

What to Show

The "Peer Review Some Draft Reports" page in the ADI Learning Hub

What to Look For

Make sure students are reaching consensus about what rating to give each criterion. Also, be sure that students are giving one another feedback that is specific, detailed, and helpful when they think an aspect of a report needs to be improved.

15. Tell the students, "Okay, I'm going to give you 15 minutes to read the report I gave you and to fill out the peer review guide. Go ahead and get started."
16. Give the students 15 minutes to work. As they work, move around from team to team to check in and see how things are going, answer questions, and offer advice.
17. After 15 minutes pass, tell the students, "Okay, time is up. Please give me the report and the peer review guide that you filled out."
18. Collect the draft reports and the PRGs, or collect the workbooks if students are using them.
19. Say, "Okay, I am now going to give you a different report to read and a new peer review guide to fill out." Give out one more report to each triad. Make sure that the report you give a triad was not written by one of the students in that triad. Give each triad one new peer review guide to fill out as a team.
20. Say, "Okay, I'm going to give you 15 minutes to read this new report and to fill out the peer review guide. Go ahead and get started."
21. Give the students 15 minutes to work. As they work, move around from team to team to check in and see how things are going, answer questions, and offer advice.
22. After 15 minutes pass, tell the students, "Okay, time is up. Please give me the report and the peer review guide that you filled out."
23. Repeat this process for the remaining draft reports. Each triad should one have one more report to review (for a total of three reviews).
24. Collect the last draft reports and PRGs. Set all the draft reports on a table in separate stacks. Each stack should have a draft report on top (with the identification symbol facing up so students can find their reports) and the PRG on the bottom.

Teaching Tip

The peer review process is one of the best ways for students to learn how to write. When students read and review different reports, they have an opportunity to (1) see examples of texts written by others for the same purpose, (2) discuss what counts as quality and why some reports are stronger than others, (3) discuss ways to strengthen a report, and (4) pick up things that they can do to improve their reports. This helps the students learn new ways to organize and present information, which will help them write better on subsequent reports.

Attending to Equity

One helpful scaffold is to give students sentence stems for what to write in their feedback to the author. Ideal feedback is respectful, specific about what needs to be changed, and helpful in that it suggests how the author should go about making that change. You can review the writing of the emerging multilingual students in your class if they have limited English language proficiency, but make sure that all emerging multilingual students participate in the peer review process. You will see tremendous growth in their writing skills as they review their peers' writing and discuss what makes some reports stronger than others.

Activity C	Revise the Reports
Time Needed for the Activity	25 minutes
Grouping	Individuals

What to Do

1. Students should work on their own for this activity.
2. Show the "Reflect on the Feedback" page on a screen using the Presentation mode in the Learning Hub so everyone in the classroom can see it.
3. Say, "Please come pick up your reports from over here on this table." Point to the table.
4. Give the students a few minutes to pick up their reports and return to their seats.
5. Say, "Please take a few minutes to read over the peer review guide. You should use it to figure out what you need to change in your report and how you will change it."
6. Allow the students to read the PRG.
7. Show the "Create a Final Report" page on a screen using the Presentation mode in the Learning Hub so everyone in the classroom can see it.
8. Say, "Okay, let's revise our reports. I would like you to rewrite your report. You can use your draft report as a starting point but use the feedback on the peer review guide to make it better."
9. Ask the students, "Do you have any questions about what you need to do?"
10. Answer any questions that come up.
11. Tell the students, "Okay, let's write." Allow about 20 minutes for the students to revise their reports.
12. After about 20 minutes, give the following directions, "Okay, time's up. I will now come around and collect your first draft, final report, and peer review guide."
13. Collect all the first drafts, PRGs, and final reports.
14. Use the "Teacher Score" columns in the PRG to grade the final report. Be sure to give the students feedback about their writing in the "Teacher Comments" section.

What to Show First

The "Reflect on the Feedback" page in the ADI Learning Hub

What to Show Second

The "Create a Final Report" page in the ADI Learning Hub

Attending to Equity

Students don't always use the feedback they receive from others to improve the quality of their reports. Make sure you move from student to student and encourage them to make changes based on the feedback. If students are not sure how to revise their reports, model how to use feedback. For example, you might say, "Okay, based on the peer review guide, I need to do more to describe what I did to collect data. I bet if I add information about ________, readers will have a better idea of what I did."

Teaching Tip

The peer review guide (PRG) is designed to be used with any ADI investigation, thus allowing you to use the same scoring rubric throughout the entire year. This is beneficial for several reasons. First, the criteria for what counts as a high-quality report do not change from investigation to investigation. Students, therefore, quickly learn what is expected from them when they write a report, and you do not have to spend valuable class time explaining the various components of the PRG each time students write a report. Second, the PRG makes it clear which components of a report need to be improved next time because the grade is not based on a holistic evaluation of the report. The PRG format allows students to see which aspects of their writing are strong enough and which aspects need improvement. Finally, and perhaps most importantly, the PRG provides you with a standardized measure of student performance that can be compared over multiple reports across semesters, thus allowing you to track improvement over time.

Teacher Notes

Fruit Fly Body Color and Wing Shape ✓ TASK

The Phenomenon

Drosophila melanogaster is a species of fly in the family Drosophilidae. The species is known generally as the common fruit fly or vinegar fly. Take a few minutes to watch a video that shows what these fruit flies look like and their life cycle. As you watch the video, write down some things you notice about the fruit flies.

Things I **noticed**...

The Task

The video you watched showed the life cycle of *Drosophila melanogaster* and how to tell the difference between male and female flies. The image below shows a photograph of a fruit fly (on the left), and an illustration of a female fruit fly and a male fruit fly as seen from above. Individual flies with the same eye color, wing shape, and body color as the flies in the image below are called "wild type" fruit flies.

***Drosophila melanogaster* with wild type traits**

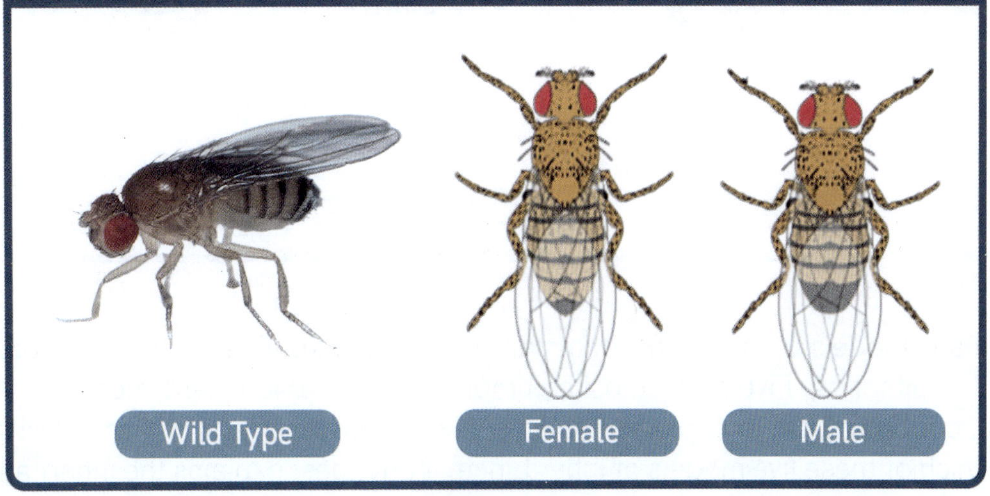

Not all fruit flies look the same. There can be a great deal of variation in the traits of fruit flies, such as eye color, body color, and wing shape. For example, some flies are born with a light-colored body. The image below shows a photograph of a fruit fly with a light-colored body (on the left), and an illustration of a female fruit fly and a male fruit fly with a light-colored body as seen from above. This trait is called "yellow body".

Drosophila melanogaster with the yellow body trait

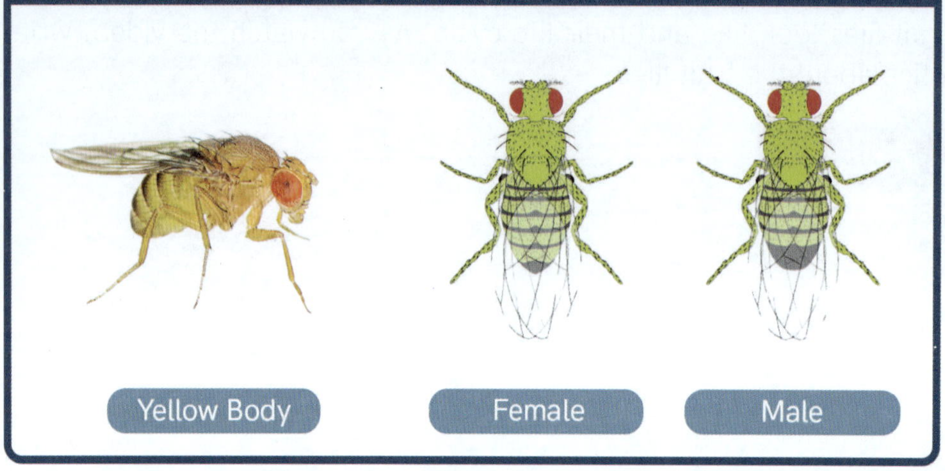

Some flies are born with wings that curl up from their body. The image below shows a photograph of a fruit fly with curled wings (on the left), and an illustration of a female fruit fly and a male fruit fly with curled wings as seen from above. This trait is called "curly wings".

Drosophila melanogaster with the curly wings trait

There are many models of inheritance that can be used to explain how traits are passed down from parent to offspring. Five of the most common models are dominant-recessive, incomplete dominance, codominance, multiple alleles, and sex-linked. Your goal in this investigation is to figure out which of these five models of inheritance, if any, best explains the inheritance pattern of the yellow body trait and curly wings trait in fruit flies. To accomplish this goal, you will use an online simulation that allows you to breed fruit flies with specific traits to see how a trait is passed down from parent to offspring. The guiding question of this investigation is:

Which model or models of inheritance best explain how the yellow body trait and the curly wing trait in fruit flies are passed down from parent to offspring?

Initial Ideas

Before you start this investigation, take a few minutes to think about how traits are inherited and what you know about these five different inheritance models. Then draw a model in the space below representing your ideas about how traits are inherited. Be sure to include labels to help explain your thinking.

Some **ideas** that I have...

Share your model with the rest of your group. Once everyone has had a chance to share their ideas, list some questions that your group has about this phenomenon.

Some **questions** we have now...

Fruit Fly Body Color and Wing Shape

IDEAS

Some Ideas You Can Use: Chromosomes, Genes, Proteins, and Traits

Chromosomes are found in cells. The image below shows the chromosomes that are found in the cells of a female fruit fly and a male fruit fly. Note that each fruit fly has a total of eight chromosomes. Both female and male fruit flies have two copies of Chromosome 2, two copies of Chromosome 3, and two copies of Chromosome 4. Female fruit flies, however, have two copies of the X chromosome whereas male fruit flies have one X chromosome and Y chromosome. Each of these chromosomes consists of a single very long **DNA molecule**.

Chromosomes of *Drosophila melanogaster*

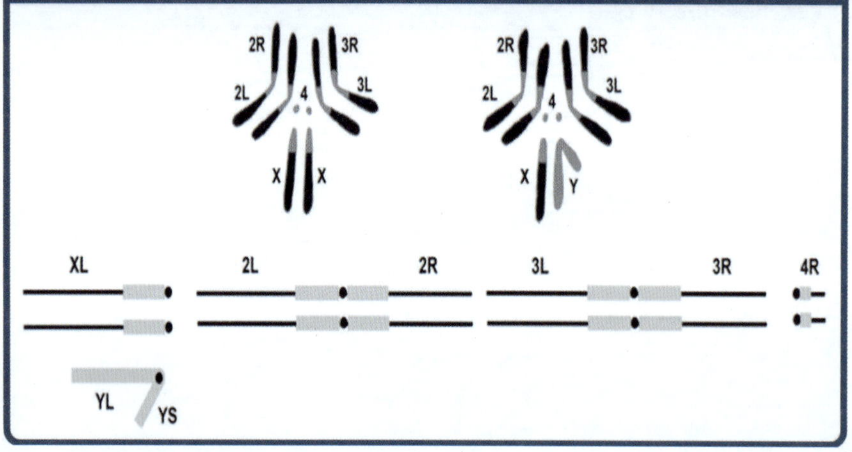

Chromosomes are passed down from parents to offspring as a result of sexual reproduction. Each offspring will inherit two copies of each of their different chromosomes. These two copies are often described as a chromosome pair. In the case of fruit flies, offspring inherit two sex chromosomes and a pair of chromosomes 2-4 from its parents. One chromosome of each pair comes from the female parent and the second chromosome comes from the male parent. The

Chromosome 3 being passed down from parents to offspring

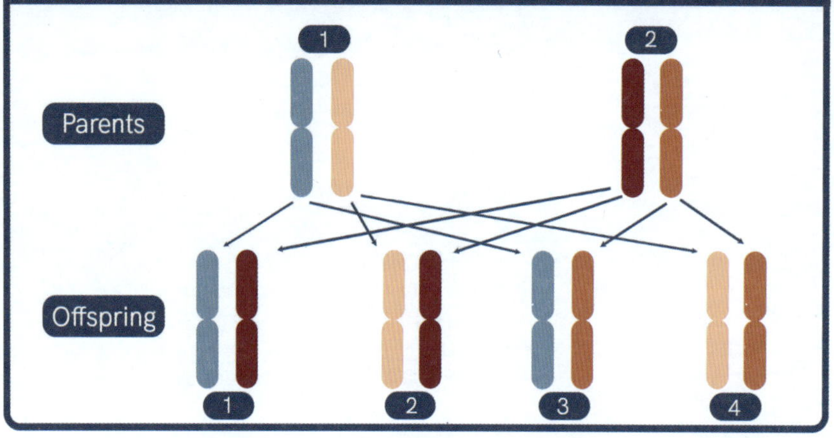

ARGUMENT-DRIVEN INQUIRY

image below shows an example of how a pair of fruit flies (parents) pass down one of their two copies of Chromosome 3 to each of their four (1-4) offspring. Although each offspring inherits a pair of Chromosome 3, each offspring can inherit a different combination.

Some **useful ideas** from what I read...

Some Ideas You Can Use: Models of Inheritance

Scientists have developed several different models of inheritance to explain how traits are passed down from parents to offspring. Five of the most common models are dominant-recessive, incomplete dominance, codominance, multiple alleles, and sex-linked. All these models share three basic assumptions. First, inherited genes determine specific traits. A **gene** is simply a region in a DNA molecule that encodes the information for making a specific protein (and the absence or presence of a protein determines the trait of an individual). Second, there are alternative versions of each gene (called **alleles**). These alternative versions of the same gene are responsible for the variation we see in traits. Second, an organism will inherit two copies of each gene because organisms inherit two copies of each of their chromosomes from their parents. What makes these five models of inheritance different from one another is the nature of the interactions that occur between alleles once they are passed down from parents to offspring, the number of alleles that are associated with a gene, and which chromosome carries the gene for a trait.

The **dominant-recessive model** of inheritance suggests that when an individual inherits two alleles, and the two alleles differ, one is fully expressed and determines the nature of a trait (this version of the gene is called the dominant allele) while the other one has no noticeable effect (this version of the gene is called the recessive allele). The image on the next page shows an example of how this model of inheritance can explain how two fruit flies, each with a different trait, can have four offspring, all with the same trait as one of the parents. In this image, one parent has the "wild type" body color trait, and the other parent has the "ebony" body color trait. Since one parent can only pass down the dominant "wild type" (+) allele and the other parent can only pass down a recessive "ebony" (-) allele, all four children inherited one copy of each allele. The "+-" genotype results in the "wild type" phenotype, which is the trait that is fully expressed by the dominant allele.

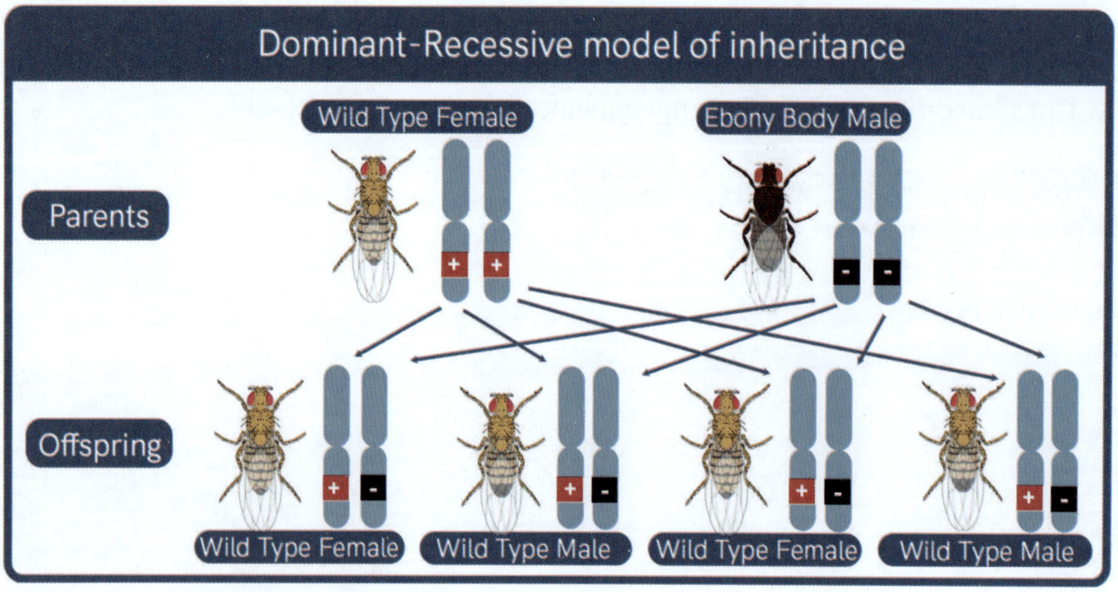

The **incomplete dominance model** of inheritance suggests that the interaction that occurs between two different alleles results in a hybrid with an appearance somewhere between the phenotypes of the two parental varieties. The image below shows an example of how this model of inheritance can explain how two parents with different traits can have four offspring who all express a third trait that is different from both parents. In this image, one parent has the "wild type" bristle (+) trait, which are long, and the other parent has the "shaven" bristle (-) trait, which are missing. The parent with "wild type" bristle has two copies of the "+" allele and the parent with no bristles has two copies of the "-". Since one parent can only pass down the incomplete dominant "+" allele and the other parent can only pass down the incomplete recessive "-" allele, all four children inherited one copy of each allele and have the "+-" genotype. The "+-" genotype results in "stuble" phenotype, which are short bristles. The "stuble" phenotype is a blending of long and missing bristle alleles.

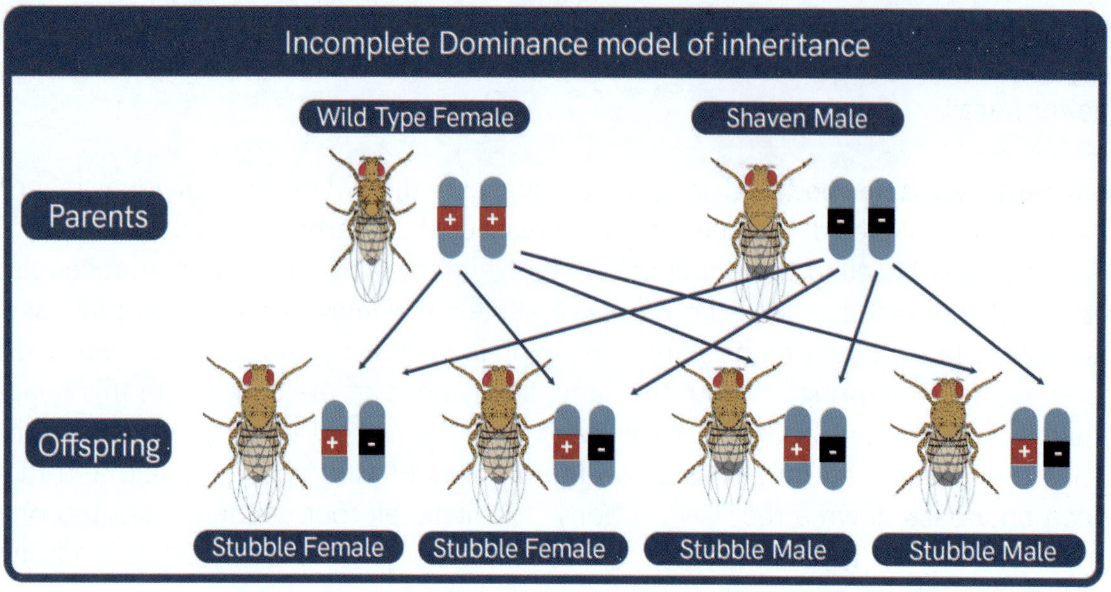

The **codominance model** of inheritance is similar to the incomplete dominance model, but in the codominance model both alleles affect the individual's phenotype in separate and distinguishable ways. The image below shows an example of how this model of inheritance can explain how two parents with different traits can have four offspring who all express a third trait that is different from both parents. In the image, one parent has the "wild type" or "long and straight" bristle trait (+) and the other parent has the "forked" or "split" bristle trait (+). Since one parent can only pass down the codominant "long and straight" allele and the other parent can only pass down the codominant "forked" allele, all four offspring inherited one copy of each allele and have the "++" genotype. The "++" genotype results in the "singed" bristle phenotype. The singed bristle phenotype is an expression of both the "long and straight" bristles and the "forked" bristles in the same individual. As a result, some of the bristles on these flies will be long and straight and some of the bristles will be forked.

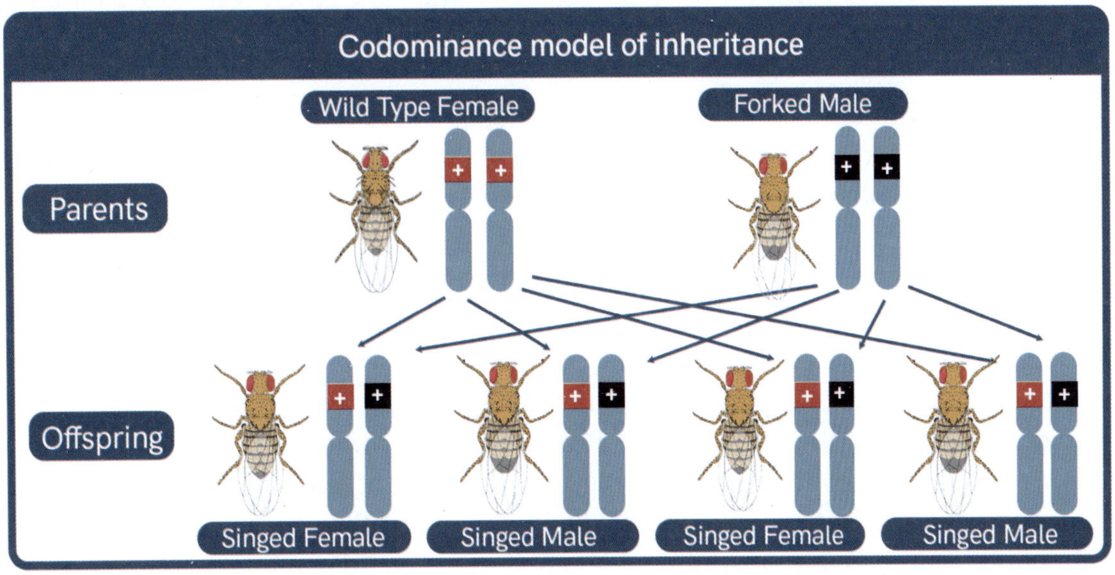

The **multiple allele model** of inheritance simply means that there are more than two versions of a gene for a given trait within a population, and each allele can be dominant, recessive, incompletely dominant, or codominant to the other alleles. The image on the next page shows an example of how this model of inheritance can explain how two parents with different traits can have offspring with traits that are different from the parents. In this image, one parent has the "wild type" or "red eyes" trait and the other parent has the "garnet eyes" trait. The "wild type" parent carries the dominant "red" allele (+) and the recessive "purple" allele. The "garnet eye" parent carries one "purple" allele and one "brown" allele. The "purple" and "brown" alleles are recessive to the "red" allele but incomplete dominant to each other. Flies that inherit both a "purple" and "brown" allele, as a result, have eyes that are a blend of purple and brown or "garnet". One parent can pass down the dominant "red" allele or the recessive "purple" allele and the other parent can pass down the recessive "brown" allele or the recessive "purple" allele. Two offspring in the example inherited one copy of the dominant "red" allele so they both have red eyes. One offspring inherits a "purple" allele from one parent and a "brown" allele from the other parent, so it has garnet eyes. Finally, the last offspring inherited two copies of the "purple" allele, one from each parent, so it has the purple eye trait.

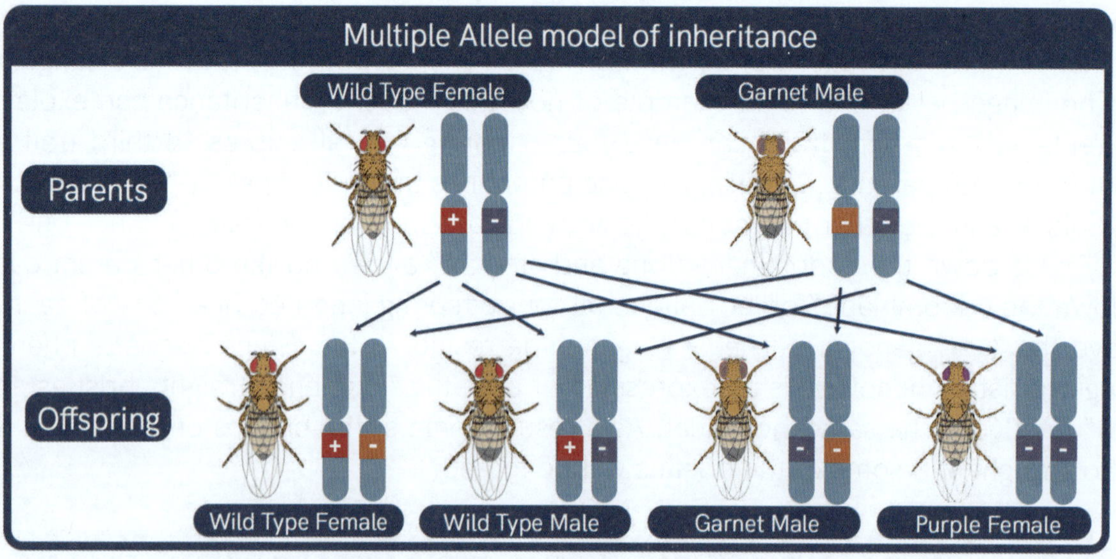

Finally, the **sex-linked model** of inheritance only applies to genes located on the sex chromosomes. Females and males differ in the number of genes they inherit when the gene is found on the sex chromosome; one sex inherits two copies of the gene, and the other sex inherits only one. The image below shows an example of how this model of inheritance can explain how a "white eye" female and "wild type" or "red eye" male can have two female offspring with red eyes and two male offspring with white eyes. In this image, the female parent has two copies of the recessive "white" allele, and the male parent has one copy of the dominant "red" allele (because males only inherit one copy of the sex chromosome where the gene is located). Because the male parent can only pass down the dominant "red" allele and the female parent can only pass down the recessive "white" allele, all the female children inherited at least one copy of the dominant "red" allele. The male offspring, in contrast, can only inherit the recessive "white" allele from the female parent. All the male offspring, as a result, have white eyes.

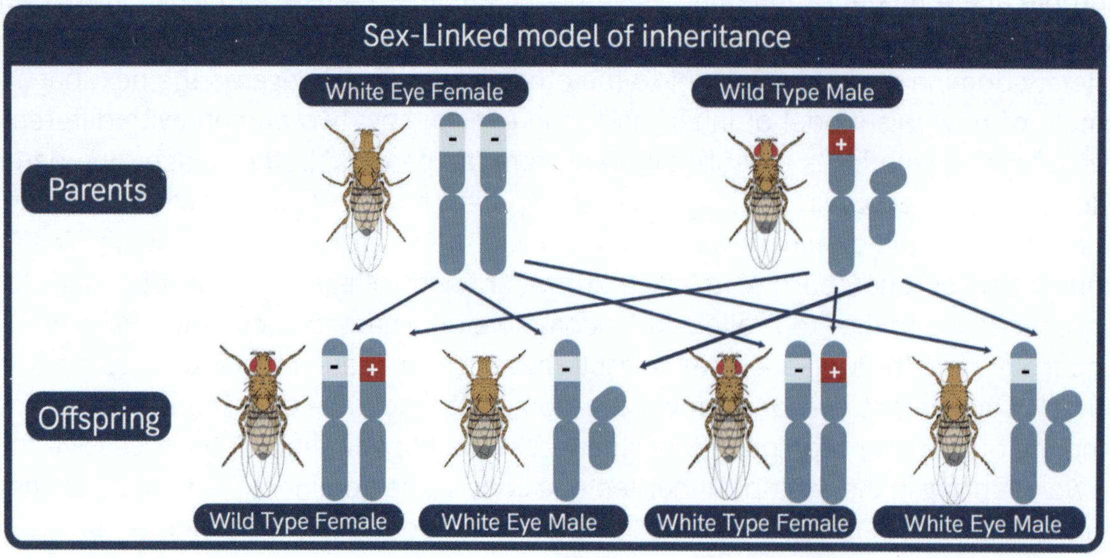

Some **useful ideas** from what I read...

Some Ideas You Can Use: Using Patterns to Identify Cause-and-Effect Relationships

A pattern is anything that happens in a predictable way. Biologists can use patterns to help identify a cause-and-effect relationship, such as how two alleles interact to produce a specific trait in an organism, because they can predict the percentage of offspring that will inherit a specific trait given each model of inheritance. For example, if a trait such as bristle length in fruit flies follows the incomplete dominance pattern, then when a "wild type" fly with long bristles is mated with a "shaven" fly, which is a fly that lacks bristles, we can predict that all the offspring of this mating will have "stubble" or short bristles. Similarly, biologists can figure out what alleles an organism carries, even if they cannot see them, by looking for patterns in the traits of their offspring. For example, if a "wild type" fly is mated with an "ebony" body fly, biologists can figure out if the "wild type" fly also carried an "ebony" body allele. If the "wild type" fly does not carry an ebony allele, then all of its offspring will be wild type; however, if some of its offspring are born with an ebony body, then it must have carried an allele for the ebony body trait. Biologists use Punnett squares to make predictions about the likelihood of different traits showing up in the next generation based on a specific model of inheritance and the possible alleles carried by the parents (see image below).

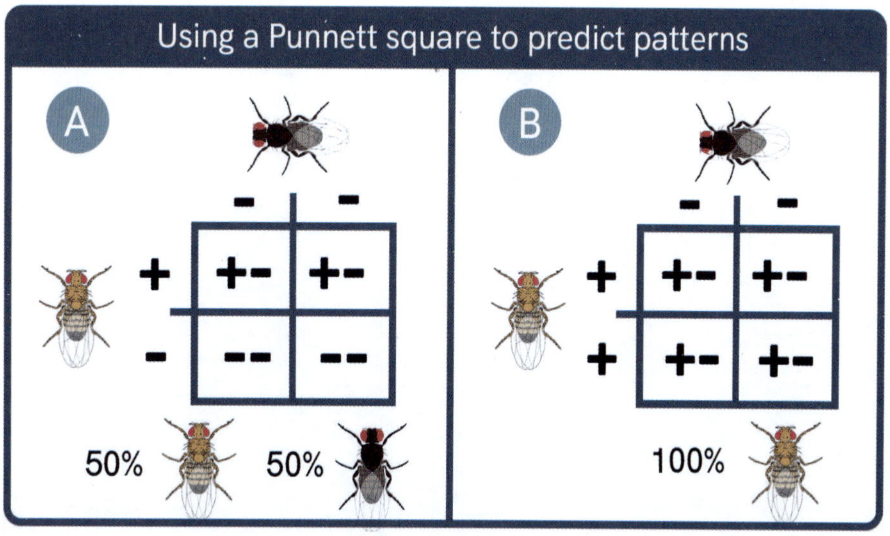

In this investigation you will use a computer-based simulation to breed fruit flies so you can figure out how the yellow body color and curly wing traits are inherited in fruit flies. This simulation, which is shown in the image below, allows you to pick flies with specific traits. You can also pick the sex of each fly. You can then mate the flies and see the traits of all the resulting offspring. Each mating will result in about 1000 offspring, which is more than the 500 eggs that a female fruit fly will typically lay after mating, so it easier to determine if a pattern you uncover is meaningful or if it is only due to chance.

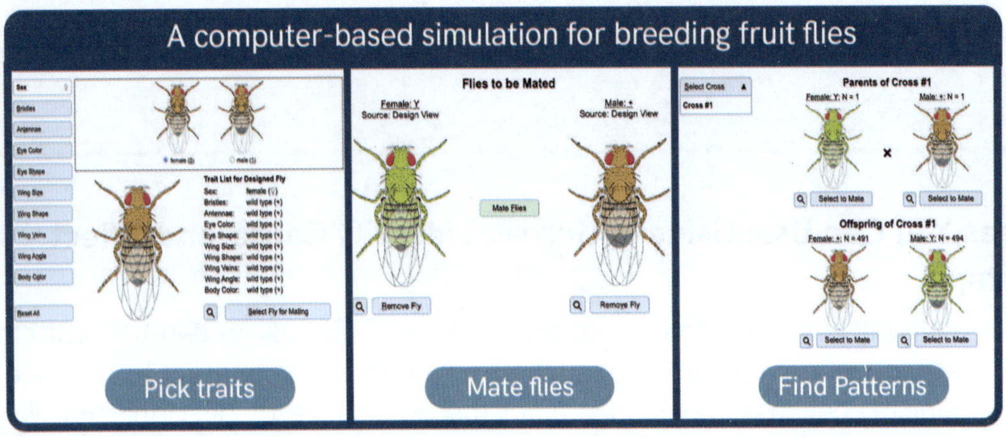

Some **useful ideas** from what I read...

Fruit Fly Body Color and Wing Shape

PLAN

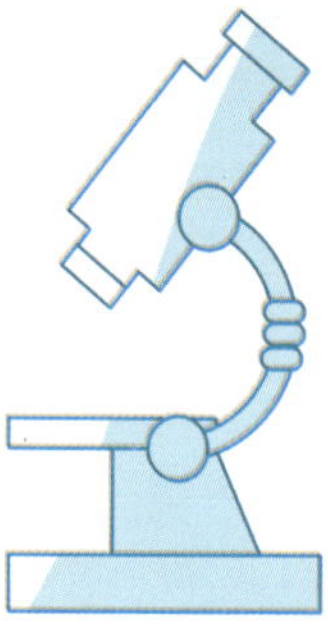

Plan Your Investigation

Prepare a plan for your investigation by filling out the chart below.

Question...

Hypothesis A | Hypothesis B | Hypothesis C

If... | **If...** | **If...**

The Test

And...

Predicted Result A | Predicted Result B | Predicted Result C

Then... | **Then...** | **Then...**

I approve of this investigation plan.

Teacher's signature ____________________ Date ____________

Fruit Fly Body Color and Wing Shape

DO

Collect Data

Keep a record of what you observe or measure during your investigation.

My **observations or measurements...**

Analyze Your Data

Use the space below to create any tables or graphs that will help you make sense of the data you collected so you can answer the guiding question.

Fruit Fly Body Color and Wing Shape

SHARE

Draft Argument

Develop an argument on a whiteboard. It should include:

1. A *claim* that answers the guiding question,
2. *Evidence* in support of the claim that provides an analysis of the data and a description of what the analysis means,
3. A *justification of the evidence* that explains why your group thinks the evidence is important.

Our Guiding Question:	
Claim:	
Evidence:	Justification of the Evidence:

Argumentation Session

Share your argument with your classmates. In the space below, be sure to keep track of any ideas that you can use to revise your argument and make it better.

Some possible ways to **improve** our argument...

Fruit Fly Body Color and Wing Shape REFLECT

Reflective Discussion

In the space below, you can keep track of any ideas from the discussion that you think are important or will be useful in the future.

Some important **ideas**...

Fruit Fly Body Color and Wing Shape

REPORT

Draft Report

Prepare a report to share what you figured out during your investigation.

Introduction

We have been studying ______________________________ in class. Before we started this investigation, we explored ______________________________

____

____

We noticed ______________________________

____

____

Our goal for this investigation was to figure out ______________________________

____

____

The guiding question was ______________________________

____

Method

To answer this question, ______________________________

____

____

____

____

____

____

____

____

____

____

Argument

We figured out ______________________________

____

____

____

The ______________ below includes information about ______________

____

____

____

This analysis suggests ______________________________

____

____

____

This evidence is based on several important scientific ideas. The first one is ______________

____

____

____

____

____

Antibiotic Resistance in Bacteria Lesson Plan

OVERVIEW

Description	Students figure out why some bacteria populations develop antibiotic resistance more quickly than other populations.
Guiding Question	Why do some populations of bacteria develop antibiotic resistance faster than others?
What Students Figure Out	A fast growth rate coupled with high mutation leads to faster evolution in bacteria.

STAGES

This investigation has seven stages. Each stage includes multiple activities. The flowchart below shows what happens during each stage and how much time is needed to complete each one.

Task (Day 1)
1 **20 minutes**
Students are introduced to a phenomenon and a guiding question to answer. They also share their initial ideas.

Ideas (Day 1)
2 **30 minutes**
Students read about DCI, a CC, and an SEP that they can use to help them figure out the phenomenon.

Plan (Day 2)
3 **15 minutes**
Students formulate a plan collecting and analyzing the data they need to answer the guiding question.

Do (Day 2)
4 **30 minutes**
Students carry out their plan, make sense of the data they collected, and answer the guiding question.

Share (Day 3)
5 **45 minutes**
Students develop, share, and revise an evidence-based argument about what they figured out.

Reflect (Day 4)
6 **20 minutes**
Students discuss a DCI and/or a CC and reflect on the SEPs they used during the investigation.

Report (Day 4)
7 **25 minutes**
Students write investigation reports to share what they figured out during the investigation.

Report (Day 5)
7 **45 minutes**
Students peer review the reports written by other students and offer feedback to improve.

Report (Day 6)
7 **30 minutes**
Students revise their reports and then submit a final draft to their teacher for evaluation.

STANDARDS ALIGNMENT

NEXT GENERATION SCIENCE STANDARDS

Performance Expectation

1. Construct an explanation based on evidence that the process of evolution primarily results from four factors: the potential for a species to increase in number, the heritable genetic variation of individuals in a species due to mutation and sexual reproduction, competition for limited resources, and the proliferation of those organisms that are better able to survive and reproduce in the environment (HS-LS4-2).
2. Construct an explanation based on evidence for how natural selection leads to the adaptation of populations (HS-LS4-4).

Core Ideas(s)	Crosscutting Concept(s)	Practices
Natural Selection (LS4.B) Adaptation (LS4.C)	Cause and Effect (CC2) Systems and System Models (CC4)	Asking Questions (SEP1) Developing and Using Models (SEP2) Planning and Carrying Out Investigations (SEP3) Analyzing and Interpreting Data (SEP4) Using Mathematics and CT (SEP5) Constructing Explanations (SEP6) Engaging in Argument from Evidence (SEP7) Obtaining, Evaluating, and Communicating Information (SEP8)

WIDA ENGLISH LANGUAGE DEVELOPMENT STANDARDS FOR SCIENCE

Explain	Argue
Interpret scientific explanations (ELD-SC.9-12.Explain.Interpretive) Construct scientific explanations (ELD-SC.9-12.Explain.Expressive)	Interpret scientific arguments (ELD-SC.9-12.Argue.Interpretive) Construct scientific arguments (ELD-SC.9-12.Argue.Expressive)

COMMON CORE ENGLISH LANGUAGE ARTS

Reading	Writing	Speaking and Listening
Key ideas and details (RST.9-10.1 and RST.9-10.2) Craft and structure (RST.9-10.4) Integration of knowledge and ideas (RST.9-10.7, RST.9-10.8, and RST.9-10.9) Range of reading (RST.9-10.10)	Text types and purposes (WHST.9-10.1 and WHST.9-10.2) Production and distribution of writing (WHST.9-10.4, WHST.9-10.5, and WHST.9-10.6)	Comprehension and collaboration (SL.9-10.1 and SL.9-10.3) Presentation of knowledge and ideas (SL.9-10.4 and SL.9-10.6)

THINGS TO KNOW BEFORE STARTING THE LESSON

The Core Ideas

Evolution describes a change in the traits of a population over time. One mechanism through which evolution occurs is natural selection. Natural selection is the process by which organisms with more advantageous traits for their environment survive and reproduce, causing a larger share of organisms to have that trait and therefore be better adapted to their environment. Natural selection is a consequence of variations of traits within a population, selective pressure within an environment, traits that can be passed onto a new generation via reproduction, and the proliferation of organisms that are better able to reproduce in that environment. However, natural selection can only occur if there is variation in the genetic information between organisms in a population as well as variation in the expression of that genetic information.

Natural selection, and therefore evolution, relies on reproduction. Reproduction is the process by which a new organism is created from one or more existing organisms. Asexual reproduction occurs when a single organism divides into two separate organisms. These new organisms receive an exact copy of the organism's original genetic information. However, this genetic information can change over time due to mutations, or random mistakes that occur during replication. Sexual reproduction, on the other hand, describes when two organisms each contribute half of their genetic information to a single offspring. In sexual reproduction, genetic information can vary due to different combinations of genetic information as well as due to mutations.

The Crosscutting Concepts

A system is a group of components that interact with each other to achieve some function. Because some systems are too complex to work with directly, scientists create models of the system so that they can test these models and use them to identify cause-and-effect relationships. However, because these models simplify complex systems, they inherently have a limited reliability and precision.

The Science and Engineering Practices

A practice of science is to **ask and refine questions** that lead to descriptions and explanations of how the natural world works. Another practice of science is to **use and construct models** as helpful tools for representing ideas and explanations. These tools include diagrams, drawings, physical replicas, mathematical representations, analogies, and computer simulations. Scientists **plan and carry out investigations** in the field or a laboratory, working collaboratively and individually. Their investigations are systematic and require clarifying what counts as data and identifying variables or parameters. Scientific investigations produce **data that must be analyzed and interpreted**. Because data patterns and trends are not always obvious, scientists use a range of tools and practices to identify the significant features and patterns in the data. Scientists also use **mathematics and computational thinking** for a range of tasks such as constructing simulations; statistically analyzing data; and recognizing, expressing, and applying quantitative relationships. The products of doing science are **explanations**. Scientific explanations often specify variables and are used by scientists to describe and predict different natural phenomena. **Scientific argumentation**, or the process of sharing, supporting, critiquing, and revising claims based on evidence is the process that scientists use to determine if an explanation is valid or acceptable. Finally, scientists must be able to **communicate clearly and persuasively** the ideas and methods they generate. Critiquing and communicating ideas individually and in groups is a critical professional activity.

PREPARATION

This lesson can be delivered paper-based, paperless, or remote. Use the paper-based delivery option of the investigation if students do not have 1:1 access to a computer and they are in the same classroom or physical space during the lesson. Use the paperless delivery option of the investigation if students have 1:1 access to a computer and they are in the same classroom or physical space during the lesson. Finally, if students are not in the same classroom or physical space during the lesson and have 1:1 access to a computer, use the remote delivery option.

Materials Needed

The materials needed by the teacher and each student in the class for each of the three modalities are listed in the table below.

Teacher Notes

Delivery Mode	Each Student Will Need	The Teacher Will Need
Paper-Based	• Copies of each student handout for the investigation (or a workbook) • A pencil	• A teacher account in the ADI Learning Hub • Access to a computer with high-speed internet and a browser (such as Chrome, Safari, Firefox, or Edge) • A projector or smart board (for displaying a computer screen)
Paperless	• A student account in the ADI Learning Hub • Access to a computer with high-speed internet and a browser (such as Chrome, Safari, Firefox, or Edge)	• A teacher account in the ADI Learning Hub • Access to a computer with high-speed internet and a browser (such as Chrome, Safari, Firefox, or Edge)
Remote	• A student account in the ADI Learning Hub • Access to a computer with high-speed internet and a browser (such as Chrome, Safari, Firefox, or Edge)	• A teacher account in the ADI Learning Hub • Access to a computer with high-speed internet and a browser (such as Chrome, Safari, Firefox, or Edge)

Additional Materials and Preparation Needed for In-Person Delivery

This in-person investigation will require students to have access to a Chromebook, desktop, laptop, iPad, or other electronic device in order to use a simulation. The simulation is embedded in the Learning Hub, so students do not need access to external sites. If students do not have access to the Learning Hub, they can use the simulation by visiting *Evolution of Antibiotic Resistance* in NetLogo. The teacher can determine if each student will use an electronic device or if each group will share.

How the Simulation Works

This simulation replicates the development of antibiotic resistance in a population of bacteria when they are placed in a giant petri dish with different doses of antibiotics. The antibiotic concentrations are set at low, medium, high, and peak levels from two sides to the center of the model. The simulation starts with the bacteria having a resistance level between one and two. Bacteria are color coded based on the antibiotic resistance level. Gray bacteria have low resistance, green bacteria have medium resistance, yellow bacteria have high resistance, and red bacteria are immune (which are called superbugs). The bacteria reproduce asexually, and users may adjust the growth rates of the various strains of bacteria. At a probability set by the mutation rate, the offsprings' resistance to antibiotics may increase or decrease within the mutation range.

Time passes in the simulation as ticks. At each tick, all bacteria in the petri dish will move a small distance in a random direction, and the nutrients needed for the bacteria to survive will be refilled in an area. It is possible for the nutrients in an area to run out and lead to the death of the bacteria in the area.

Lastly, the simulation has two views. The bacteria view gives a more realistic view of the petri dish, but the icon view allows the simulation to run faster.

How to Use the Simulation

The simulation interface is shown in the image on the next page.

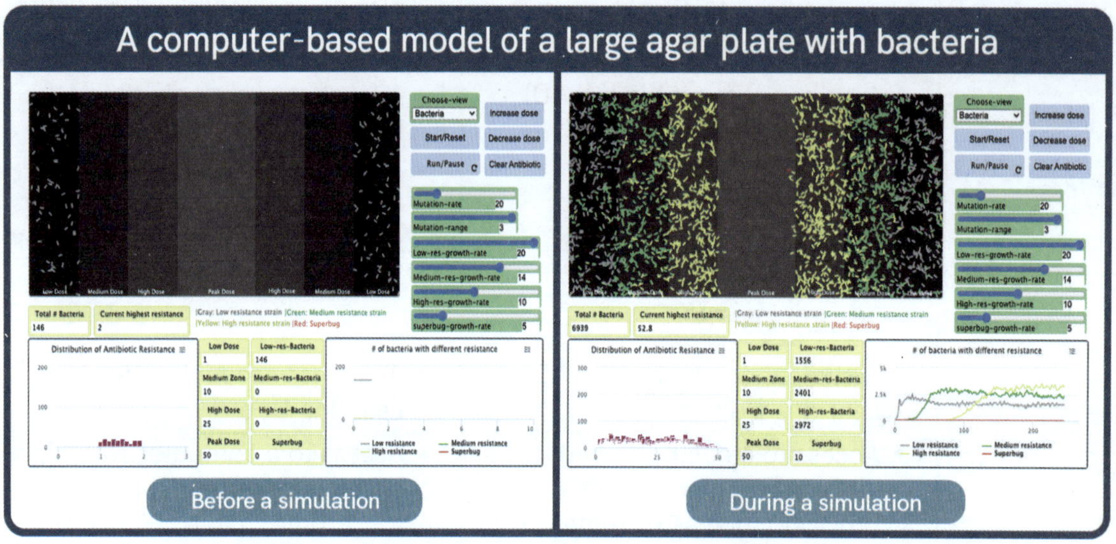

To start the simulation, click on the **Start/Reset** button to add bacteria to the petri dish. Click on the Run/Pause button to start the simulation. Users can change the dosage of the antibiotic on the petri dish while the simulation runs by clicking on the **Increase Dose** button, the **Decrease Dose** button, or the **Clear Antibiotic** button. Click on the **Run/Pause** button to pause the simulation. Users can then use the slider Mutation-rate to change how often mutations occur in the population of bacteria and the slider **Mutation-range** to adjust how much a mutation affects antibiotic resistance. Users can also change the growth rate of bacteria with low resistance, medium resistance, high resistance, and immunity to the antibiotic using the sliders. Additionally, the yellow meters provide information about the current antibiotic dosage in each section of the petri dish and information about the bacteria in the dish. One graph allows users to see the current distribution of antibiotic-resistant levels within the entire population. The other graph shows how the number of bacteria within each level of antibiotic resistance changes over time. If users would like to collect a new round of data, they can click on the **Start/Reset** button to reset the simulation with bacteria that have different characteristics.

Safety Precautions

Remind students to follow all normal lab safety rules.

Laboratory Waste Disposal

There is no laboratory waste generated during this investigation.

FACILITATION NOTES

The facilitation notes for this investigation describe how to deliver this lesson in person. They are written specifically for the paper-based mode of delivery. These same facilitation notes, however, can also be used for a paperless delivery of the lesson by simply substituting the directions that mention paper with the entry into the ADI Learning Hub. If you are teaching this lesson using the remote mode of delivery, you only need to assign due dates and prerequisites for each stage. The students can then simply follow the directions built into the ADI Learning Hub as they move through the investigation.

ADI Go Digital

The digital version of this lesson is available in the **ADI Learning Hub**. Be sure to check out the built-in teacher presentation mode.

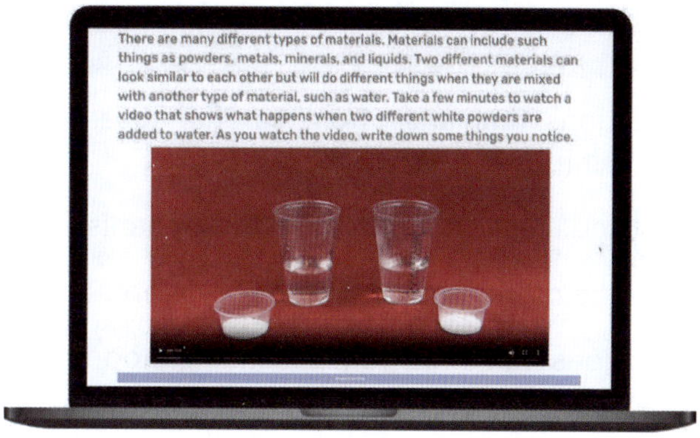

Teacher Notes

Antibiotic Resistance in Bacteria Lesson Plan

Stage 1 TASK

Total Time for the Stage	45 minutes
Grouping	Individual; Teams

Learning Goals

Students should make progress on their ability to do the following by the end of this stage of the lesson:

Science

1. Ask questions that arise from careful observation of phenomena to clarify and/or seek additional information (SEP1).
2. Develop and use a model to provide mechanistic accounts of phenomena (SEP2).
3. Evaluate the merits and limitations of different models of the same phenomenon (SEP2).
4. Ask questions that arise from examining models to clarify and/or seek additional information and relationships (SEP 1).
5. Ask questions to clarify and refine a model (SEP1).

English Language Development

None

English Language Arts

1. Initiate and participate effectively in small-group collaborative discussions with diverse partners by building on others' ideas and expressing their own clearly (SL.9-10.1).
2. Present information clearly, concisely, and logically such that listeners can follow the line of reasoning and the organization, substance, and style are appropriate to the purpose, audience, and task (SL.9-10.4).
3. Adapt speech to a variety of contexts and tasks, demonstrating command of formal English when indicated or appropriate (SL.9-10.6).

Activities within the Stage

Activity A	The Phenomenon
Time Needed for the Activity	5 minutes
Grouping	Teams

What to Do

1. Have students work in their assigned teams for this activity.
2. Begin the activity by giving out a copy of the Task handout to each student.
3. Show "The Phenomenon" page on a screen using the Presentation mode in the Learning Hub so everyone in the classroom can see it.
4. Read the introduction of the phenomenon out loud and then start the video. You can also have students read the introduction of the phenomenon and watch the video on their own devices.
5. Give the students a few moments to write down their observations on their handouts before moving on to the next activity.

What to Give Out

Task handout for **Antibiotic Resistance in Bacteria**

What to Show

"The Phenomenon" page in the ADI Learning Hub

Teaching Tip

Instead of assigning fixed roles to students by name at the beginning of the investigation, give each member of a group a different number so you can easily assign and change roles as needed during the various investigation stages. For example, you might give directions such as "1s will share out for your group" or "3s, will you please come back here and pick up these materials for your group?" or "2s and 3s, you will present the argument for your group while 1s and 4s travel to the other groups to offer feedback."

Attending to Equity

Be sure to intentionally group emerging multilingual students at times with peers who know the same languages as they do and at other times with peers whose English language development is slightly more advanced. Thoughtful grouping that varies throughout an investigation allows emerging multilingual students to benefit from the different linguistic resources of their peers.

Activity B	Share What You Noticed
Time Needed for the Activity	5 minutes
Grouping	Teams

What to Do

1. Show the "Share What You Noticed" page on a screen using the Presentation mode in the Learning Hub so everyone in the classroom can see it.
2. Say, "Okay, please share what you noticed with the others at your table."
3. Give the students about two minutes to share at their tables.
4. Say, "Okay, let's take a few minutes to share with the class. Team X (call on a team at random), what did you talk about?"
5. Allow Team X to share something they noticed. If it is related to natural selection or reproduction, say, "That is really interesting. We are actually going to examine that in more detail during this investigation." If what is shared is not related to natural selection or reproduction, say, "That is really interesting. Team Y (call on another team at random), what did you notice?" and allow that team to share. Follow this pattern until a team mentions something related to why the *E. coli* bacteria developed resistance to antibiotics so quickly. If no team mentions this, say, "You know what I found interesting? I saw that the *E. coli* bacteria developed antibiotic resistance very quickly, and I wondered if other bacteria could develop antibiotic resistance at that rate. Did anyone notice that as well?"
6. Move on to the next activity.

What to Show

The "Share What You Noticed" page in the ADI Learning Hub

What to Look For

As soon as a group mentions something related to why the *E. coli* bacteria developed resistance to antibiotics so quickly, move on to the next activity.

Attending to Equity

Be sure to provide students with opportunities to talk in small groups during this activity. Small-group structures offer all students, including emerging multilingual students, a chance to engage in sense-making with their peers and offers them space to use their verbal and nonverbal (e.g., writing, drawing pictures, and using gestures) resources to express their ideas. Small-group structures also give students a chance to learn from other students' uses of these resources.

Activity C	What You Need to Figure Out
Time Needed for the Activity	5 minutes
Grouping	Teams

What to Do

1. Show the "What You Need to Figure Out" page on a screen using the Presentation mode in the Learning Hub so everyone in the classroom can see it.
2. Read the task out loud and have students follow along on their handout, on their devices, or on the screen.
3. Allow students to ask clarifying questions about the task or the guiding question.
4. Once everyone understands what they will need to figure out during the investigation, move on to the next activity.

What to Show

The "What You Need to Figure Out" page in the ADI Learning Hub

Teaching Tip

Have students highlight, circle, or place a star next to the guiding question. Marking the guiding question helps students find it later during the Plan stage.

Teacher Notes

Activity D	Share Your Ideas
Time Needed for the Activity	15 minutes
Grouping	Teams

What to Do

1. Show the "Share Your ideas" page on a screen using the Presentation mode in the Learning Hub so everyone in the classroom can see it.
2. Read the directions on the handout out loud and then give students five minutes to draw their initial representations of the science.
3. Say, "Okay, will everyone share their representations with the rest of your team? Please make sure everyone has time to talk. Also, be sure to ask one another questions about the scientific ideas that are included in each representation."
4. Give the students about 8-9 minutes to talk and share. Be sure to walk around and listen to the discussions.
5. Say, "Okay, I heard some interesting ideas as I walked around the room."
6. Move on to the next stage.

What to Show

The "Share Your Ideas"page in the ADI Learning Hub

What to Look For

Each student's initial representation should include a combination of symbols, diagrams, graphs, or words. It should also clearly show how a scientific idea can be used to help complete the task and answer the guiding question.

Attending to Equity

When students are tasked with expressing their ideas, be sure to encourage them to do so through multiple modes (speaking, writing, drawings, and gestures) and to use both content-specific and everyday registers. This process helps students fully express themselves, allowing peers to evaluate, question, and build on ideas. Also, make sure that every student in every team has a chance to share their initial representation.

Activity E	More Questions
Time Needed for the Activity	15 minutes
Grouping	Teams

What to Do

1. Show the "More Questions" page on a screen using the Presentation mode in the Learning Hub so everyone in the classroom can see it.
2. Say, "Okay, I heard a bunch of interesting questions as I was walking around the room. Will you work with your team to identify two or three questions that you think are important to learn more about that came up as you all discussed the different representations you made? You can write these questions on your handout."
3. Give students about five minutes to talk and write down questions. Be sure to walk around and listen to the discussions.
4. Say, "Okay, let's take a few minutes to share. Team X (call on a team at random), will you share one question you all have?"
5. Have each team share a question. Then say, "I want to share information about some useful scientific ideas with you. I bet this information will not only answer many of these questions but will also help you complete the task and answer the guiding question of the investigation."
6. Move on to the next stage.

What to Show

The "More Questions" page in the ADI Learning Hub

What to Look For

Each student should have 2-3 questions written on their handout.

Teaching Tip

When students are asked to create representations of the underlying scientific ideas, share these ideas with their peers, and then identify things they want to know more about, it sparks their curiosity and helps make the investigation meaningful. Also, identifying questions before reading (which students do during the Ideas stage) is an outstanding pre-reading activity that helps improve comprehension.

Antibiotic Resistance in Bacteria Lesson Plan

Stage 2 IDEAS

Total Time for the Stage	30 minutes
Grouping	Teams

Learning Goals

Students should make progress on their ability to do the following by the end of this stage of the lesson:

Science

1. Ask questions that arise from examining theory to clarify and/or seek additional information and relationships (SEP1).
2. Critically read scientific literature adapted for classroom use to determine the central ideas or conclusions and/or to obtain scientific and/or technical information to summarize complex concepts, processes, or information presented in a text by paraphrasing them in simpler but still accurate terms (SEP8).
3. Integrate sources of information presented in different media or formats (e.g., visually and quantitatively) as well as in words to address a scientific question (SEP8).
4. Gather, read, and evaluate scientific and/or technical information from multiple authoritative sources, assessing the evidence and usefulness of each source (SEP8).
5. Describe how natural selection occurs only if there is both variation in the genetic information between organisms in a population and variation in the expression of that genetic information that leads to differences in performance among individuals (LS4.B).
6. Identify how evolution is a consequence of four factors: the potential for species to increase in number, the genetic variation of individuals in a species, competition for an environment's limited resources, and the ensuing proliferation of the organisms that are better able to survive and reproduce in that environment (LS4.C).
7. Describe how natural selection leads to adaptation, or to a population dominated by organisms that are anatomically, behaviorally, and physiologically well suited to survive and reproduce in a specific environment (LS4.C).
8. Explain how organisms in a population may have an advantageous heritable trait, which leads to an increase in the proportion of individuals in future generations that have the trait (LS4.C).
9. Use empirical evidence to differentiate between cause and correlation in order to make claims about specific causes and effects (CC2).
10. Use models to simulate systems and interactions within and between systems at different scales (CC4).

English Language Development

1. Paraphrase central ideas in complex evidence, concepts, processes, and information to help explain how or why a phenomenon occurs (ELD-SLC.9-12.Explain.Interpretive).

English Language Arts

1. Determine the central ideas or conclusions of a text; trace the text's explanation or depiction of a complex process, phenomenon, or concept; provide an accurate summary of the text (RST.9-10.2).
2. Determine the meaning of symbols, key terms, and other domain-specific words and phrases as they are used in a specific scientific or technical context (RST.9-10.4).
3. Read and comprehend scientific/technical texts (RST.9-10.10).
4. Initiate and participate effectively in small-group collaborative discussions with diverse partners by building on others' ideas and expressing their own clearly (SL.9-10.1).

Teacher Notes

Activities

Activity A	Read About a Core Idea
Time Needed for the Activity	10 minutes
Grouping	Teams

What to Do

1. Have students work in their assigned teams during this stage. Give out a copy of the Ideas handout to each student.
2. The first activity provides students with information about evolution via natural selection.
3. Show the "Read About a Core Idea You Can Use" page on a screen using the Presentation mode in the Learning Hub so everyone in the classroom can see it. You can also click on the images to make them full screen and easier for students to see.
4. Have students read the text on their own, either on their handouts or on the screen.
5. Give the students about three minutes to discuss the text with the others in their teams and develop a shared list of ideas that they think are important or useful.
6. Ask a team to share an idea they thought was useful. Have the remaining teams share a different idea until there are no more ideas to share.
7. Move on to the next activity.

Teaching Tip

Students do not need to do a close read of this text or completely understand all the ideas in it before moving on to the next activity. They will be encouraged to revisit this text later in the investigation when creating their arguments and writing their reports. All students need to be able to do during this activity is to work with the other members of their group to identify one or two ideas that they think are important to keep in mind or are potentially helpful.

What to Give Out

Ideas handout for **Antibiotic Resistance in Bacteria**

What to Show

The "Read About a Core Idea You Can Use" page in the ADI Learning Hub

What to Look For

Students should write down some of the following ideas in the box labeled, "Some things I know from what I read" on their handouts:

- Evolution is the change in the traits of a population over time.
- Natural selection is a mechanism through which evolution can occur.
- Evolution via natural selection is a consequence of four factors: variations in a trait within a population, selective pressure in the environment that prevents some individuals from reproducing, the individuals able to reproduce passing their traits onto the next generation, and the proliferation of those organisms that are better able to reproduce in that environment.
- Natural selection leads to the development of specific adaptations, or to a population dominated by organisms that are anatomically, behaviorally, and physiologically well suited to reproduce within a specific environment.
- Natural selection can only occur if there is both variation in the genetic information of organisms in a population and variation in the expression of that genetic information.

Activity B	Read About a Core Idea
Time Needed for the Activity	10 minutes
Grouping	Teams

What to Do

1. The second activity provides students with information about reproduction in bacteria.
2. Show the "Read About Another Core Idea You Can Use" page on a screen using the Presentation mode in the Learning Hub so everyone in the classroom can see it. You can also click on the images to make them full screen and easier for students to see.
3. Have students read the text on their own, either on their handouts or on the screen.
4. Give the students about three minutes to discuss the text with the others in their teams and develop a shared list of ideas that they think are important or useful.
5. Ask a team to share an idea they thought was useful. Have the remaining teams share a different idea until there are no more ideas to share.
6. Move on to the next activity.

What to Show

The "Read About Another Core Idea You Can Use" page in the ADI Learning Hub

What to Look For

Students should write down some of the following ideas in the box labeled, "Some things I know from what I read" on their handouts:

- Reproduction is the process by which one or more existing organisms create a new organism.
- Asexual reproduction occurs when a single organism divides into two new organisms.
- Organisms born via asexual reproduction receive an exact copy of its parents' genetic material.
- Sexual reproduction occurs when two organisms donate half of their genetic material to their offspring.
- Bacteria reproduce asexually, and the time it takes for bacteria to reproduce is called its growth rate.
- The growth rate of bacteria varies by species and is dependent on environmental conditions such as food, space, and other resources.
- Bacteria duplicate their genetic information and some cellular structures and then split into two new organisms in what is known as binary fission.
- Bacteria experience changes in their genetic information due to mutations, or random changes that occur during the replication of their genetic material.
- The mutation rate for an organism describes how often a mutation occurs when an organism duplicates their genetic information.

Attending to Equity

There are many supports for helping students comprehend what they read already embedded into this activity (e.g., activating prior knowledge, providing a shared experience, making connections, synthesizing, and talking with peers). So, you might not need to provide much extra support. If you are concerned about some students not understanding what they are reading because of their scores on past reading comprehension tests, you can read it out loud as they follow along. As you read the text out loud, be sure to stop at each important idea and ask the students to put a star (or other annotation) next to it in the margin of their handouts. They can then discuss these ideas in their small groups.

Teacher Notes

Activity C	Read About a Final Core Idea
Time Needed for the Activity	10 minutes
Grouping	Teams

What to Do

1. The third activity provides students with information about using models to represent a system.
2. Show the "Read About a Final Core Idea You Can Use" page on a screen using the Presentation mode in the Learning Hub so everyone in the classroom can see it. You can also click on the images to make them full screen and easier for students to see.
3. Have students read the text on their own, either on their handouts or on the screen.
4. Give the students about three minutes to discuss the text with the others in their teams and develop a shared list of ideas that they think are important or useful.
5. Ask a team to share an idea they thought was useful. Have the remaining teams share a different idea until there are no more ideas to share.
6. Move on to the next activity

What to Show

The "Read About a Final Core Idea You Can Use" page in the ADI Learning Hub

What to Look For

Students should write down some of the following ideas in the box labeled, "Some things I know from what I read" on their handouts:

- Models are used to represent an object, event, or process.
- Models can be physical, visual, or a computer-based simulation.
- Models can be used to identify cause-and-effect relationships.
- Models have limited reliability and precision due to the approximations and assumptions used to create the model.

Attending to Equity

To support emerging multilingual students during this activity, be sure to create a word bank on the wall of high-frequency, high-need concrete vocabulary that is needed to communicate scientific ideas. The word bank highlights terms needed for basic communication in an academic context. It also helps with the spelling. You can also encourage your emerging multilingual students to:

1. Use strategic learning techniques such as concept mapping, drawing, memorizing, comparing, contrasting, and reviewing.
2. Decode (sound out) words using skills such as recognizing sound-letter relationships and identifying cognates, affixes, roots, and base words.
3. Use visual and contextual support and support from peers and teachers to enhance and confirm understanding of the texts.

Antibiotic Resistance in Bacteria Lesson Plan

Stage 3 PLAN

Total Time for the Stage	15 minutes
Grouping	Teams

Learning Goals

Students should make progress on their ability to do the following by the end of this stage of the lesson:

Science

1. Evaluate a question to determine if it is testable and relevant (SEP1).
2. Plan an investigation individually and collaboratively to produce data to serve as the basis for evidence as part of building and revising models (SEP3).
3. Consider possible variables or effects and evaluate the confounding investigation's design to ensure variables are controlled (SEP3).
4. Ask questions to determine relationships, including quantitative relationships, between independent and dependent variables (SEP1).
5. Decide on types, quantity, and accuracy of data needed and consider limitations on the precision of the data (e.g., number of trials, cost, risk, and time), and refine the design accordingly (SEP3).
6. Select appropriate tools to collect, record, and analyze data (SEP3).

English Language Development

1. Define investigable questions on observations, information, and/or data about a phenomenon (ELD-SLC.9-12.Explain.Interpretive).
2. Develop reasoning to illustrate and/or predict the relationships between variables (ELD-SLC.9-12.Explain.Expressive).

English Language Arts

1. Initiate and participate effectively in small-group collaborative discussions with diverse partners by building on others' ideas and expressing their own clearly (SL.9-10.1).
2. Present information clearly, concisely, and logically such that listeners can follow the line of reasoning and the organization, substance, and style are appropriate to the purpose, audience, and task (SL.9-10.4).
3. Adapt speech to a variety of contexts and tasks, demonstrating command of formal English when indicated or appropriate (SL.9-10.6).

Activities

Activity A	Some Materials You Can Use
Time Needed for the Activity	5 minutes
Grouping	Teams

What to Do

1. Show the "Some Materials You Can Use" page on a screen using the Presentation mode in the Learning Hub so everyone in the classroom can see it. You can also click on the images to make them full screen and easier for students to see.
2. Say, "Okay, it sounds like we now have enough information to plan an investigation that will help us figure out why some bacteria populations become resistant to antibiotics more quickly than others."
3. Say, "Before we get started, I want to show you some materials that we are going to use during our investigation. Will you all please look at the screen so I can show you the simulation we will be using?"
4. Direct students to look at the screen where the simulation is displayed.
5. Ask, "What do you think we could use this simulation for?"
6. Allow students to share ideas. Then say, "We can use this simulation to test what causes some bacteria populations to acquire antibiotic resistance faster than others."
7. Say, "Okay, that is all the stuff we can use. Will you all head back to your teams so we can plan our investigation?"
8. Move on to the next activity.

What to Set Up

Pull up the simulation in the Learning Hub on a screen so that all students can see it. If you do not have a Learning Hub subscription, you can access the simulation by visiting *Evolution of Antibiotic Resistance in NetLogo.*

What to Show

The "Create a Final Report" page in the ADI Learning Hub

Teaching Tip

This activity helps students see for themselves all the materials they will be able to use during the hands-on part of the investigation instead of just reading about them or looking at pictures. It is important to show each material to the students and then point out how they can use it during the investigation so they can participate in developing a plan for the investigation during the next activity.

Activity B	Make a Plan
Time Needed for the Activity	10 minutes
Grouping	Teams

What to Do

1. Have students work in their assigned teams during this activity. Give a copy of the Plan handout to each student.
2. Show the "Make a Plan" page on a screen using the Presentation mode in the Learning Hub so everyone in the classroom can see it.
3. Say, "Work as a team to create a plan for your investigation by filling out the chart on your handout. As part of your plan, be sure to include which tools and techniques you will use to solve problems."
4. Give the students at least 10 minutes to complete this task. As students work in small teams to create their investigation proposals, be sure to move from team to team to answer questions and offer advice.
5. When a team thinks that they have a good plan, read it over. If you agree, sign off at the bottom of their handouts to let them know that their plan has been approved. If you do not agree, tell them what they will need to add or change before you will approve it. Once they have made the changes you can sign their handouts.
6. Once all the teams are done, you can move on to the next activity. You can also use an approved investigation plan as an "exit ticket" out of class.

What to Give Out

Plan handout for **Antibiotic Resistance in Bacteria**

What to Show

The "Make a Plan" page in the ADI Learning Hub

What to Look For

The investigation plan should, at a minimum, include:

- The question they are trying to answer.
- What conditions they will test using the simulation.
- What data they will collect from the simulation.
- How they will analyze the data they collect.
- How they will use their analysis to answer the guiding question.

Attending to Equity

To support emerging multilingual students during this stage, be sure to create a word bank on the wall of high-frequency, high-need concrete vocabulary that is needed to communicate scientific processes and ideas. The word bank highlights terms needed for basic communication in an academic context. It also helps with the spelling. Also, you can encourage your emerging multilingual students to use modes (speaking, writing, drawings, and gestures) and to use both content-specific and everyday registers to discuss and create their plans. This process helps students fully express themselves, allowing peers to evaluate, question, and build on ideas.

Teaching Tip

One way to do a quick check-in with a group during this activity is to use the "two questions, a nudge, and a goal" technique. This technique includes the following steps:

1. Sit down with a group of students.
2. Ask, "Would one of you tell me a little about what you all have been doing or thinking about?" (Question 1)
3. A student answers on behalf of the group.
4. Ask, "Would anyone else like to add anything to that?" (Question 2)
5. A different student provides additional information on behalf of the group.
6. Ask, "Have you thought about [a hint or something you want to highlight]?" (A nudge)
7. Say, "Okay, I'm going to check on the other groups. When I get back, I want to see [something that you want the group to accomplish]." (A goal)

When you have completed the check-in with one group, move on to the next group.

Teacher Notes

Stage 4 DO

Total Time for the Stage	45 minutes
Grouping	Teams

Learning Goals

Students should make progress on their ability to do the following by the end of this stage of the lesson:

Science

1. Conduct an investigation individually and collaboratively to produce data to serve as the basis for evidence (SEP3).
2. Analyze data using tools and/or technologies to make valid and reliable scientific claims (SEP4).
3. Consider limitations of data analysis (e.g., measurement error and sample selection) when analyzing and interpreting data (SEP4).
4. Evaluate the impact of new data on a working explanation and/or model of a proposed process or system (SEP4).
5. Apply ratios, rates, percentages, and unit conversions in the context of complicated measurement problems involving quantities with derived or compound units (SEP5).
6. Make a quantitative and/or qualitative claim regarding the relationship between dependent and independent variables (SEP6).
7. Construct and revise an explanation based on valid and reliable evidence obtained from a variety of sources and the assumption that theories and laws that describe the natural world operate today as they did in the past and will continue to do so in the future (SEP6).
8. Apply scientific ideas, principles, and/or evidence to provide an explanation of phenomena taking into account possible unanticipated effects (SEP6).

English Language Development

1. Paraphrase central ideas in complex evidence, concepts, processes, and information to help explain how or why a phenomenon occurs (ELD-SLC.9-12.Explain.Interpretive).
2. Evaluate the extent to which reasoning, theory, and/or models link evidence to claims and support conclusions (ELD-SLC.9-12.Explain.Interpretive).
3. Develop reasoning to illustrate the relationships between variables (ELD-SLC.9-12.Explain.Expressive).

English Language Arts

1. Initiate and participate effectively in small-group collaborative discussions with diverse partners by building on others' ideas and expressing their own clearly (SL.9-10.1).
2. Present information clearly, concisely, and logically such that listeners can follow the line of reasoning and the organization, substance, and style are appropriate to the purpose, audience, and task (SL.9-10.4).
3. Adapt speech to a variety of contexts and tasks, demonstrating command of formal English when indicated or appropriate (SL.9-10.6).

Activities

Activity A	Carry Out Your Plan
Time Needed for the Activity	25 minutes
Grouping	Teams

What to Do

1. Students should work in their assigned teams for this activity. Begin this activity by giving out a copy of the Do handout to each student.
2. Show the "Carry Out Your Plan" page on a screen using the Presentation mode in the Learning Hub so everyone in the classroom can see it.
3. Say, "Okay, it sounds like we are now ready to carry out our plans and collect the data or observations we need to answer the guiding question. You can get the materials and tools you need to use from this table." Point to the table where you have placed them.
4. Ask, "Does anyone have any questions about these materials or tools?"
5. Answer any questions that come up.
6. Say, "As you work, be sure to record your observations and measurements on your handout so you can keep track of everything you do. Does anyone have any final questions?"
7. Answer any questions that come up.
8. Say, "Okay, go ahead and get started. You have the next 25 minutes to work. I'll be available to help if any team needs me."
9. Give the students at least 25 minutes to carry out their plans and solve the problem. As students work in small teams, be sure to move from team to team to answer questions and offer advice.

What to Set Up

Provide each student, or group of students, with access to a computer so they can use the computer-based model. Students can access the model in the Learning Hub using their accounts. If students do not have a Learning Hub account, they can access the simulation by going to *Evolution of Antibiotic Resistance in* NetLogo.

What to Give Out

Do handout for **Antibiotic Resistance in Bacteria**

What to Show

The "Carry Out Your Plan" page in the ADI Learning Hub

Teaching Tip

One way to do a quick check-in with a group during this activity is to use the "two questions, a nudge, and a goal" technique. This technique includes the following steps:

1. Sit down with a group of students.
2. Ask, "Would one of you tell me a little about what you all have been doing or thinking about?" (Question 1)
3. A student answers on behalf of the group.
4. Ask, "Would anyone else like to add anything to that?" (Question 2)
5. A different student provides additional information on behalf of the group.
6. Ask, "Have you thought about [a hint or something you want to highlight]?" (A nudge)
7. Say, "Okay, I'm going to check on the other groups. When I get back, I want to see [something that you want the group to accomplish]." (A goal)

When you have completed the check-in with one group, move on to the next group.

Attending to Equity

Be sure to intentionally group emerging multilingual students at times with peers who know the same languages as they do and at other times with peers whose English language development is slightly more advanced. Thoughtful grouping that varies during different investigation (or activities within an investigation) allows emerging multilingual students to benefit from the different linguistic resources of their peers. We also recommend assessing emerging multilingual students based on how much they contribute to the group during this activity, rather than solely on what they write or draw on their handouts.

Teacher Notes

Activity B	Make Sense of Your Data
Time Needed for the Activity	20 minutes
Grouping	Teams

What to Do

1. Students should work in their assigned teams for this activity. Begin this activity by showing the "Make Sense of Your Data" page on a screen using the Presentation mode in the Learning Hub so everyone in the classroom can see it.
2. Say, "Okay, it sounds like we are now ready to make sense of the data we collected. One way to do this is to create a table, graph, or other visualization that shows a trend or change over time, a comparison between teams, or a relationship between things you tested. You can create a table, graph, or other visualization here on your handout."
3. Point to the box on the handout.
4. Say, "Please work together in your teams to figure out the best way to make sense of your data. Remember, you are looking for any trends, changes, differences, or relationships in the data that will help you answer the guiding question. As you work, be sure to use the space on your handout to perform any calculations that you need to do and create any representations that you need. Does anyone have any last questions?"
5. Answer any questions that come up.
6. Say, "Okay, go ahead and get started. You have the rest of the period to work. I'll be available to help if any team needs me."
7. Give the students at least 15 minutes to make sense of their data. As students work together in small teams, be sure to move from team to team to answer questions and offer advice.
8. Make sure students return the materials and tools to the table where they found them before dismissing class. Cleanup should take no longer than five minutes.

What to Show

The "Make Sense of Your Data" page in the ADI Learning Hub

What to Look For

One of the most useful ways to make sense of the data during this activity is to make several graphs that compare the antibiotic resistance levels of the bacteria under different conditions.

Teaching Tip

Collecting and returning materials or tools to a central table works best if you assign the job to a specific number within each team. For example, you can say, "2s, please come up and grab the materials and tools that your team needs" at the start of the activity. Then you can say, "3s, please return the materials and tools your team used back to this table," at the end of the activity. Assigning jobs like this not only cuts down on how many students need to be up and moving around, but also allows you to hold specific students accountable for the materials and tools.

Attending to Equity

Encourage your emerging multilingual students to speak using learning strategies such as requesting assistance, employing nonverbal cues, and using synonyms and circumlocution (conveying ideas by defining or describing when exact English words are not known) during this activity. It is also helpful to point out and/or draw emerging multilingual learners' attention to elements of the English sound system in newly acquired vocabulary such as long and short vowels, silent letters, and consonant clusters as they come up naturally during this activity.

Teacher Notes

Antibiotic Resistance in Bacteria Lesson Plan

Stage 5 SHARE

Total Time Needed for the Stage	45 minutes
Grouping	Teams

Learning Goals

Students should make progress on their ability to do the following by the end of this stage of the lesson:

Science

1. Conduct an investigation individually and collaboratively to produce data to serve as the basis for evidence (SEP3).
2. Analyze data using tools and/or technologies to make valid and reliable scientific claims (SEP4).
3. Consider limitations of data analysis (e.g., measurement error and sample selection) when analyzing and interpreting data (SEP4).
4. Evaluate the impact of new data on a working explanation and/or model of a proposed process or system (SEP4).
5. Apply ratios, rates, percentages, and unit conversions in the context of complicated measurement problems involving quantities with derived or compound units (SEP5).
6. Make a quantitative and/or qualitative claim regarding the relationship between dependent and independent variables (SEP6).
7. Construct and revise an explanation based on valid and reliable evidence obtained from a variety of sources and the assumption that theories and laws that describe the natural world operate today as they did in the past and will continue to do so in the future (SEP6).
8. Apply scientific ideas, principles, and/or evidence to provide an explanation of phenomena taking into account possible unanticipated effects (SEP6).

English Language Development

1. Identify appropriate and sufficient evidence from data, models, and/or information from investigations of a phenomenon (ELD-SLC.9-12.Argue.Interpretive).
2. Compare reasoning and claims based on evidence from competing arguments (ELD-SLC.9-12.Argue.Interpretive).
3. Defend or refute a claim based on data and evidence (ELD-SLC.9-12.Argue.Expressive).
4. Establish and maintain an appropriate tone and stance (ELD-SLC.9-12.Argue.Expressive).
5. Signal logical relationships among reasoning, evidence, data, and/or models when making and defending a claim, counterclaim, and/or rebuttal (ELD-SLC.9-12.Argue.Expressive).

English Language Arts

1. Initiate and participate effectively in small-group collaborative discussions with diverse partners by building on others' ideas and expressing their own clearly (SL.9-10.1).
2. Integrate multiple sources of information presented in diverse media or formats (e.g., visually, quantitatively and orally) evaluating the credibility and accuracy of each source (SL.9-10.2).
3. Evaluate a speaker's reasoning and use of evidence and rhetoric, identifying any fallacious reasoning or exaggerated or distorted evidence (SL.9-10.3).

(continued)

4. Present information clearly, concisely, and logically such that listeners can follow the line of reasoning, and the organization, substance, and style are appropriate to the purpose, audience, and task (SL.9-10.4).
5. Make strategic use of media (e.g., textual, graphical and visual) in presentations to enhance understanding of findings, reasoning, and evidence and to add interest (SL.9.10.5).
6. Adapt speech to a variety of contexts and tasks, demonstrating command of formal English when indicated or appropriate (SL.9-10.6).

Activities

Activity A	Create a Draft Argument
Time Needed for the Activity	15 minutes
Grouping	Teams

What to Do

1. Students should work in their assigned teams for this activity. Begin this activity by giving out a copy of the Share handout to each student.
2. Show the "Create a Draft Argument" page on a screen using the Presentation mode in the Learning Hub so everyone in the classroom can see it.
3. Say, "Okay, it sounds like we have all solved the task. I would now like you to create a draft argument to share what you figured out with the other teams. Your argument should include everything in this image here." Point to the image of a science-based argument projected on the screen. Be sure to go over what should be included in each component of the argument and where students can find the information they need.
4. Say, "I would like you to make your draft argument on a whiteboard. You can use the materials on this table to make your argument." Point to the table where you have placed the materials they need.
5. Ask, "Does anyone have any questions?"
6. Answer any questions that come up.
7. Say, "I'm going to give you exactly 15 minutes to get this done. Your draft argument doesn't need to be perfect. We just need something down on the whiteboards so we can share our ideas."
8. Give each team of students 15 minutes to create a draft argument. As students work in small teams, be sure to move from team to team to answer questions and offer advice.

What to Set Up

Place the following items on a table so students can use them during this stage:

- 24″ x 36″ whiteboard
- 3-4 dry-erase pens in different colors
- Eraser or cloth towel

What to Give Out

Share handout for **Antibiotic Resistance in Bacteria**

What to Show

The "Create a Draft Argument" page in the ADI Learning Hub

What to Look for

Students should include core ideas that they read about during the Ideas stage in their justification.

Teaching Tip

Be sure to plant some different ideas with each group about how to make a strong argument as you move from group to group to check in, answer questions, and offer advice. This strategy helps to ensure that there will be some really strong examples of the evidence and justification components of the argument for students to see and discuss during the next activity. For example, you could help two groups have a really strong evidence component by telling one group, "I think a graph would be really helpful here" and telling another group, "I think your evidence would be stronger if you explained, in words, what this graph shows." You could help a third group have a really strong justification by saying, "Let's brainstorm some ideas that will help you defend your choice of evidence," or "Let's go back to our Ideas handout and see if there are any concepts we can include."

It is important you make sure that all the arguments are different at this point in the investigation by planting different ideas with different groups, because the students will have nothing to critique or discuss during the next activity if they all have the same argument. The goal during this activity is to set the stage for the next activity by ensuring that the various components of a strong argument can be found on different whiteboards within the class. These strong components will then be taken up by all the groups because they reflect good ideas. Be sure to give students a voice and choice during this activity, so they can make mistakes and learn from them.

Attending to Equity

It is often helpful to assign and change roles at least once during this activity. For example, you might assign the 1s to be "in charge of writing" at the beginning of the activity. You can then easily switch roles about five minutes into the activity by saying, "1s, please hand the pens to the 2s. 2s, you are now in charge of writing for the group." By changing roles, you can ensure that all students have a chance to participate and contribute their ideas to the creation of a draft argument.

ADI Go Digital

The **ADI Learning Hub** has tons of features that allow students to share their thinking with each other when they are doing the investigation paperless or remote.

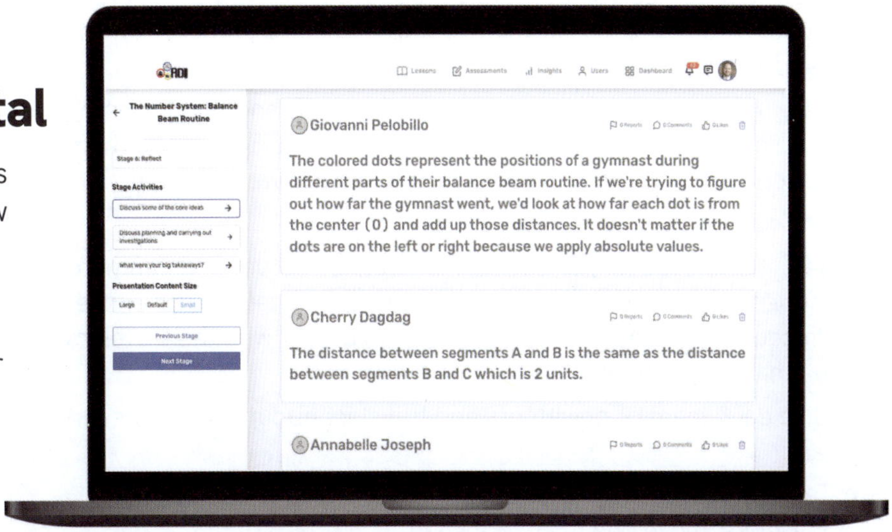

Activity B	Present Your Draft Argument
Time Needed for the Activity	15 minutes
Grouping	Teams

What to Do

1. Students should work in their assigned teams for this activity.
2. Show the "Present Your Draft Argument" page on a screen using the Presentation mode in the Learning Hub so everyone in the classroom can see it.
3. Say, "We are now going to share our arguments. Please set up your whiteboards so everyone can see them." Allow students time to set up their whiteboards. They should be set up on stands or chairs, so they are facing toward the outside of the room.
4. Say, "Okay, before we get started, I want to explain what we are going to do next. I'm going to ask some of you to present your arguments to your classmates. If you are presenting your argument, your job is to share your team's claim, evidence, and justification of the evidence. The rest of you will be travelers. If you are a traveler, your job is to listen to the presenters, ask them questions if you do not understand something, and then offer them some suggestions about ways to make their argument better. Do you have any questions about these instructions?"
5. Answer any questions that come up.
6. Say, "As you talk to one another, make sure you write down any good ideas or suggestions that are important in the space on your handout, so you don't forget about them." Point to the box labeled, "Some possible ways to improve our argument" on the handout.
7. Say, "After we have a chance to learn from one another, I'm going to give you some time to revise your arguments and make them better. Do you have any questions about what we are doing?"
8. Answer any questions that come up.
9. Say, "Okay. Let's get started. Travelers, move one team to the left. Don't move to the next team until I tell you to move. Presenters, go ahead and share your argument with the travelers when they get there."
10. Give each team of presenters and travelers about 3-4 minutes to talk.
11. Tell the students, "Okay. Let's move on to the next argument. Travelers, move one team to the left."
12. Again, give each team of presenters and travelers about 3-4 minutes to talk.

What to Show

The "Present Your Draft Argument" page in the ADI Learning Hub

What to Look for

The critiques and feedback that students give one another during this activity should:

- Be shared in a way that is kind (respectful),
- Provide information about what, exactly, needs to be changed or improved (specific), and
- Offer a suggestion about how to make something better rather than just saying something is wrong (helpful).

In addition, students should be focusing their critiques on the science included in the arguments, not the appearance of the arguments.

(continued)

13. Repeat this process until each team has had their argument read and critiqued three times. Then tell the students to return to their workstations.
14. Be sure to move between workstations during this activity to keep everyone involved and model what it means to communicate and critique scientific ideas and reasoning.

Attending to Equity

Be sure to encourage students to provide feedback that is respectful, specific, and helpful during this activity. It is important to encourage student-to-student talk because it helps students develop speaking and listening skills and enables them to make sense of what and how they know. It also provides a context for emerging multilingual students to use both productive (speaking and writing) and receptive (listening and reading) language and learn science through meaningful interactions and scientific sense-making.

Teaching Tip

There are two types of formats for sharing and critiquing arguments. The first format is called a *gallery walk.* In this format, each group sets up their argument at their workstation so others can see it. The entire group then moves to a different workstation. While at a different workstation, the group is expected to read the argument at that workstation (which was created by a different group) and offer critiques and suggestions for improvement. The students should move to a different workstation after a few minutes to read and critique an argument written by a different group. Students should also be encouraged to keep a record of good ideas or potential ways to improve their own arguments as they move from workstation to workstation to offer critiques and feedback. After repeating this process in the gallery walk 3-4 times, every group will have had an opportunity to read and critique three or four different arguments and receive feedback from three or four different groups.

The second format is called a *modified gallery walk.* In the modified gallery walk format, one or two members of the group stay at their workstation to share their group's ideas (these students are called presenters) while the other members of their group go to different groups, one group at a time, to listen to and critique the arguments developed by their classmates (these students are called travelers). The travelers should visit at least three different workstations during the argumentation session. We also recommend that the presenters keep a record of the critiques made by their classmates and any suggestions for improvement. The travelers should also be encouraged to keep a record of good ideas or potential ways to improve their own arguments as they travel from group to group.

Attending to Equity

Another good way to have students, especially emerging multilingual learners, give one another feedback is to have them use four different color sticky notes to make comments. Give each group a single pink, blue, yellow, and green sticky note. Then direct the students to use the pink sticky note to suggest "something to add" to the argument, the blue sticky note to suggest "something to change" in the argument, the yellow sticky note to highlight "something they like" in the argument, and the green sticky note to ask a question about "something they don't understand". This strategy is useful because everyone in the classroom will be actively engaged during this activity, and everyone will have a chance to see different arguments. This format, however, does not give students a chance to support, critique, and challenge the ideas, solutions, scientific reasoning of other groups through talk because two different groups are never at the same workstation at the same time.

Activity C	Revise Your Argument
Time Needed for the Activity	15 minutes
Grouping	Teams

What to Do

1. Students should work in their assigned teams for this activity.
2. Show the "Revise Your Argument" page on a screen using the Presentation mode in the Learning Hub so everyone in the classroom can see it.
3. Say, "I'm now going to give you all about 10 minutes to revise your argument. Please return to your original teams."
4. Wait for the students to move back into their original teams.
5. Say, "Okay, take a few minutes to talk in your teams and determine what you want to change to make your argument better. Once you have decided what to change, go ahead and make the changes to your whiteboard."
6. Ask, "Do you have any questions about what you need to do?"
7. Answer any questions that come up.
8. Tell the students, "Okay. Let's get started."
9. Give the students 10 minutes to work in their teams on their arguments. As they work, move from team to team to check in, ask probing questions, and offer a suggestion if a team gets stuck.
10. When students are finished with their final arguments, have them take a picture of it or have them put the boards in a safe spot so they can refer to the picture or the boards during the Report stage.

What to Show

The "Revise Your Argument" page in the ADI Learning Hub

What to Look For

A strong argument includes the following information:

- A claim that states, "A fast growth rate coupled with high mutation leads to faster evolution in bacteria."
- Several graphs that compare the antibiotic resistance levels of the bacteria under different conditions.
- An explanation of the graphs.
- An explanation of the chosen evidence that uses core ideas from the investigation.

Teaching Tip

It is important to give students time to meet with their original groups so they can discuss what they learned by interacting with individuals from the other groups. At this point in the process, students often realize that their initial problem-solving model was flawed in some way. It is therefore important to encourage students to "redo the science" as needed. At the end of this activity, each group should have a final argument that is much better than their original one.

Antibiotic Resistance in Bacteria Lesson Plan

Stage 6 REFLECT

Total Time for the Stage	25 minutes
Grouping	Whole class, Teams

Learning Goals

Students should make progress on their ability to do the following by the end of this stage of the lesson:

Science

1. Ask questions that arise from careful observation of phenomena, or unexpected results, to clarify and/or seek additional information as needed (SEP1).
2. Describe how natural selection occurs only if there is both variation in the genetic information between organisms in a population and variation in the expression of that genetic information that leads to differences in performance among individuals (LS4.B).
3. Identify how evolution is a consequence of four factors: the potential for species to increase in number, the genetic variation of individuals in a s pecies, competition for an environment's limited resources, and the ensuing proliferation of the organisms that are better able to survive and reproduce in that environment (LS4.C).
4. Describe how natural selection leads to adaptation, or to a population dominated by organisms that are anatomically, behaviorally, and physiologically well suited to survive and reproduce in a specific environment (LS4.C).
5. Explain how organisms in a population may have an advantageous heritable trait, which leads to an increase in the proportion of individuals in future generations that have the trait (LS4.C).
6. Use empirical evidence to differentiate between cause and correlation in order to make claims about specific causes and effects (CC2).
7. Use models to simulate systems and interactions within and between systems at different scales (CC4).

English Language Development

1. Paraphrase central ideas in complex evidence, concepts, processes, and information to help explain how or why a phenomenon occurs (ELD-SLC.9-12.Explain.Interpretive).
2. Use scientific explanations about a phenomenon in a different context or situation (ELD-SLC.9-12.Explain.Expressive).

English Language Arts

1. Participate effectively in small-group collaborative discussions with diverse partners on texts by building on others' ideas and expressing their own clearly and persuasively (SL.9-10.1).

Activities

Activity A	Discuss a Core Idea
Time Needed for the Activity	10 minutes
Grouping	Whole class/Teams

What to Do

1. Students should be in small teams for this activity. However, they do not need to be in the same teams as they were during previous stages.
2. Begin this activity by giving out a copy of the Reflect handout to each student.
3. Show the "Discuss a Core Idea" page on a screen using the Presentation mode in the Learning Hub so everyone in the classroom can see it.
4. Say, "Okay. Now that we have all revised our arguments and made them better, let's take some time to discuss some of the core ideas you used during this investigation and to justify the evidence you used in your arguments, so we are all on the same page. Look at the picture on the screen."
5. Pause for about 30 seconds.
6. Say, "Take a few minutes to talk in your teams and see if you can use what you know about natural selection and reproduction in bacteria to explain what is happening in the picture."
7. Give the students at least three minutes to talk in their small teams.
8. Say, "Okay. Let's share with the whole class. Team A (or whichever team you want to share first), what did you talk about?"
9. Allow the first team to share.
10. Say, "Interesting. Would anyone like to add to, challenge, or refine the ideas shared by Team A?"
11. Allow students to respond.
12. Say, "Okay. Team B (or whichever team you want to share second), what did you talk about?"
13. Allow the second team to share.

(continued)

What to Give Out

Reflect handout for **Antibiotic Resistance in Bacteria**

What to Show

The "Discuss a Core Idea" page in the ADI Learning Hub

Explanation of the Image

The image that the students discuss during this activity shows the four factors that affect natural selection in a population of beetles.

What to Look For During the Discussion

Students who understand the content will likely mention some, if not all, of the following ideas when providing an explanation for the image:

- Evolution is the change in the traits of a population over time.
- Natural selection is a mechanism through which evolution can occur.
- Evolution via natural selection is a consequence of four factors: variations in a trait within a population, selective pressure in the environment that prevents some individuals from reproducing, the individuals able to reproduce passing

(continued)

14. Say, "Interesting. Would anyone like to add to, challenge, or refine the ideas shared by Team B?"
15. Allow students to respond.
16. Keep up this pattern of questioning until all teams have shared. Be sure to keep encouraging students to talk about the content until they reach a correct and consensual explanation for what is shown in the image. If they are unable to reach a consensus or reach consensus using an inaccurate idea, be sure to provide an accurate explanation for the image before moving on to the next activity.

their traits onto the next generation, and the proliferation of those organisms that are better able to reproduce in that environment.

- Natural selection leads to the development of specific adaptations, or to a population dominated by organisms that are anatomically, behaviorally, and physiologically well suited to reproduce within a specific environment.
- Natural selection can only occur if there is both variation in the genetic information of organisms in a population and variation in the expression of that genetic information.

Teaching Tip

This activity is a discussion, not a presentation about what the students "should have seen" or "should have learned" during the investigation. The more students talk during this activity, the more meaningful the experience will be for them and the more you can learn about their thinking. Be sure to use talk goals and moves to help facilitate the discussion. Below are some examples of talk goals that you can use to help facilitate the discussion and some questions that you can ask the students to help accomplish each goal. These talk goals and questions are only suggestions.

Goal 1: Encourage students to share, expand on, or clarify an idea.

- What do you see going on here?
- What do you think is happening?
- How might we explain that?
- Can you tell me more about that?
- What do you mean by that?
- Can you give me an example of that?
- Can you say that in another way?

Goal 2: Encourage students to support an idea.

- Why do you think that?
- What is your evidence for that?
- How did you arrive at that conclusion?
- What makes you say that?
- How do you know that?

Goal 3: Encourage students to think with others.

- Would anyone like to add to that idea?
- What should we keep or change about that idea?
- Can anyone else give me some examples of that?
- Can anyone take that idea and push it a little further?
- Does anyone want to respond to that idea?
- Do you agree or disagree with that idea, and why?
- What do we think about that idea?
- Does that fit with what we have figured out so far?
- Is there another way to do it?

Goal 4: Encourage students to think about what they know and don't know yet.

- I'm confused now. What have we figured out so far?
- What do we need to figure out now?
- So, what do we agree on right now, and what are we still trying to figure out?
- What are we sure and not so sure about?
- What questions do you have?

Activity B	Discuss the Practices
Time Needed for the Activity	10 minutes
Grouping	Whole class/Teams

What to Do	What to Show
1. Show the first page of "Discuss the Practices" on a screen using the Presentation mode in the Learning Hub so everyone in the classroom can see it. 2. Say, "Okay. Now that we have talked about some of the core ideas that we used during our investigation, let's take a few minutes to talk about the science practices that we used during the investigation. An example of a scientific practice used was analyzing and interpreting data. On the screen there is a question that I want you all to think about: What other things did you do during your investigation that you feel were good examples of a scientific practice or attempting to figure out something in a scientific way?" 3. Pause for about 30 seconds. 4. Say, "Take a few minutes to talk about this question in your teams." 5. Give the students at least three minutes to talk in their small teams. 6. Say, "Okay. Let's share with the whole class. Team A (or whichever team you want to share first), what did you talk about?" 7. Allow the first team to share. 8. Say, "Okay. Team B (or whichever team you want to share second), what did you talk about?" 9. Allow the second team to share. 10. Continue asking teams until every team has a chance to share. 11. Show the second page of "Discuss the Practices" on a screen using the Presentation mode in the Learning Hub so everyone in the classroom can see it. 12. Say, "Here is another question that I want you all to think about: What things did you do during your investigation that were not good examples of a scientific practice or not very scientific?" 13. Pause for about 30 seconds.	**What to Show First** The first page of "Discuss the Practices" in the ADI Learning Hub **What to Show Second** The second page of "Discuss the Practices" in the ADI Learning Hub **What to Show Third** The third page of "Discuss the Practices" in the ADI Learning Hub

(continued)

14. Say, "Take a few minutes to talk about this question in your team."
15. Give the students at least three minutes to talk in their small teams.
16. Say, "Okay. Let's share with the whole class. Team A (or whichever team you want to share first), what did you talk about?"
17. Allow the first team to share.
18. Say, "Okay. Team B (or whichever team you want to share second), what did you talk about?"
19. Allow the second team to share.
20. Continue asking teams until every team has a chance to share.
21. Show the third page of "Discuss the Practices" on a screen using the Presentation mode in the Learning Hub so everyone in the classroom can see it.
22. Say, "Here is a final question that I want you all to think about: What should we all do during our next investigation to make sure we are using scientific practices and being scientific in the way we try to figure something out?"
23. Pause for about 30 seconds.
24. Say, "Take a few minutes to talk about this question in your team."
25. Give the students at least three minutes to talk in their small teams.
26. Say, "Okay. Let's share with the whole class. Team A (or whichever team you want to share first), what did you talk about?"
27. Allow the first team to share.
28. Say, "Okay. Team B (or whichever team you want to share second), what did you talk about?"
29. Allow the second team to share.
30. Continue asking teams until every team has a chance to share.
31. Once everyone has shared, offer any suggestions or advice about what they can do to improve their next investigation and then move on to the next stage.

Teaching Tip

A great way to wrap up this activity is for the students to contribute to a list of class norms or "gotta-haves" to keep in mind when planning and carrying out an investigation in science. To create this list, each group can suggest for a potential norm or "gotta-have" to add to the list at the end of this activity, then the class votes on the one that they think is most important. The most important one gets added to the list. The list then serves as reminder for students during the Plan and Do stages of subsequent investigations. Over time, the list will grow and each addition to the list will be more nuanced than the last as students develop expertise with scientific processes.

Attending to Equity

Be sure to give all students, especially emerging multilingual students, enough time to talk in small groups before asking anyone to share with the entire class during this activity. Small-group structures offer all students, including emerging multilingual students, a chance to engage in sense-making with their peers as well as space to use their verbal and nonverbal (e.g., writing, drawing pictures, and using gestures) resources to express their ideas. Small-group structures also give students a chance to learn from other students' uses of these resources. When students are tasked with expressing their ideas, be sure that they know it is okay for them to do so in a mode or combination of modes (speaking, writing, drawings, and gestures).

Antibiotic Resistance in Bacteria Lesson Plan

Stage 7 REPORT

Total Time Needed for the Stage	100 minutes (over three days)
Grouping	Individual, Teams

Learning Goals

Students should make progress on their ability to do the following by the end of this stage of the lesson:

Science

1. Construct an explanation based on evidence that the process of evolution primarily results from four factors: the potential for a species to increase in number, the heritable genetic variation of individuals in a species due to mutation and sexual reproduction, competition for limited resources, and the proliferation of those organisms that are better able to survive and reproduce in the environment (HS-LS4-2).
2. Construct an explanation based on evidence for how natural selection leads to the adaptation of populations (HS-LS4-4).
3. Make and defend a claim based on evidence about the natural world that reflects scientific knowledge and student-generated evidence (SEP7).
4. Construct a written argument based on evidence (SEP7).
5. Evaluate the claims, evidence, and/or reasoning to determine the merits of arguments (SEP7).
6. Respectfully give critiques on scientific arguments by and challenging ideas and conclusions (SEP7).
7. Communicate scientific information or ideas in multiple formats including graphically and textually (SEP8).

English Language Development

1. Identify appropriate and sufficient evidence from data, models, and/or information from investigations of a phenomenon (ELD-SLC.9-12.Argue.Interpretive).
2. Compare reasoning and claims based on evidence from competing arguments (ELD-SLC.9-12.Argue.Interpretive).
3. Defend or refute a claim based on data and evidence (ELD-SLC.9-12.Argue.Expressive).
4. Establish and maintain an appropriate tone and stance (ELD-SLC.9-12.Argue.Expressive).
5. Signal logical relationships among reasoning, evidence, data, and/or models when making and defending a claim (ELD-SLC.9-12.Argue.Expressive).

English Language Arts

1. Write routinely over extended time frames (time for reflection and revision) and shorter time frames for a range of discipline-specific tasks, purposes, and audiences (WHST.9-10.10).
2. Write arguments focused on discipline-specific content (WHST.9-10.1).
3. Write informative/explanatory texts, including the narration of scientific procedures/experiments or technical processes (WHST.9-10.2).
4. Produce clear and coherent writing in which the development, organization, and style are appropriate to task, purpose, and audience (WHST.9-10.4).

(continued)

5. Strengthen writing as needed by revising, editing, rewriting, and focusing on addressing what is most significant for a specific purpose and audience (WHST.9-10.5).
6. Determine the central ideas or conclusions of a text; trace the text's explanation or depiction of a complex process, phenomenon, or concept; provide an accurate summary of the text (RST.9-10.2).
7. Determine the meaning of symbols, key terms, and other domain-specific words and phrases as they are used in a specific scientific or technical context (RST.9-10.4).
8. Analyze the structure of the relationships among concepts in a text, including relationships among key terms (RST.9-10.5).
9. Assess the extent to which the reasoning and evidence in a text support the author's claim (RST.9-10.8).
10. Compare and contrast findings presented in a text to those from other sources (including their own experiments), noting when the findings support or contradict other accounts (RST.9-10.9).

Activities

Activity A	Draft Report
Time Needed for the Activity	20 minutes
Grouping	Individual

What to Do

1. Have students work on their own during this activity. Give a copy of the Report handout to each student.
2. Show the "Create a Draft Report" page on a screen using the Presentation mode in the Learning Hub so everyone in the classroom can see it.
3. Say, "You are now going to write an investigation report to share what you figured out. The first part of the report is called the 'Introduction'. In this section, you should explain to the reader what you were investigating, why you were investigating it, and what question you were trying to answer. All this information can be found on your Task handout."
4. Hold up the Task handout.
5. Ask the students, "Do you have any questions about what you need to do?"
6. Answer any questions that come up.
7. Say, "Here are some sentence starters to help you begin writing."
8. Hold up the Report handout and point to the sentence starters.
9. Tell the students, "Okay, let's write."
10. Give the students five minutes to write the "Introduction" section of the report. As they work, move from student to student to check in, ask probing questions, and offer a suggestion if a student gets stuck.

(continued)

What to Give Out

Report handout for **Antibiotic Resistance in Bacteria**

What to Show

The "Create a Draft Report" page in the ADI Learning Hub

11. Say, "Okay, let's move on to the second part of the report. The second part of the report is called the 'Method'. In this section, you should explain to the reader what you did during the investigation to complete the task and answer the guiding question. All this information can be found in the Plan handout. Remember that you all planned and carried out different investigations, so do not assume that the reader will know what you did."
12. Ask the students, "Do you have any questions about what you need to do?"
13. Answer any questions that come up.
14. Say, "Here are some sentence starters to help you begin writing."
15. Hold up the Report handout and point to the sentence starters.
16. Tell the students, "Okay, let's write."
17. Give the students five minutes to write the "Method" section of the report. As they work, move from student to student to check in, ask probing questions, and offer a suggestion if a student gets stuck.
18. Say, "Okay, let's move on to the last part of the report. The third part of the report is called the 'Argument'. In this section, you should explain to the reader what you figured out and how you know you are correct. All this information can be found on your Do handout and on your team's whiteboard."
19. Hold up the Do handout and point to the boards (or remind them that they have a picture of their board).
20. Ask the students, "Do you have any questions about what you need to do?"
21. Answer any questions that come up.
22. Say, "Here are some sentence starters to help you begin writing."
23. Hold up the Report handout and point to the sentence starters.
24. Tell the students, "Okay, let's write."
25. Give the students 10 minutes to write the "Argument" section of the report. As they work, move from student to student to check in, ask probing questions, and offer a suggestion if a student gets stuck.
26. Collect all the draft reports before moving on to the next activity. Tell students to mark their draft reports with a symbol that they can use to identify it later. They should NOT write their names on their draft reports.

Teaching Tip

Writing is an important component of this model. When a student writes the report on their own, it not only helps the student learn to write in the context of science (an important literacy skill) but also allows that student to write to learn (develop a better understanding of scientific concepts by writing about them). There are many writing supports (e.g., sentence starters) already embedded into this activity, so you might not need to provide much extra support for your students.

This activity can be done completely in class or started in class and then finished as homework. Do not assign the entire activity as homework unless you know that every student in your class has dedicated time, space, and resources to write outside of class and everyone understands and is comfortable with the report writing process.

If you are concerned about students feeling overwhelmed, you can develop a class "gotta-have-it" list of information that needs to be included in each section of the report. Students can then refer to this list as they write each section.

Don't worry if students do not produce a perfect report at this point. All the reports will be reviewed and revised before they are turned in to you. The reports should therefore be viewed as a starting point. Be sure to encourage students to "just get something down on paper" so "we can work on it together."

Attending to Equity

Emerging multilingual students' access to instructional tasks that require complex thinking, such as writing a report, is enhanced when linguistic complexity and instructional support match their current level of language proficiency. Therefore, it is important to scaffold the writing of the reports for the emerging multilingual students based on their current English language proficiency level. The use of sentence starters to model written language production skills can help emerging multilingual students develop English language skills because the development of academic language and academic content knowledge are interrelated processes. However, it is important that such scaffolds be used purposefully and removed when no longer needed.

ADI Go Digital

Use the **ADI Learning Hub** to go paperless. When you create a class and add students to it, students can complete all the stages right in the system.

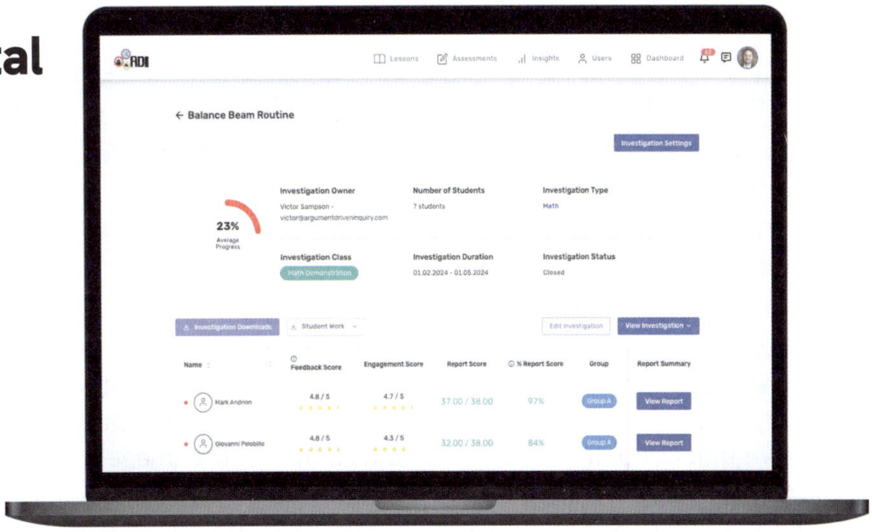

Activity B	Peer Review the Reports
Time Needed for the Activity	45 minutes
Grouping	Teams

What to Do

1. Students should work in teams of three for this activity. These can be completely different teams.
2. Show the "Peer Review Some Draft Reports" page on a screen using the Presentation mode in the Learning Hub so everyone in the classroom can see it.
3. Say, "We are now going to review our reports to find ways to make them better."
4. If possible, use a document camera to project the peer review guide (PRG) on a screen.
5. Say, "We are going to use this peer review guide to give each other feedback." Point to the image.
6. Say, "I'm going to ask you to work in teams of three to do this. I will give each team a draft report to read. You three will then read the report together. Once you are done reading the report, I want you to answer each of the questions on the peer review guide." Point to the review questions on the image of the PRG.
7. Tell the students, "You can check No, Almost, or Yes after each question." Point to the checkboxes on the image of the PRG.
8. Say, "This will be your rating for this part of the report. Make sure you agree on the rating you give the author. If you mark No or Almost, then you need to tell the author what he or she needs to do to get a Yes." Point to the space for the reviewer feedback on the image of the peer review guide.
9. Say, "It is really important for you to give the authors feedback that is helpful. That means you need to tell them exactly what they need to do to make their report better."
10. Ask the students, "Do you have any questions about what you need to do?"
11. Answer any questions that come up.
12. Tell the students, "Please get into your teams of three."
13. Allow the students time to move into their triads.
14. Say, "Okay, I'm now going to give you one report to read." Give out one draft report to each triad. Make sure that the report you give a triad was not written by one of the students in that triad. Give each triad one PRG to fill out.

What to Give Out

Peer Review Guide (one per group per review, so for this investigation you need the same number of peer review guides as the number of draft reports that will be reviewed)

What to Show

The "Peer Review Some Draft Reports" page in the ADI Learning Hub

What to Look For

Make sure students are reaching consensus about what rating to give each criterion. Also, be sure that students are giving one another feedback that is specific, detailed, and helpful when they think an aspect of a report needs to be improved.

(continued)

15. Tell the students, "Okay, I'm going to give you 15 minutes to read the report I gave you and to fill out the peer review guide. Go ahead and get started."
16. Give the students 15 minutes to work. As they work, move around from team to team to check in and see how things are going, answer questions, and offer advice.
17. After 15 minutes pass, tell the students, "Okay, time is up. Please give me the report and the peer review guide that you filled out."
18. Collect the draft reports and the PRGs, or collect the workbooks if students are using them.
19. Say, "Okay, I am now going to give you a different report to read and a new peer review guide to fill out." Give out one more report to each triad. Make sure that the report you give a triad was not written by one of the students in that triad. Give each triad one new peer review guide to fill out as a team.
20. Say, "Okay, I'm going to give you 15 minutes to read this new report and to fill out the peer review guide. Go ahead and get started."
21. Give the students 15 minutes to work. As they work, move around from team to team to check in and see how things are going, answer questions, and offer advice.
22. After 15 minutes pass, tell the students, "Okay, time is up. Please give me the report and the peer review guide that you filled out."
23. Repeat this process for the remaining draft reports. Each triad should one have one more report to review (for a total of three reviews).
24. Collect the last draft reports and PRGs. Set all the draft reports on a table in separate stacks. Each stack should have a draft report on top (with the identification symbol facing up so students can find their reports) and the PRG on the bottom.

Teaching Tip

The peer review process is one of the best ways for students to learn how to write. When students read and review different reports, they have an opportunity to (1) see examples of texts written by others for the same purpose, (2) discuss what counts as quality and why some reports are stronger than others, (3) discuss ways to strengthen a report, and (4) pick up things that they can do to improve their reports. This helps the students learn new ways to organize and present information, which will help them write better on subsequent reports.

Attending to Equity

One helpful scaffold is to give students sentence stems for what to write in their feedback to the author. Ideal feedback is respectful, specific about what needs to be changed, and helpful in that it suggests how the author should go about making that change. You can review the writing of the emerging multilingual students in your class if they have limited English language proficiency, but make sure that all emerging multilingual students participate in the peer review process. You will see tremendous growth in their writing skills as they review their peers' writing and discuss what makes some reports stronger than others.

Activity C	Revise the Reports
Time Needed for the Activity	25 minutes
Grouping	Individuals

What to Do

1. Students should work on their own for this activity.
2. Show the "Reflect on the Feedback" page on a screen using the Presentation mode in the Learning Hub so everyone in the classroom can see it.
3. Say, "Please come pick up your reports from over here on this table." Point to the table.
4. Give the students a few minutes to pick up their reports and return to their seats.
5. Say, "Please take a few minutes to read over the peer review guide. You should use it to figure out what you need to change in your report and how you will change it."
6. Allow the students to read the PRG.
7. Show the "Create a Final Report" page on a screen using the Presentation mode in the Learning Hub so everyone in the classroom can see it.
8. Say, "Okay, let's revise our reports. I would like you to rewrite your report. You can use your draft report as a starting point but use the feedback on the peer review guide to make it better."
9. Ask the students, "Do you have any questions about what you need to do?"
10. Answer any questions that come up.
11. Tell the students, "Okay, let's write." Allow about 20 minutes for the students to revise their reports.
12. After about 20 minutes, give the following directions, "Okay, time's up. I will now come around and collect your first draft, final report, and peer review guide."
13. Collect all the first drafts, PRGs, and final reports.
14. Use the "Teacher Score" columns in the PRG to grade the final report. Be sure to give the students feedback about their writing in the "Teacher Comments" section.

What to Show First

The "Reflect on the Feedback" page in the ADI Learning Hub

What to Show Second

The "Create a Final Report" page in the ADI Learning Hub

Attending to Equity

Students don't always use the feedback they receive from others to improve the quality of their reports. Make sure you move from student to student and encourage them to make changes based on the feedback. If students are not sure how to revise their reports, model how to use feedback. For example, you might say, "Okay, based on the peer review guide, I need to do more to describe what I did to collect data. I bet if I add information about ________, readers will have a better idea of what I did."

Teaching Tip

The peer review guide (PRG) is designed to be used with any ADI investigation, thus allowing you to use the same scoring rubric throughout the entire year. This is beneficial for several reasons. First, the criteria for what counts as a high-quality report do not change from investigation to investigation. Students, therefore, quickly learn what is expected from them when they write a report, and you do not have to spend valuable class time explaining the various components of the PRG each time students write a report. Second, the PRG makes it clear which components of a report need to be improved next time because the grade is not based on a holistic evaluation of the report. The PRG format allows students to see which aspects of their writing are strong enough and which aspects need improvement. Finally, and perhaps most importantly, the PRG provides you with a standardized measure of student performance that can be compared over multiple reports across semesters, thus allowing you to track improvement over time.

Teacher Notes

Antibiotic Resistance in Bacteria

TASK

The Phenomenon

Doctors prescribe antibiotics to people with a bacterial infection. An example of a bacteria that causes an infection in humans is called *E. coli* (see image at right). Unfortunately, more and more types of bacteria are developing resistance to certain types of antibiotics. Take a few minutes to watch a video that shows how a population of *E. coli* can develop resistance to a type of antibiotic called ciprofloxacin. As you watch the video, write down some things you notice in the space below.

A cluster of E.coli bacteria

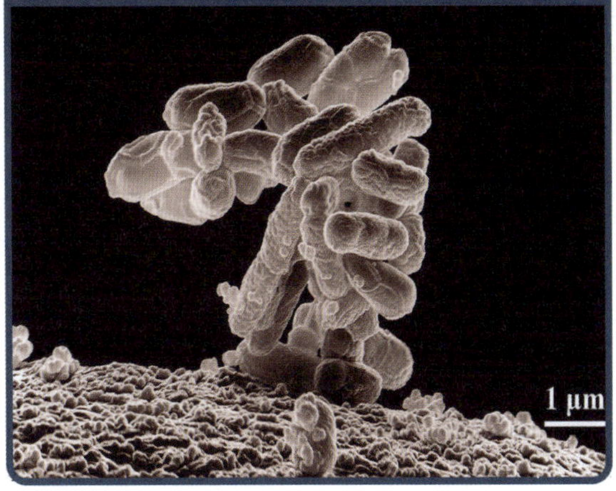

Things I **noticed...**

The Task

The video you watched showed how a population of *E. coli* developed resistance to the ciprofloxacin in only 12 days. Some populations of bacteria, however, take much longer to develop resistance to antibiotics, and other populations never do. Your goal in this investigation is to figure out how the interaction of different environmental conditions and specific characteristics, such as growth rate and mutation rate, affect how quickly a population of bacteria can change from being antibiotic sensitive to antibiotic resistant. The guiding question of this investigation is:

Why do some populations of bacteria develop antibiotic resistance faster than others?

Initial Ideas

Before you start this investigation, take a few minutes to think about how the bacteria in the video developed antibiotic resistance. Then, draw a model in the space below that shows what you think happened. Be sure to include labels to help explain your thinking.

Some **ideas** that I have...

Share your model with the rest of your group. Once everyone has had a chance to share their ideas, list some questions that your group has about this phenomenon.

Some **questions** we have now...

Antibiotic Resistance in Bacteria

IDEAS

Some Ideas You Can Use: Evolution by Means of Natural Selection

Evolution can be defined as a change in the traits of a population over time. One mechanism or means of evolution is a process called **natural selection**. Evolution by means of natural selection is a consequence of the interaction of four factors: (1) a variation in a trait within a population, (2) a selective pressure in the environment, such as a predator, lack of food, a disease, or a lack of mates, that prevents some individuals from being able to reproduce, (3) the individuals that are able to reproduce passing their traits on to the next generation, and (4) the proliferation of those organisms that are better able to reproduce in that environment. These four factors are illustrated in the image below using a hypothetical beetle population as an example.

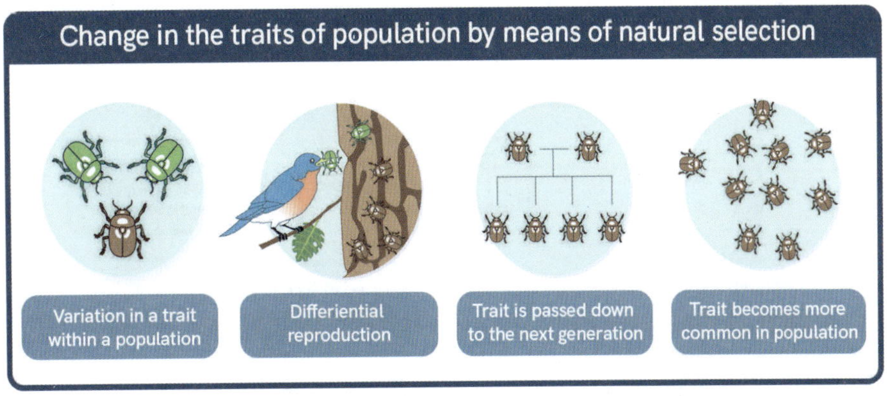

When a population evolves by means of natural selection, traits that support successful reproduction in a specific environment become more common; those that do not become less common. Natural selection, in other words, leads to the development of specific **adaptations**, that is, to a population dominated by organisms that are anatomically, behaviorally, and physiologically well suited to reproduce within a specific environment. Natural selection, however, can only occur if there is both (1) variation in the genetic information between organisms in a population and (2) variation in the expression of that genetic information, which leads to a variation in one or more traits. Without this **variation**, the process of natural selection cannot happen.

Some **useful ideas** from what I read...

Some Ideas You Can Use: Reproduction in Bacteria

A characteristic of all living things is the ability to reproduce. **Reproduction** is the process by which a new organism is created from one or more existing organisms. There are two main forms of reproduction: asexual and sexual. **Asexual reproduction** occurs when a single organism divides into two separate organisms. The new organism that arises through asexual reproduction receives a complete copy of its genetic information from its one parent. **Sexual reproduction**, in contrast, is the creation of an offspring from two existing parent organisms. To create a new offspring through sexual reproduction, an organism produces a specialized reproductive cell called a gamete. A **gamete** contains half the genetic information that is needed to create a new organism. A new organism is then created when two gametes fuse together. The image below is a simplified model of asexual and sexual reproduction.

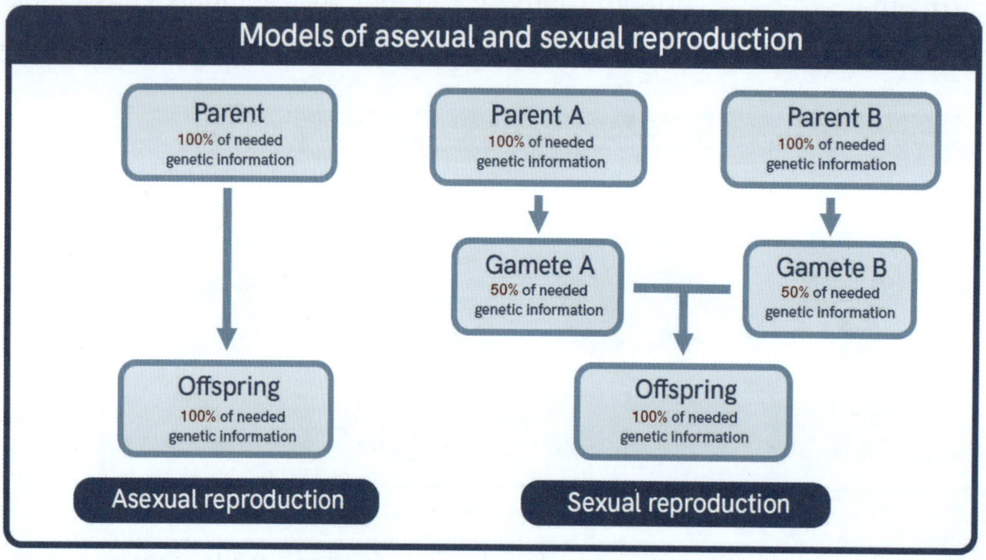

Bacteria reproduce asexually. The time it takes for a species of bacteria to reproduce is called its **growth rate**. The growth rate of bacteria is typically very fast, with many species able to reproduce in a few hours; for example, a common bacterium like *E. coli* can reproduce every 120 minutes under optimal conditions (see image below). The growth rate of bacteria varies by species and is also dependent on environmental conditions. Bacteria reproduce faster when they have access to unlimited food, space, and other resources and slower when they do not.

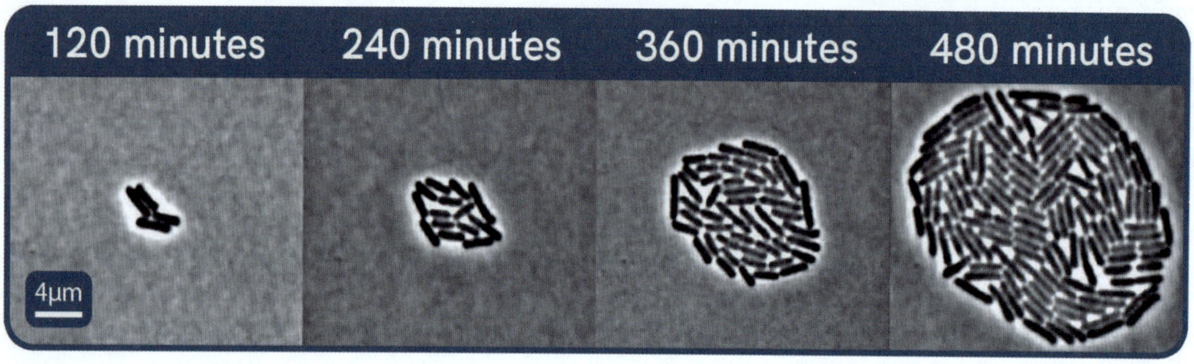

Bacteria can reproduce quickly because they only need to duplicate their genetic information and some other important cellular structures and then split into two. This process is called **binary fission**. A downside of binary fission, however, is each offspring is a clone of its parent, which results in no variation in genetic information. Offspring that are produced through sexual reproduction, in contrast, are a combination of the genetic information from two parents, which leads to much more variation between individuals in a population. The image below shows how much genetic information is shared between individuals over three generations because of asexual and sexual reproduction. Asexual reproduction only requires one individual (the parent) that splits into, and becomes, the offspring (two, in the case of binary fission). So, after two reproduction events, there will be a total of four individuals all of whom share 100% of the genetic information. Sexual reproduction, in contrast, requires two individuals per generation to create offspring. If each reproduction event produces one offspring, there will be five individuals alive in two generations. None of the individuals in the population will have 100% of their genetic information in common.

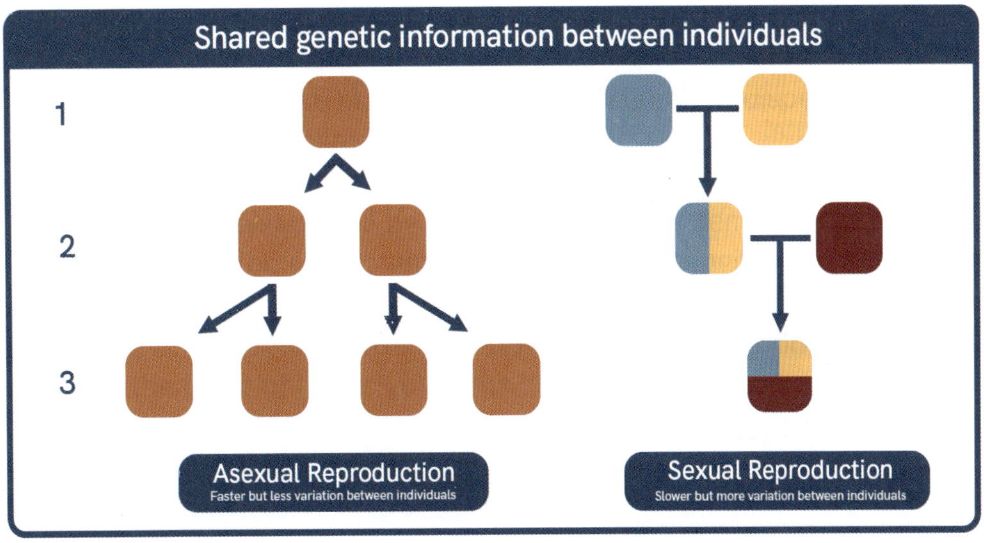

One of the only sources of genetic variation in a population of bacteria is through mistakes in DNA replication. Although DNA replication is tightly regulated and remarkably accurate, errors do occur and result in mutations. A **mutation** is simply a change in the genetic code of a gene from one generation to the next. Mutations are a source of genetic variation because even small changes in the DNA sequence of gene can lead to the production of a completely different protein, and it is the presence or absence of specific proteins that determines a specific trait. The two images in the picture on the next page illustrate how a single mutation in a gene can lead to the production of a different protein. The top part of the image shows how an existing, or "normal" gene codes for a specific protein. The bottom part shows what happens when an extra base is inserted into that gene due to a mistake in DNA replication. The mutation in the gene results in a different molecule of mRNA being produced, which in turn, codes for the production of a protein that is very different from the protein that is produced by the original gene.

How muations in DNA can lead to the production of different proteins

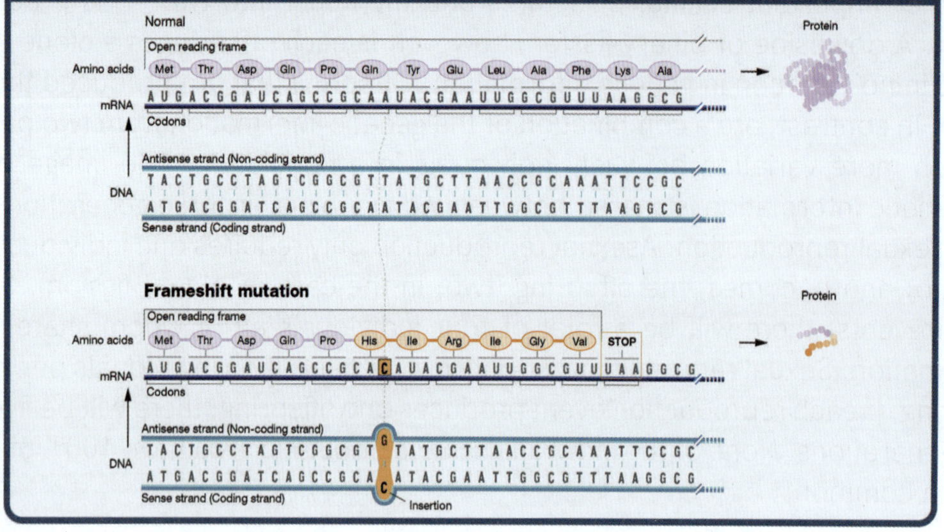

How often a mutation occurs when bacteria duplicate their DNA in order to reproduce is called the **mutation rate**. The mutation rate of bacteria is typically around 0.003 mutations per generation. However, mutation rates vary depending on the type of bacteria and environmental factors such as the presence of mutagens. Mutagens are chemicals or radiations that increase the frequency of mutation by damaging DNA.

Some **useful ideas** from what I read...

Some Ideas You Can Use: Using Models to Identify Cause-and Effect Relationships

Scientists often use models to help identify potential **cause-and-effect relationships**. A **model** can be physical, mathematical, or a computer-based simulation. You will use a computer-based simulation to model the development of antibiotic resistance in a population of bacteria. This simulation has been coded so it replicates what the scientists in the video did during their demonstration. The model includes a unit of time called a "tick". You can assume that each tick represents one hour. The **parts** of the model include an agar plate with sections that contain different doses of antibiotic, a population of bacteria, and a way to see the bacteria evolve without having to wait 12 days for the population to develop antibiotic resistance like in the video. This model, as a result, will enable you to figure how the interaction of the environment and specific bacteria characteristics, such as growth rate and mutation rate, affect how quickly a population of bacteria can change from being antibiotic sensitive to antibiotic resistant.

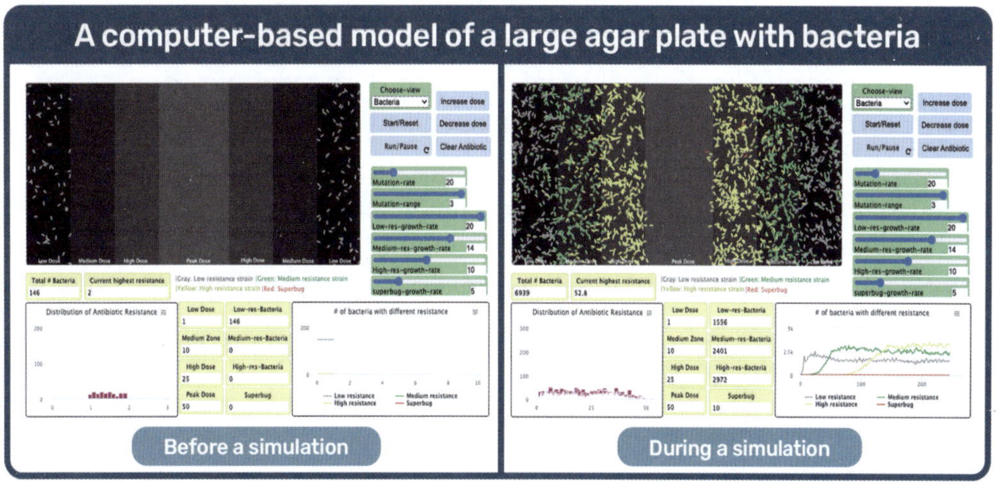

You can change an **input** of the model by changing either the environment or the bacteria. The simulated agar plate is always divided into seven sections. Two sections will contain a low dose antibiotic, two will contain a medium dose, two will contain a high dose, and one section will contain a maximum, or peak, dose of antibiotic. You can, however, increase the dose across all seven of the sections at the same time by clicking on the "increase dose" button, decrease it by clicking on the "decrease dose" button, or remove all the antibiotics from all seven sections by clicking on the "clear antibiotic" button.

The population of bacteria in the model start with the ability to survive a very low dose of antibiotic (indicated as resistant level 1 or 2 in the model). These bacteria are show as a gray color. The bacteria reproduce asexually. At a probability defined by the mutation rate, the offsprings' resistance may increase or decrease within the mutation range (up or down jumps 0-3). The bacteria will change their color to green when they are able survive a medium dose of antibiotic, yellow when they can survive the high dose, and red when they can survive the highest dose of antibiotic. All bacteria in the model also move a small distance in a random direction with each tick. You can change the mutation rate, the mutation range (the maximum possible change in antibiotic resistant level when a mutation occurs), and the growth rate of different strains of bacteria (how many ticks pass before they reproduce). These different inputs will allow you identify and test potential variables that would indicate a cause-and effect relationship.

The **output** in this model is the antibiotic resistance of the bacteria. The antibiotic resistance of the bacteria can be tracked in several different ways. You can track how many total bacteria are low resistant, medium resistant, high resistant, or completely resistant (which in the model is called a superbug) strain. You can also track the highest resistance level of any one bacterium in the population. There are also graphs that show the distribution of bacteria at each antibiotic resistance value and the number of bacteria in each strain of "antibiotic resistance level" (low, medium, high, superbug) changes over time. You will need to decide the best way to measure for the effect you are interested in tracking.

As you use the simulation, keep in mind that models like this can be useful for identifying possible cause-and-effect relationships but **they have limited precision and reliability** due to the assumptions and approximation inherent to them. For example, one assumption in this model is that we can control the growth rate of a specific strain of bacteria. An approximation is the fact the nutrients in the agar that the bacteria feed on are replenished with each tick, which does not happen in real bacterial culture. Models like these, as result, are good for testing ideas and to see what might happen and not for generating the empirical evidence that is needed to differentiate between cause and correlation, and make claims about specific causes and effects.

Some **useful ideas** from what I read...

Antibiotic Resistance in Bacteria

PLAN

Plan Your Investigation

Prepare a plan for your investigation by filling out the chart below.

I am trying to answer the following **question**...

I will use the following **observations or measurements**...

I will **collect** these observations or measurements by...

I will **analyze** the observations or measurements that I collect by...

I approve of this investigation plan.

Teacher's signature

Date

Antibiotic Resistance in Bacteria

DO

Collect Data

Keep a record of what you observe or measure during your investigation.

My **observations or measurements...**

Make Sense of Your Data

Use the space below to create your model so you can answer the guiding question.

My **model...**

Antibiotic Resistance in Bacteria

SHARE

Draft Argument

Develop an argument on a whiteboard. It should include:

1. A *claim* that answers the guiding question,
2. *Evidence* in support of the claim that provides an analysis of the data and a description of what the analysis means,
3. A *justification of the evidence* that explains why your group thinks the evidence is important.

Our Guiding Question:

Claim:

Evidence:	Justification of the Evidence:

Argumentation Session

Share your argument with your classmates. In the space below, be sure to keep track of any ideas that you can use to revise your argument and make it better.

Some possible ways to **improve** our argument...

Antibiotic Resistance in Bacteria REFLECT

Reflective Discussion

In the space below, you can keep track of any ideas from the discussion that you think are important or will be useful in the future.

Some important **ideas**...

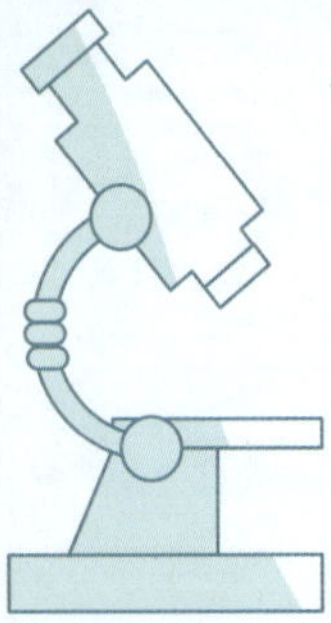

Antibiotic Resistance in Bacteria

REPORT

Draft Report

Prepare a report to share what you figured out during your investigation.

Introduction

We have been studying ____________________ in class. Before we started this investigation, we explored ____________________

We noticed ____________________

Our goal for this investigation was to figure out ____________________

The guiding question was ____________________

Method

To answer this question, ____________________

Argument

The model below shows ____________________

My evidence for this model is ____________________

This evidence is based on several important scientific ideas. The first one is ____________________

Cichlids of Lake Malawi Lesson Plan

OVERVIEW

Description	Students figure out why Lake Malawi is home to so many different species of cichlids.
Guiding Question	Why are there so many species of cichlids in Lake Malawi?
What Students Figure Out	The speciation of cichlids is driven by sexual selection with females choosing mates based on specific rituals, appearances, and timing.

STAGES

This investigation has seven stages. Each stage includes multiple activities. The flowchart below shows what happens during each stage and how much time is needed to complete each one.

Task (Day 1)
1 — **20 minutes**
Students are introduced to a phenomenon and a guiding question to answer. They also share their initial ideas.

Ideas (Day 1)
2 — **30 minutes**
Students read about DCI, a CC, and an SEP that they can use to help them figure out the phenomenon.

Plan (Day 2)
3 — **15 minutes**
Students formulate a plan collecting and analyzing the data they need to answer the guiding question.

Do (Day 2)
4 — **30 minutes**
Students carry out their plan, make sense of the data they collected, and answer the guiding question.

Share (Day 3)
5 — **45 minutes**
Students develop, share, and revise an evidence-based argument about what they figured out.

Reflect (Day 4)
6 — **20 minutes**
Students discuss a DCI and/or a CC and reflect on the SEPs they used during the investigation.

Report (Day 4)
7 — **25 minutes**
Students write investigation reports to share what they figured out during the investigation.

Report (Day 5)
7 — **45 minutes**
Students peer review the reports written by other students and offer feedback to improve.

Report (Day 6)
7 — **30 minutes**
Students revise their reports and then submit a final draft to their teacher for evaluation.

STANDARDS ALIGNMENT

NEXT GENERATION SCIENCE STANDARDS

Performance Expectation

1. Apply concepts of statistics and probability to support explanations that organisms with an advantageous heritable trait tend to increase in proportion to organisms lacking this trait (HS-LS4-3).
2. Evaluate the evidence supporting claims that changes in environmental conditions may result in increases in the number of individuals of some species, the emergence of new species over time, and the extinction of other species (HS-LS4-5).

Core Ideas(s)	Crosscutting Concept(s)	Practices
Natural Selection (LS4.B) Adaptation (LS4.C)	Systems and System Models (CC4)	Asking Questions (SEP1) Developing and Using Models (SEP2) Planning and Carrying Out Investigations (SEP3) Analyzing and Interpreting Data (SEP4) Using Mathematics and CT (SEP5) Constructing Explanations (SEP6) Engaging in Argument from Evidence (SEP7) Obtaining, Evaluating, and Communicating Information (SEP8)

WIDA ENGLISH LANGUAGE DEVELOPMENT STANDARDS FOR SCIENCE

Explain	Argue
Interpret scientific explanations (ELD-SC.9-12.Explain.Interpretive) Construct scientific explanations (ELD-SC.9-12.Explain.Expressive)	Interpret scientific arguments (ELD-SC.9-12.Argue.Interpretive) Construct scientific arguments (ELD-SC.9-12.Argue.Expressive)

COMMON CORE ENGLISH LANGUAGE ARTS

Reading	Writing	Speaking and Listening
Key ideas and details (RST.9-10.1 and RST.9-10.2) Craft and structure (RST.9-10.4) Integration of knowledge and ideas (RST.9-10.7, RST.9-10.8, and RST.9-10.9) Range of reading (RST.9-10.10)	Text types and purposes (WHST.9-10.1 and WHST.9-10.2) Production and distribution of writing (WHST.9-10.4, WHST.9-10.5, and WHST.9-10.6)	Comprehension and collaboration (SL.9-10.1 and SL.9-10.3) Presentation of knowledge and ideas (SL.9-10.4 and SL.9-10.6)

THINGS TO KNOW BEFORE STARTING THE LESSON

The Core Ideas

According to the phylogenetic species definition, a species is the smallest set of organisms that share an ancestor and can be distinguished from other such sets of organisms. A new species is created in what is known as a speciation event. A speciation event describes the point in time as well as the location where a new species splits off from an existing species. This event occurs because a small group of organisms within a larger population developed one or more characteristics that changed how the small group interacted with the larger population, usually making mating unlikely or unsuccessful.

These differing characteristics usually develop due to natural selection, sexual selection, or some combination of the two. Natural selection is the process by which organisms with traits that are advantageous to surviving and reproducing in their environment pass on these traits to their offspring, making this trait more common among the population. On the other hand, sexual selection is the process by which organisms with traits that are advantageous for attracting a mate are more likely to reproduce, causing this trait to become more common within the population. Natural and sexual selection both rely on variation in the genetic information between organisms in a population and variation in the expression of said genetic information in order to occur.

The Crosscutting Concepts

A system is a group of components that interact with one another to achieve some function. Because some systems are too complex to work with directly, scientists create models of the system so that they can test the system and learn more about how the system's components work together. However, because these models simplify complex systems, they inherently have a limited reliability and precision.

The Science and Engineering Practices

A practice of science is to **ask and refine questions** that lead to descriptions and explanations of how the natural world works. Another practice of science is to **use and construct models** as helpful tools for representing ideas and explanations. These tools include diagrams, drawings, physical replicas, mathematical representations, analogies, and computer simulations. Scientists **plan and carry out investigations** in the field or a laboratory, working collaboratively and individually. Their investigations are systematic and require clarifying what counts as data and identifying variables or parameters. Scientific investigations produce **data that must be analyzed and interpreted**. Because data patterns and trends are not always obvious, scientists use a range of tools and practices to identify the significant features and patterns in the data. Scientists also use **mathematics and computational thinking** for a range of tasks such as constructing simulations; statistically analyzing data; and recognizing, expressing, and applying quantitative relationships. The products of doing science are **explanations**. Scientific explanations often specify variables and are used by scientists to describe and predict different natural phenomena. **Scientific argumentation**, or the process of sharing, supporting, critiquing, and revising claims based on evidence is the process that scientists use to determine if an explanation is valid or acceptable. Finally, scientists must be able to **communicate clearly and persuasively** the ideas and methods they generate. Critiquing and communicating ideas individually and in groups is a critical professional activity.

PREPARATION

This lesson can be delivered paper-based, paperless, or remote. Use the paper-based delivery option of the investigation if students do not have 1:1 access to a computer and they are in the same classroom or physical space during the lesson. Use the paperless delivery option of the investigation if students have 1:1 access to a computer and they are in the same classroom or physical space during the lesson. Finally, if students are not in the same classroom or physical space during the lesson and have 1:1 access to a computer, use the remote delivery option.

Materials Needed

The materials needed by the teacher and each student in the class for each of the three modalities are listed in the table below.

Teacher Notes

Delivery Mode	Each Student Will Need	The Teacher Will Need
Paper-Based	• Copies of each student handout for the investigation (or a workbook) • A pencil	• A teacher account in the ADI Learning Hub • Access to a computer with high-speed internet and a browser (such as Chrome, Safari, Firefox, or Edge) • A projector or smart board (for displaying a computer screen)
Paperless	• A student account in the ADI Learning Hub • Access to a computer with high-speed internet and a browser (such as Chrome, Safari, Firefox, or Edge)	• A teacher account in the ADI Learning Hub • Access to a computer with high-speed internet and a browser (such as Chrome, Safari, Firefox, or Edge)
Remote	• A student account in the ADI Learning Hub • Access to a computer with high-speed internet and a browser (such as Chrome, Safari, Firefox, or Edge)	• A teacher account in the ADI Learning Hub • Access to a computer with high-speed internet and a browser (such as Chrome, Safari, Firefox, or Edge)

Additional Materials and Preparation Needed for In-Person Delivery

This in-person investigation will require students to have access to a Chromebook, desktop, laptop, iPad, or other electronic device in order to use a simulation. The simulation is embedded in the Learning Hub, so students do not need access to external sites. If students do not have access to the Learning Hub, they can use the simulation by visiting *Speciation* by NetLogo. The teacher can determine if each student will use an electronic device or if each group will share.

How the Simulation Works

This simulation replicates the coloration evolution of a fish population in a lake. Users create a boundary to divide the population, differentiate the environment, as well as adjust various factors, and then observe what happens to the fish color distribution over time. The fish in the simulation are given a random age from 0 to 9. Each fish executes the same set of rules for moving, growth and death, reproduction, and survival. The fish swim in the simulated lake. When a fish encounters a boundary, it resets its direction and moves away. The fish's age increases every hypothetical year. Any fish older than 20 years dies. In each "year", each female chooses a male from the adjacent area as a mate and produces an offspring. The female prefers a mate with a similar color. The color of the offspring is the average of its parents' colors unless there is a mutation that affects the offspring's coloring. The more that the fish's color differs from the color of the water, the more likely it is to become the prey of a visual predator. Finally, if the number of fish exceeds the maximum number that the lake can support, every fish has a 50% chance of dying regardless of its gender, age, or color.

How to Use the Simulation

The simulation interface is shown in the image below.

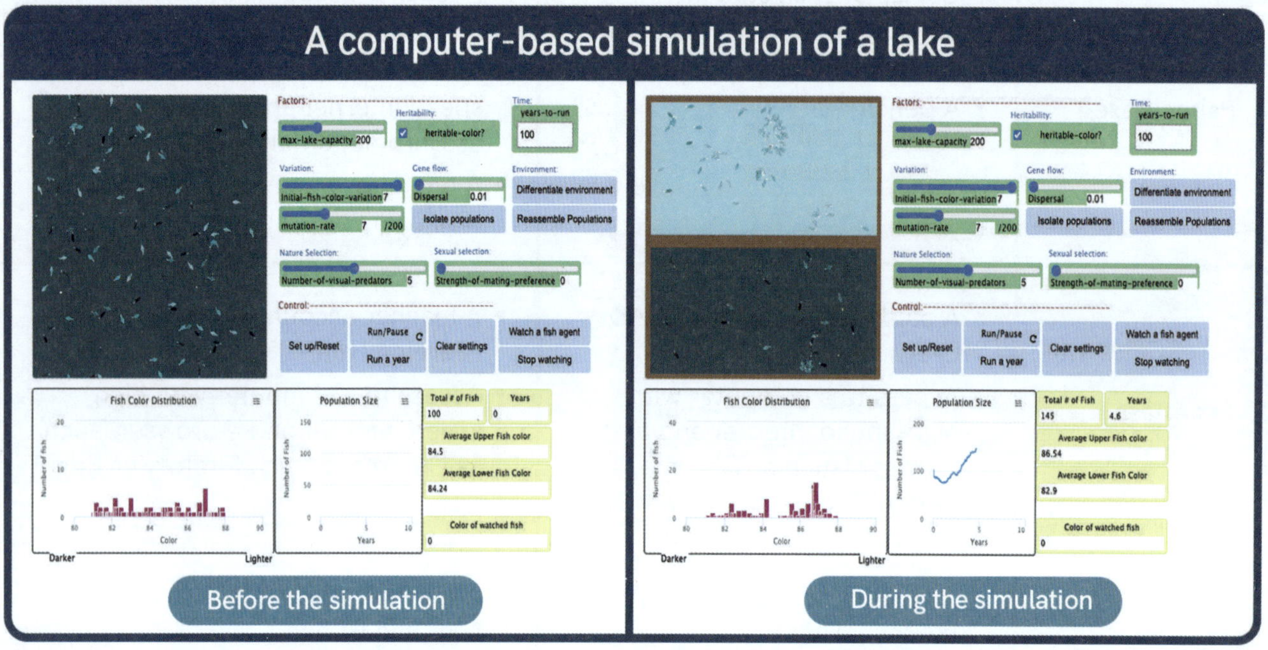

To start the simulation, click on the **SETUP** button and then click on the **RUN/PAUSE** button to start it. When the simulation starts, users will see a population of fish swimming in a lake, represented by a dark blue background. While running the simulation, users can click on the:

1. **ISOLATE POPULATIONS** button to create a boundary that divides the simulation window into upper and lower areas.
2. **DIFFERENTIATE ENVIRONMENT** button to set different environment colors for the upper and lower areas.
3. **REASSEMBLE POPULATION** button to remove the boundary and/or change the background to its original color.

To adjust the different factor(s), click on the **RUN/PAUSE** button to pause the simulation. Users can the change the following conditions using the green sliders:

1. **FISH-COLOR-VARIATION** sets how much variation there is in the color of fish.
2. **HERITABLE-COLOR?** makes the color trait inheritable.
3. **MUTATION-RATE** sets how often there is a random change in the fish color trait.
4. **MAX-LAKE-CAPACITY** sets the maximum number of fish that can be in the lake.
5. **DISPERSAL** controls the fish's ability to move to different parts of the lake.
6. **NUMBER-OF-VISUAL-PREDATORS** sets how many predators feed on the fish.
7. **STRENGTH-OF-MATING-PREFERENCE** determines if the female will chose a male with similar color or mate at random.

Once the conditions are set, click on the **SETUP** button and then click on **RUN/PAUSE** button to start the simulation. Users can the use the meters to track the number of fish in the environment, the average color of fish in different sections of the lake, and the numbers of years the simulation has run. The graphs allow users to track how the size of the fish population changes over time and the current distribution of fish color.

Safety Precautions

Remind students to follow all normal lab safety rules.

Laboratory Waste Disposal

There is no laboratory waste generated during this investigation.

FACILITATION NOTES

The facilitation notes for this investigation describe how to deliver this lesson in person. They are written specifically for the paper-based mode of delivery. These same facilitation notes, however, can also be used for a paperless delivery of the lesson by simply substituting the directions that mention paper with the entry into the ADI Learning Hub. If you are teaching this lesson using the remote mode of delivery, you only need to assign due dates and prerequisites for each stage. The students can then simply follow the directions built into the ADI Learning Hub as they move through the investigation.

ADI Go Digital

The digital version of this lesson is available in the **ADI Learning Hub**. Be sure to check out the built-in teacher presentation mode.

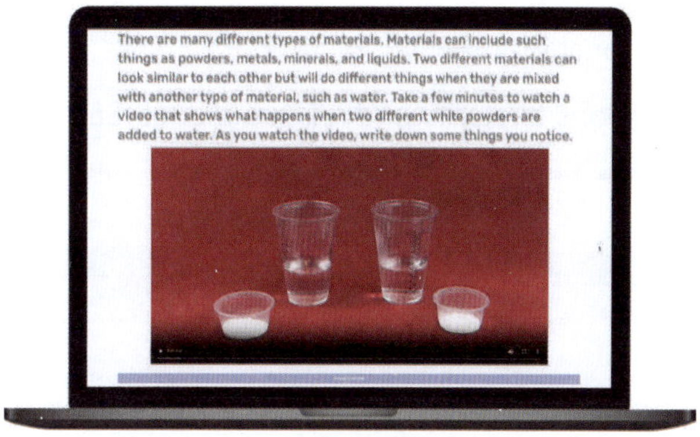

Teacher Notes

Cichlids of Lake Malawi Lesson Plan

Stage 1 TASK

Total Time for the Stage	45 minutes
Grouping	Individual; Teams

Learning Goals

Students should make progress on their ability to do the following by the end of this stage of the lesson:

Science

1. Ask questions that arise from careful observation of phenomena to clarify and/or seek additional information (SEP1).
2. Develop and use a model to provide mechanistic accounts of phenomena (SEP2).
3. Evaluate the merits and limitations of different models of the same phenomenon (SEP2).
4. Ask questions that arise from examining models to clarify and/or seek additional information and relationships (SEP 1).
5. Ask questions to clarify and refine a model (SEP1).

English Language Development

None

English Language Arts

1. Initiate and participate effectively in small-group collaborative discussions with diverse partners by building on others' ideas and expressing their own clearly (SL.9-10.1).
2. Present information clearly, concisely, and logically such that listeners can follow the line of reasoning and the organization, substance, and style are appropriate to the purpose, audience, and task (SL.9-10.4).
3. Adapt speech to a variety of contexts and tasks, demonstrating command of formal English when indicated or appropriate (SL.9-10.6).

Activities within the Stage

Activity A	The Phenomenon
Time Needed for the Activity	5 minutes
Grouping	Teams

What to Do

1. Have students work in their assigned teams for this activity.
2. Begin the activity by giving out a copy of the Task handout to each student.
3. Show "The Phenomenon" page on a screen using the Presentation mode in the Learning Hub so everyone in the classroom can see it.
4. Read the introduction of the phenomenon out loud and then start the video. You can also have students read the introduction of the phenomenon and watch the video on their own devices.
5. Give the students a few moments to write down their observations on their handouts before moving on to the next activity.

What to Give Out

Task handout for **Cichlids in Lake Malawi**

What to Show

"The Phenomenon" page in the ADI Learning Hub

Teaching Tip

Instead of assigning fixed roles to students by name at the beginning of the investigation, give each member of a group a different number so you can easily assign and change roles as needed during the various investigation stages. For example, you might give directions such as "1s will share out for your group" or "3s, will you please come back here and pick up these materials for your group?" or "2s and 3s, you will present the argument for your group while 1s and 4s travel to the other groups to offer feedback."

Attending to Equity

Be sure to intentionally group emerging multilingual students at times with peers who know the same languages as they do and at other times with peers whose English language development is slightly more advanced. Thoughtful grouping that varies throughout an investigation allows emerging multilingual students to benefit from the different linguistic resources of their peers.

Activity B	Share What You Noticed
Time Needed for the Activity	5 minutes
Grouping	Teams

What to Do

1. Show the "Share What You Noticed" page on a screen using the Presentation mode in the Learning Hub so everyone in the classroom can see it.
2. Say, "Okay, please share what you noticed with the others at your table."
3. Give the students about two minutes to share at their tables.
4. Say, "Okay, let's take a few minutes to share with the class. Team X (call on a team at random), what did you talk about?"
5. Allow Team X to share something they noticed. If it is related to speciation, natural selection, or sexual selection, say, "That is really interesting. We are actually going to examine that in more detail during this investigation." If what is shared is not related to speciation, natural selection, or sexual selection, say, "That is really interesting. Team Y (call on another team at random), what did you notice?" and allow that team to share. Follow this pattern until a team mentions something related to why there are so many species of cichlids in Lake Malawi. If no team mentions this, say, "You know what I found interesting? I saw that there are hundreds of species of cichlids in Lake Malawi. Did anyone notice that as well?"
6. Move on to the next activity.

What to Show

The "Share What You Noticed" page in the ADI Learning Hub

What to Look For

As soon as a group mentions something related to why there are so many species of cichlids in Lake Malawi, move on to the next activity.

Attending to Equity

Be sure to provide students with opportunities to talk in small groups during this activity. Small-group structures offer all students, including emerging multilingual students, a chance to engage in sense-making with their peers and offers them space to use their verbal and nonverbal (e.g., writing, drawing pictures, and using gestures) resources to express their ideas. Small-group structures also give students a chance to learn from other students' uses of these resources.

Activity C	What You Need to Figure Out
Time Needed for the Activity	5 minutes
Grouping	Teams

What to Do

1. Show the "What You Need to Figure Out" page on a screen using the Presentation mode in the Learning Hub so everyone in the classroom can see it.
2. Read the task out loud and have students follow along on their handout, on their devices, or on the screen.
3. Allow students to ask clarifying questions about the task or the guiding question.
4. Once everyone understands what they will need to figure out during the investigation, move on to the next activity.

What to Show

The "What You Need to Figure Out" page in the ADI Learning Hub

Teaching Tip

Have students highlight, circle, or place a star next to the guiding question. Marking the guiding question helps students find it later during the Plan stage.

Teacher Notes

Activity D	Share Your Ideas
Time Needed for the Activity	15 minutes
Grouping	Teams

What to Do

1. Show the "Share Your ideas" page on a screen using the Presentation mode in the Learning Hub so everyone in the classroom can see it.
2. Read the directions on the handout out loud and then give students five minutes to draw their initial representations of the science.
3. Say, "Okay, will everyone share their representations with the rest of your team? Please make sure everyone has time to talk. Also, be sure to ask one another questions about the scientific ideas that are included in each representation."
4. Give the students about 8-9 minutes to talk and share. Be sure to walk around and listen to the discussions.
5. Say, "Okay, I heard some interesting ideas as I walked around the room."
6. Move on to the next stage.

What to Show

The "Share Your Ideas"page in the ADI Learning Hub

What to Look For

Each student's initial representation should include a combination of symbols, diagrams, graphs, or words. It should also clearly show how a scientific idea can be used to help complete the task and answer the guiding question.

Attending to Equity

When students are tasked with expressing their ideas, be sure to encourage them to do so through multiple modes (speaking, writing, drawings, and gestures) and to use both content-specific and everyday registers. This process helps students fully express themselves, allowing peers to evaluate, question, and build on ideas. Also, make sure that every student in every team has a chance to share their initial representation.

Activity E	More Questions
Time Needed for the Activity	15 minutes
Grouping	Teams

What to Do

1. Show the "More Questions" page on a screen using the Presentation mode in the Learning Hub so everyone in the classroom can see it.
2. Say, "Okay, I heard a bunch of interesting questions as I was walking around the room. Will you work with your team to identify two or three questions that you think are important to learn more about that came up as you all discussed the different representations you made? You can write these questions on your handout."
3. Give students about five minutes to talk and write down questions. Be sure to walk around and listen to the discussions.
4. Say, "Okay, let's take a few minutes to share. Team X (call on a team at random), will you share one question you all have?"
5. Have each team share a question. Then say, "I want to share information about some useful scientific ideas with you. I bet this information will not only answer many of these questions but will also help you complete the task and answer the guiding question of the investigation."
6. Move on to the next stage.

What to Show

The "More Questions" page in the ADI Learning Hub

What to Look For

Each student should have 2-3 questions written on their handout.

Teaching Tip

When students are asked to create representations of the underlying scientific ideas, share these ideas with their peers, and then identify things they want to know more about, it sparks their curiosity and helps make the investigation meaningful. Also, identifying questions before reading (which students do during the Ideas stage) is an outstanding pre-reading activity that helps improve comprehension.

Cichlids of Lake Malawi Lesson Plan

Stage 2 IDEAS

Total Time for the Stage	30 minutes
Grouping	Teams

Learning Goals

Students should make progress on their ability to do the following by the end of this stage of the lesson:

Science

1. Ask questions that arise from examining theory to clarify and/or seek additional information and relationships (SEP1).
2. Critically read scientific literature adapted for classroom use to determine the central ideas or conclusions and/or to obtain scientific and/or technical information to summarize complex concepts, processes, or information presented in a text by paraphrasing them in simpler but still accurate terms (SEP8).
3. Integrate sources of information presented in different media or formats (e.g., visually and quantitatively) as well as in words to address a scientific question (SEP8).
4. Gather, read, and evaluate scientific and/or technical information from multiple authoritative sources, assessing the evidence and usefulness of each source (SEP8).
5. Describe how natural selection occurs only if there is both variation in the genetic information between organisms in a population and variation in the expression of that genetic information, leading to differences in performance among individuals (LS4.B).
6. Explain how traits that positively affect an organism's survival are more likely to be reproduced, and thus become more common in the population (LS4.B).
7. Describe how natural selection leads to adaptation, or to a population dominated by organisms that are anatomically, behaviorally, and physiologically well suited to survive and reproduce in a specific environment (LS4.C).
8. Identify how changes in an environment have contributed to the expansion of some species, the emergence of new distinct species, and the decline and extinction of other species (LS4.C).
9. Explain how species become extinct because they can no longer survive and reproduce in their altered environment (LS4.C).
10. Use models to simulate systems and interactions within and between systems at different scales (CC4).

English Language Development

1. Paraphrase central ideas in complex evidence, concepts, processes, and information to help explain how or why a phenomenon occurs (ELD-SLC.9-12.Explain.Interpretive).

ARGUMENT-DRIVEN INQUIRY.

English Language Arts

1. Determine the central ideas or conclusions of a text; trace the text's explanation or depiction of a complex process, phenomenon, or concept; provide an accurate summary of the text (RST.9-10.2).
2. Determine the meaning of symbols, key terms, and other domain-specific words and phrases as they are used in a specific scientific or technical context (RST.9-10.4).
3. Read and comprehend scientific/technical texts (RST.9-10.10).
4. Initiate and participate effectively in small-group collaborative discussions with diverse partners by building on others' ideas and expressing their own clearly (SL.9-10.1).

Teacher Notes

Activities

Activity A	Read About a Core Idea
Time Needed for the Activity	10 minutes
Grouping	Teams

What to Do

1. Have students work in their assigned teams during this stage. Give out a copy of the Ideas handout to each student.
2. The first activity provides students with information about species and speciation.
3. Show the "Read About a Core Idea You Can Use" page on a screen using the Presentation mode in the Learning Hub so everyone in the classroom can see it. You can also click on the images to make them full screen and easier for students to see.
4. Have students read the text on their own, either on their handouts or on the screen.
5. Give the students about three minutes to discuss the text with the others in their teams and develop a shared list of ideas that they think are important or useful.
6. Ask a team to share an idea they thought was useful. Have the remaining teams share a different idea until there are no more ideas to share.
7. Move on to the next activity.

Teaching Tip

Students do not need to do a close read of this text or completely understand all the ideas in it before moving on to the next activity. They will be encouraged to revisit this text later in the investigation when creating their arguments and writing their reports. All students need to be able to do during this activity is to work with the other members of their group to identify one or two ideas that they think are important to keep in mind or are potentially helpful.

What to Give Out

Ideas handout for **Cichlids of Lake Malawi**

What to Show

The "Read About a Core Idea You Can Use" page in the ADI Learning Hub

What to Look For

Students should write down some of the following ideas in the box labeled, "Some things I know from what I read" on their handouts:

- According to the phenotypic species definition, a species is a group of organisms that look the same.
- According to the biological species definition, a species is a group of organisms that can interbreed with each other in nature.
- According to the phylogenetic species definition, a species is the smallest set of organisms that share an ancestor and can be distinguished from other such sets.
- Speciation describes when a new species is created.
- A speciation event is a point in time and location where a new species splits off from an existing one.

Activity B	Read About a Core Idea
Time Needed for the Activity	10 minutes
Grouping	Teams

What to Do

1. The second activity provides students with information about natural selection and sexual selection.
2. Show the "Read About Another Core Idea You Can Use" page on a screen using the Presentation mode in the Learning Hub so everyone in the classroom can see it. You can also click on the images to make them full screen and easier for students to see.
3. Have students read the text on their own, either on their handouts or on the screen.
4. Give the students about three minutes to discuss the text with the others in their teams and develop a shared list of ideas that they think are important or useful.
5. Ask a team to share an idea they thought was useful. Have the remaining teams share a different idea until there are no more ideas to share.
6. Move on to the next activity.

What to Show

The "Read About Another Core Idea You Can Use" page in the ADI Learning Hub

What to Look For

Students should write down some of the following ideas in the box labeled, "Some things I know from what I read" on their handouts:

- A speciation event can occur due to natural selection, sexual selection, or a combination of the two processes.
- Natural selection is the process by which organisms in a population become more suited for their environment because they inherit traits that are more advantageous for their survival.
- Natural selection is a consequence of the amount of variation within a population, a selective pressure in the environment that results in some individuals surviving to reproduce while others do not, and the inheritability of the traits that support survival.
- Sexual selection describes when traits become more prominent within a population due to how these traits help an organism attract a mate.
- Natural and sexual selection can only occur if there is both variation in the genetic information between organisms in a population and variation in the expression of that genetic information.

Attending to Equity

There are many supports for helping students comprehend what they read already embedded into this activity (e.g., activating prior knowledge, providing a shared experience, making connections, synthesizing, and talking with peers). So, you might not need to provide much extra support. If you are concerned about some students not understanding what they are reading because of their scores on past reading comprehension tests, you can read it out loud as they follow along. As you read the text out loud, be sure to stop at each important idea and ask the students to put a star (or other annotation) next to it in the margin of their handouts. They can then discuss these ideas in their small groups.

Activity C	Read About a Final Core Idea
Time Needed for the Activity	10 minutes
Grouping	Teams

What to Do

1. The third activity provides students with information about using models to represent a system.
2. Show the "Read About a Final Core Idea You Can Use" page on a screen using the Presentation mode in the Learning Hub so everyone in the classroom can see it. You can also click on the images to make them full screen and easier for students to see.
3. Have students read the text on their own, either on their handouts or on the screen.
4. Give the students about three minutes to discuss the text with the others in their teams and develop a shared list of ideas that they think are important or useful.
5. Ask a team to share an idea they thought was useful. Have the remaining teams share a different idea until there are no more ideas to share.
6. Move on to the next activity

What to Show

The "Read About a Final Core Idea You Can Use" page in the ADI Learning Hub

What to Look For

Students should write down some of the following ideas in the box labeled, "Some things I know from what I read" on their handouts:

- Models are used to represent an object, event, or process.
- Models can be physical, visual, or a computer-based simulation.
- Models have limited reliability and precision due to the approximations and assumptions used to create the model.

Attending to Equity

To support emerging multilingual students during this activity, be sure to create a word bank on the wall of high-frequency, high-need concrete vocabulary that is needed to communicate scientific ideas. The word bank highlights terms needed for basic communication in an academic context. It also helps with the spelling. You can also encourage your emerging multilingual students to:

1. Use strategic learning techniques such as concept mapping, drawing, memorizing, comparing, contrasting, and reviewing.
2. Decode (sound out) words using skills such as recognizing sound-letter relationships and identifying cognates, affixes, roots, and base words.
3. Use visual and contextual support and support from peers and teachers to enhance and confirm understanding of the texts.

Cichlids of Lake Malawi Lesson Plan

Stage 3 PLAN

Total Time for the Stage	15 minutes
Grouping	Teams

Learning Goals

Students should make progress on their ability to do the following by the end of this stage of the lesson:

Science

1. Evaluate a question to determine if it is testable and relevant (SEP1).
2. Plan an investigation individually and collaboratively to produce data to serve as the basis for evidence as part of building and revising models (SEP3).
3. Consider possible variables or effects and evaluate the confounding investigation's design to ensure variables are controlled (SEP3).
4. Ask questions to determine relationships, including quantitative relationships, between independent and dependent variables (SEP1).
5. Decide on types, quantity, and accuracy of data needed and consider limitations on the precision of the data (e.g., number of trials, cost, risk, and time), and refine the design accordingly (SEP3).
6. Select appropriate tools to collect, record, and analyze data (SEP3).

English Language Development

1. Define investigable questions on observations, information, and/or data about a phenomenon (ELD-SLC.9-12.Explain.Interpretive).
2. Develop reasoning to illustrate and/or predict the relationships between variables (ELD-SLC.9-12.Explain.Expressive).

English Language Arts

1. Initiate and participate effectively in small-group collaborative discussions with diverse partners by building on others' ideas and expressing their own clearly (SL.9-10.1).
2. Present information clearly, concisely, and logically such that listeners can follow the line of reasoning and the organization, substance, and style are appropriate to the purpose, audience, and task (SL.9-10.4).
3. Adapt speech to a variety of contexts and tasks, demonstrating command of formal English when indicated or appropriate (SL.9-10.6).

Activities

Activity A	Some Materials You Can Use
Time Needed for the Activity	5 minutes
Grouping	Teams

What to Do

1. Show the "Some Materials You Can Use" page on a screen using the Presentation mode in the Learning Hub so everyone in the classroom can see it. You can also click on the images to make them full screen and easier for students to see.
2. Say, "Okay, it sounds like we now have enough information to plan an investigation that will help us figure out why there are so many species of cichlids in Lake Malawi."
3. Say, "Before we get started, I want to show you some materials that we are going to use during our investigation. Will you all look at the screen so I can show you what we have?"
4. Direct students to look at the screen where the simulation has been pulled up.
5. Ask, "What do you think we could use this simulation for?"
6. Allow students to share ideas. Then say, "We can use the simulation to test what conditions would cause a group of fish to evolve into different species."
7. Say, "Okay, that is all the stuff we can use. Will you all head back to your teams so we can plan our investigation?"
8. Move on to the next activity.

What to Set Up

Pull up the simulation in the Learning Hub on a screen so that all students can see it. If you do not have a Learning Hub subscription, you can access the simulation by visiting *Speciation* by NetLogo.

What to Show

The "Create a Final Report" page in the ADI Learning Hub

Teaching Tip

This activity helps students see for themselves all the materials they will be able to use during the hands-on part of the investigation instead of just reading about them or looking at pictures. It is important to show each material to the students and then point out how they can use it during the investigation so they can participate in developing a plan during the next activity.

Activity B	Make a Plan
Time Needed for the Activity	10 minutes
Grouping	Teams

What to Do

1. Have students work in their assigned teams during this activity. Give a copy of the Plan handout to each student.
2. Show the "Make a Plan" page on a screen using the Presentation mode in the Learning Hub so everyone in the classroom can see it.
3. Say, "Work as a team to create a plan for your investigation by filling out the chart on your handout. As part of your plan, be sure to include which tools and techniques you will use to solve problems."
4. Give the students at least 10 minutes to complete this task. As students work in small teams to create their investigation proposals, be sure to move from team to team to answer questions and offer advice.
5. When a team thinks that they have a good plan, read it over. If you agree, sign off at the bottom of their handouts to let them know that their plan has been approved. If you do not agree, tell them what they will need to add or change before you will approve it. Once they have made the changes you can sign their handouts.
6. Once all the teams are done, you can move on to the next activity. You can also use an approved investigation plan as an "exit ticket" out of class.

What to Give Out

Plan handout for **Cichlids of Lake Malawi**

What to Show

The "Make a Plan" page in the ADI Learning Hub

What to Look For

The investigation plan should, at a minimum, include:

- The question they are trying to answer.
- What conditions they will change in the simulation.
- What data they will collect using the simulation.
- How they will analyze the data they collect.
- How they will use their analysis to answer the guiding question.

Attending to Equity

To support emerging multilingual students during this stage, be sure to create a word bank on the wall of high-frequency, high-need concrete vocabulary that is needed to communicate scientific processes and ideas. The word bank highlights terms needed for basic communication in an academic context. It also helps with the spelling. Also, you can encourage your emerging multilingual students to use modes (speaking, writing, drawings, and gestures) and to use both content-specific and everyday registers to discuss and create their plans. This process helps students fully express themselves, allowing peers to evaluate, question, and build on ideas.

Teaching Tip

One way to do a quick check-in with a group during this activity is to use the "two questions, a nudge, and a goal" technique. This technique includes the following steps:

1. Sit down with a group of students.
2. Ask, "Would one of you tell me a little about what you all have been doing or thinking about?" (Question 1)
3. A student answers on behalf of the group.
4. Ask, "Would anyone else like to add anything to that?" (Question 2)
5. A different student provides additional information on behalf of the group.
6. Ask, "Have you thought about [a hint or something you want to highlight]?" (A nudge)
7. Say, "Okay, I'm going to check on the other groups. When I get back, I want to see [something that you want the group to accomplish]." (A goal)

When you have completed the check-in with one group, move on to the next group.

Teacher Notes

Cichlids of Lake Malawi Lesson Plan

Stage 4 DO

Total Time for the Stage	45 minutes
Grouping	Teams

Learning Goals

Students should make progress on their ability to do the following by the end of this stage of the lesson:

Science

1. Conduct an investigation individually and collaboratively to produce data to serve as the basis for evidence (SEP3).
2. Analyze data using tools and/or technologies to make valid and reliable scientific claims (SEP4).
3. Consider limitations of data analysis (e.g., measurement error and sample selection) when analyzing and interpreting data (SEP4).
4. Evaluate the impact of new data on a working explanation and/or model of a proposed process or system (SEP4).
5. Apply ratios, rates, percentages, and unit conversions in the context of complicated measurement problems involving quantities with derived or compound units (SEP5).
6. Make a quantitative and/or qualitative claim regarding the relationship between dependent and independent variables (SEP6).
7. Construct and revise an explanation based on valid and reliable evidence obtained from a variety of sources and the assumption that theories and laws that describe the natural world operate today as they did in the past and will continue to do so in the future (SEP6).
8. Apply scientific ideas, principles, and/or evidence to provide an explanation of phenomena taking into account possible unanticipated effects (SEP6).

English Language Development

1. Paraphrase central ideas in complex evidence, concepts, processes, and information to help explain how or why a phenomenon occurs (ELD-SLC.9-12.Explain.Interpretive).
2. Evaluate the extent to which reasoning, theory, and/or models link evidence to claims and support conclusions (ELD-SLC.9-12.Explain.Interpretive).
3. Develop reasoning to illustrate the relationships between variables (ELD-SLC.9-12.Explain.Expressive).

English Language Arts

1. Initiate and participate effectively in small-group collaborative discussions with diverse partners by building on others' ideas and expressing their own clearly (SL.9-10.1).
2. Present information clearly, concisely, and logically such that listeners can follow the line of reasoning and the organization, substance, and style are appropriate to the purpose, audience, and task (SL.9-10.4).
3. Adapt speech to a variety of contexts and tasks, demonstrating command of formal English when indicated or appropriate (SL.9-10.6).

Activities

Activity A	Carry Out Your Plan
Time Needed for the Activity	25 minutes
Grouping	Teams

What to Do

1. Students should work in their assigned teams for this activity. Begin this activity by giving out a copy of the Do handout to each student.
2. Show the "Carry Out Your Plan" page on a screen using the Presentation mode in the Learning Hub so everyone in the classroom can see it.
3. Say, "Okay, it sounds like we are now ready to carry out our plans and collect the data or observations we need to answer the guiding question. You can get the materials and tools you need to use from this table." Point to the table where you have placed them.
4. Ask, "Does anyone have any questions about these materials or tools?"
5. Answer any questions that come up.
6. Say, "As you work, be sure to record your observations and measurements on your handout so you can keep track of everything you do. Does anyone have any final questions?"
7. Answer any questions that come up.
8. Say, "Okay, go ahead and get started. You have the next 25 minutes to work. I'll be available to help if any team needs me."
9. Give the students at least 25 minutes to carry out their plans and solve the problem. As students work in small teams, be sure to move from team to team to answer questions and offer advice.

What to Set Up

Provide each student, or group of students, with access to a computer so they can use the computer-based model. Students can access the model in the Learning Hub using their accounts. If students do not have a Learning Hub account, they can access the simulation by going to *Speciation* by NetLogo.

What to Give Out

Do handout for **Cichlids of Lake Malawi**

What to Show

The "Carry Out Your Plan" page in the ADI Learning Hub

Teaching Tip

One way to do a quick check-in with a group during this activity is to use the "two questions, a nudge, and a goal" technique. This technique includes the following steps:

1. Sit down with a group of students.
2. Ask, "Would one of you tell me a little about what you all have been doing or thinking about?" (Question 1)
3. A student answers on behalf of the group.
4. Ask, "Would anyone else like to add anything to that?" (Question 2)
5. A different student provides additional information on behalf of the group.
6. Ask, "Have you thought about [a hint or something you want to highlight]?" (A nudge)
7. Say, "Okay, I'm going to check on the other groups. When I get back, I want to see [something that you want the group to accomplish]." (A goal)

When you have completed the check-in with one group, move on to the next group.

Attending to Equity

Be sure to intentionally group emerging multilingual students at times with peers who know the same languages as they do and at other times with peers whose English language development is slightly more advanced. Thoughtful grouping that varies during different investigation (or activities within an investigation) allows emerging multilingual students to benefit from the different linguistic resources of their peers. We also recommend assessing emerging multilingual students based on how much they contribute to the group during this activity, rather than solely on what they write or draw on their handouts.

Teacher Notes

Activity B	Make Sense of Your Data
Time Needed for the Activity	20 minutes
Grouping	Teams

What to Do

1. Students should work in their assigned teams for this activity. Begin this activity by showing the "Make Sense of Your Data" page on a screen using the Presentation mode in the Learning Hub so everyone in the classroom can see it.
2. Say, "Okay, it sounds like we are now ready to make sense of the data we collected. One way to do this is to create a table, graph, or other visualization that shows a trend or change over time, a comparison between teams, or a relationship between things you tested. You can create a table, graph, or other visualization here on your handout."
3. Point to the box on the handout.
4. Say, "Please work together in your teams to figure out the best way to make sense of your data. Remember, you are looking for any trends, changes, differences, or relationships in the data that will help you answer the guiding question. As you work, be sure to use the space on your handout to perform any calculations that you need to do and create any representations that you need. Does anyone have any last questions?"
5. Answer any questions that come up.
6. Say, "Okay, go ahead and get started. You have the rest of the period to work. I'll be available to help if any team needs me."
7. Give the students at least 15 minutes to make sense of their data. As students work together in small teams, be sure to move from team to team to answer questions and offer advice.
8. Make sure students return the materials and tools to the table where they found them before dismissing class. Cleanup should take no longer than five minutes.

What to Show

The "Make Sense of Your Data" page in the ADI Learning Hub

What to Look For

One of the most useful ways to make sense of the data during this activity is to make several graphs that compare how different conditions affected the fish population and its characteristics.

Teaching Tip

Collecting and returning materials or tools to a central table works best if you assign the job to a specific number within each team. For example, you can say, "2s, please come up and grab the materials and tools that your team needs" at the start of the activity. Then you can say, "3s, please return the materials and tools your team used back to this table," at the end of the activity. Assigning jobs like this not only cuts down on how many students need to be up and moving around, but also allows you to hold specific students accountable for the materials and tools.

Attending to Equity

Encourage your emerging multilingual students to speak using learning strategies such as requesting assistance, employing nonverbal cues, and using synonyms and circumlocution (conveying ideas by defining or describing when exact English words are not known) during this activity. It is also helpful to point out and/or draw emerging multilingual learners' attention to elements of the English sound system in newly acquired vocabulary such as long and short vowels, silent letters, and consonant clusters as they come up naturally during this activity.

Teacher Notes

Cichlids of Lake Malawi Lesson Plan

Stage 5 SHARE

Total Time Needed for the Stage	45 minutes
Grouping	Teams

Learning Goals

Students should make progress on their ability to do the following by the end of this stage of the lesson:

Science

1. Conduct an investigation individually and collaboratively to produce data to serve as the basis for evidence (SEP3).
2. Analyze data using tools and/or technologies to make valid and reliable scientific claims (SEP4).
3. Consider limitations of data analysis (e.g., measurement error and sample selection) when analyzing and interpreting data (SEP4).
4. Evaluate the impact of new data on a working explanation and/or model of a proposed process or system (SEP4).
5. Apply ratios, rates, percentages, and unit conversions in the context of complicated measurement problems involving quantities with derived or compound units (SEP5).
6. Make a quantitative and/or qualitative claim regarding the relationship between dependent and independent variables (SEP6).
7. Construct and revise an explanation based on valid and reliable evidence obtained from a variety of sources and the assumption that theories and laws that describe the natural world operate today as they did in the past and will continue to do so in the future (SEP6).
8. Apply scientific ideas, principles, and/or evidence to provide an explanation of phenomena taking into account possible unanticipated effects (SEP6).

English Language Development

1. Identify appropriate and sufficient evidence from data, models, and/or information from investigations of a phenomenon (ELD-SLC.9-12.Argue.Interpretive).
2. Compare reasoning and claims based on evidence from competing arguments (ELD-SLC.9-12.Argue.Interpretive).
3. Defend or refute a claim based on data and evidence (ELD-SLC.9-12.Argue.Expressive).
4. Establish and maintain an appropriate tone and stance (ELD-SLC.9-12.Argue.Expressive).
5. Signal logical relationships among reasoning, evidence, data, and/or models when making and defending a claim, counterclaim, and/or rebuttal (ELD-SLC.9-12.Argue.Expressive).

English Language Arts

1. Initiate and participate effectively in small-group collaborative discussions with diverse partners by building on others' ideas and expressing their own clearly (SL.9-10.1).
2. Integrate multiple sources of information presented in diverse media or formats (e.g., visually, quantitatively, and orally) evaluating the credibility and accuracy of each source (SL.9-10.2).
3. Evaluate a speaker's reasoning and use of evidence and rhetoric, identifying any fallacious reasoning or exaggerated or distorted evidence (SL.9-10.3).

(continued)

4. Present information clearly, concisely, and logically such that listeners can follow the line of reasoning, and the organization, substance, and style are appropriate to the purpose, audience, and task (SL.9-10.4).
5. Make strategic use of media (e.g., textual, graphical, and visual) in presentations to enhance understanding of findings, reasoning, and evidence and to add interest (SL.9.10.5).
6. Adapt speech to a variety of contexts and tasks, demonstrating command of formal English when indicated or appropriate (SL.9-10.6).

Activities

Activity A	Create a Draft Argument
Time Needed for the Activity	15 minutes
Grouping	Teams

What to Do

1. Students should work in their assigned teams for this activity. Begin this activity by giving out a copy of the Share handout to each student.
2. Show the "Create a Draft Argument" page on a screen using the Presentation mode in the Learning Hub so everyone in the classroom can see it.
3. Say, "Okay, it sounds like we have all solved the task. I would now like you to create a draft argument to share what you figured out with the other teams. Your argument should include everything in this image here." Point to the image of a science-based argument projected on the screen. Be sure to go over what should be included in each component of the argument and where students can find the information they need.
4. Say, "I would like you to make your draft argument on a whiteboard. You can use the materials on this table to make your argument." Point to the table where you have placed the materials they need.
5. Ask, "Does anyone have any questions?"
6. Answer any questions that come up.
7. Say, "I'm going to give you exactly 15 minutes to get this done. Your draft argument doesn't need to be perfect. We just need something down on the whiteboards so we can share our ideas."
8. Give each team of students 15 minutes to create a draft argument. As students work in small teams, be sure to move from team to team to answer questions and offer advice.

What to Set Up

Place the following items on a table so students can use them during this stage:

- 24″ x 36″ whiteboard
- 3-4 dry-erase pens in different colors
- Eraser or cloth towel

What to Give Out

Share handout for **Cichlids of Lake Malawi**

What to Show

The "Create a Draft Argument" page in the ADI Learning Hub

What to Look for

Students should include core ideas that they read about during the Ideas stage in their justification.

Teaching Tip

Be sure to plant some different ideas with each group about how to make a strong argument as you move from group to group to check in, answer questions, and offer advice. This strategy helps to ensure that there will be some really strong examples of the evidence and justification components of the argument for students to see and discuss during the next activity. For example, you could help two groups have a really strong evidence component by telling one group, "I think a graph would be really helpful here" and telling another group, "I think your evidence would be stronger if you explained, in words, what this graph shows." You could help a third group have a really strong justification by saying, "Let's brainstorm some ideas that will help you defend your choice of evidence," or "Let's go back to our Ideas handout and see if there are any concepts we can include."

It is important you make sure that all the arguments are different at this point in the investigation by planting different ideas with different groups, because the students will have nothing to critique or discuss during the next activity if they all have the same argument. The goal during this activity is to set the stage for the next activity by ensuring that the various components of a strong argument can be found on different whiteboards within the class. These strong components will then be taken up by all the groups because they reflect good ideas. Be sure to give students a voice and choice during this activity, so they can make mistakes and learn from them.

Attending to Equity

It is often helpful to assign and change roles at least once during this activity. For example, you might assign the 1s to be "in charge of writing" at the beginning of the activity. You can then easily switch roles about five minutes into the activity by saying, "1s, please hand the pens to the 2s. 2s, you are now in charge of writing for the group." By changing roles, you can ensure that all students have a chance to participate and contribute their ideas to the creation of a draft argument.

ADI Go Digital

The **ADI Learning Hub** has tons of features that allow students to share their thinking with each other when they are doing the investigation paperless or remote.

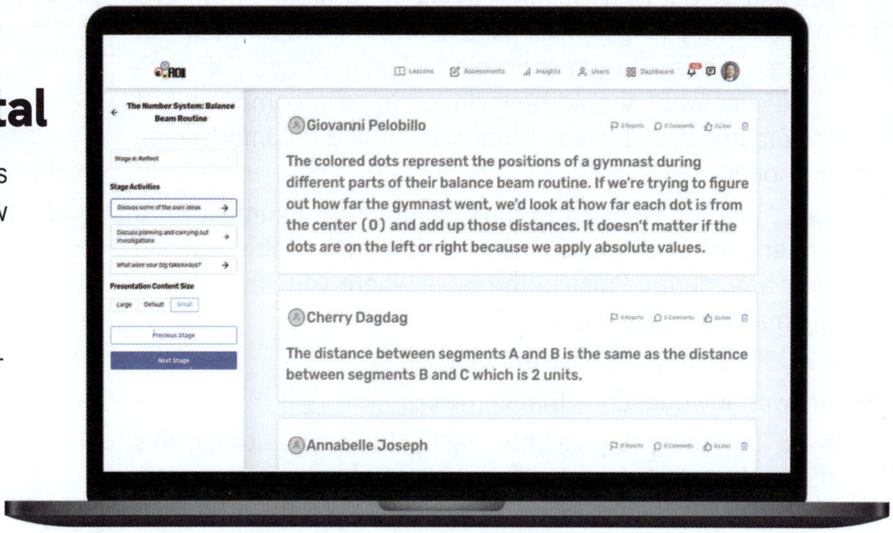

Activity B	Present Your Draft Argument
Time Needed for the Activity	15 minutes
Grouping	Teams

What to Do

1. Students should work in their assigned teams for this activity.
2. Show the "Present Your Draft Argument" page on a screen using the Presentation mode in the Learning Hub so everyone in the classroom can see it.
3. Say, "We are now going to share our arguments. Please set up your whiteboards so everyone can see them." Allow students time to set up their whiteboards. They should be set up on stands or chairs, so they are facing toward the outside of the room.
4. Say, "Okay, before we get started, I want to explain what we are going to do next. I'm going to ask some of you to present your arguments to your classmates. If you are presenting your argument, your job is to share your team's claim, evidence, and justification of the evidence. The rest of you will be travelers. If you are a traveler, your job is to listen to the presenters, ask them questions if you do not understand something, and then offer them some suggestions about ways to make their argument better. Do you have any questions about these instructions?"
5. Answer any questions that come up.
6. Say, "As you talk to one another, make sure you write down any good ideas or suggestions that are important in the space on your handout, so you don't forget about them." Point to the box labeled, "Some possible ways to improve our argument" on the handout.
7. Say, "After we have a chance to learn from one another, I'm going to give you some time to revise your arguments and make them better. Do you have any questions about what we are doing?"
8. Answer any questions that come up.
9. Say, "Okay. Let's get started. Travelers, move one team to the left. Don't move to the next team until I tell you to move. Presenters, go ahead and share your argument with the travelers when they get there."
10. Give each team of presenters and travelers about 3-4 minutes to talk.
11. Tell the students, "Okay. Let's move on to the next argument. Travelers, move one team to the left."
12. Again, give each team of presenters and travelers about 3-4 minutes to talk.

What to Show

The "Present Your Draft Argument" page in the ADI Learning Hub

What to Look for

The critiques and feedback that students give one another during this activity should:

- Be shared in a way that is kind (respectful),
- Provide information about what, exactly, needs to be changed or improved (specific), and
- Offer a suggestion about how to make something better rather than just saying something is wrong (helpful).

In addition, students should be focusing their critiques on the science included in the arguments, not the appearance of the arguments.

(continued)

13. Repeat this process until each team has had their argument read and critiqued three times. Then tell the students to return to their workstations.
14. Be sure to move between workstations during this activity to keep everyone involved and model what it means to communicate and critique scientific ideas and reasoning.

Attending to Equity

Be sure to encourage students to provide feedback that is respectful, specific, and helpful during this activity. It is important to encourage student-to-student talk because it helps students develop speaking and listening skills and enables them to make sense of what and how they know. It also provides a context for emerging multilingual students to use both productive (speaking and writing) and receptive (listening and reading) language and learn science through meaningful interactions and scientific sense-making.

Teaching Tip

There are two types of formats for sharing and critiquing arguments. The first format is called a *gallery walk.* In this format, each group sets up their argument at their workstation so others can see it. The entire group then moves to a different workstation. While at a different workstation, the group is expected to read the argument at that workstation (which was created by a different group) and offer critiques and suggestions for improvement. The students should move to a different workstation after a few minutes to read and critique an argument written by a different group. Students should also be encouraged to keep a record of good ideas or potential ways to improve their own arguments as they move from workstation to workstation to offer critiques and feedback. After repeating this process in the gallery walk 3-4 times, every group will have had an opportunity to read and critique three or four different arguments and receive feedback from three or four different groups.

The second format is called a *modified gallery walk.* In the modified gallery walk format, one or two members of the group stay at their workstation to share their group's ideas (these students are called presenters) while the other members of their group go to different groups, one group at a time, to listen to and critique the arguments developed by their classmates (these students are called travelers). The travelers should visit at least three different workstations during the argumentation session. We also recommend that the presenters keep a record of the critiques made by their classmates and any suggestions for improvement. The travelers should also be encouraged to keep a record of good ideas or potential ways to improve their own arguments as they travel from group to group.

Attending to Equity

Another good way to have students, especially emerging multilingual learners, give one another feedback is to have them use four different color sticky notes to make comments. Give each group a single pink, blue, yellow, and green sticky note. Then direct the students to use the pink sticky note to suggest "something to add" to the argument, the blue sticky note to suggest "something to change" in the argument, the yellow sticky note to highlight "something they like" in the argument, and the green sticky note to ask a question about "something they don't understand". This strategy is useful because everyone in the classroom will be actively engaged during this activity, and everyone will have a chance to see different arguments. This format, however, does not give students a chance to support, critique, and challenge the ideas, solutions, scientific reasoning of other groups through talk because two different groups are never at the same workstation at the same time.

Activity C	Revise Your Argument
Time Needed for the Activity	15 minutes
Grouping	Teams

What to Do

1. Students should work in their assigned teams for this activity.
2. Show the "Revise Your Argument" page on a screen using the Presentation mode in the Learning Hub so everyone in the classroom can see it.
3. Say, "I'm now going to give you all about 10 minutes to revise your argument. Please return to your original teams."
4. Wait for the students to move back into their original teams.
5. Say, "Okay, take a few minutes to talk in your teams and determine what you want to change to make your argument better. Once you have decided what to change, go ahead and make the changes to your whiteboard."
6. Ask, "Do you have any questions about what you need to do?"
7. Answer any questions that come up.
8. Tell the students, "Okay. Let's get started."
9. Give the students 10 minutes to work in their teams on their arguments. As they work, move from team to team to check in, ask probing questions, and offer a suggestion if a team gets stuck.
10. When students are finished with their final arguments, have them take a picture of it or have them put the boards in a safe spot so they can refer to the picture or the boards during the Report stage.

What to Show

The "Revise Your Argument" page in the ADI Learning Hub

What to Look For

A strong argument includes the following information:

- A claim that states, "The speciation of cichlids is driven by sexual selection with females choosing mates based on specific rituals, appearances, and timing."
- Several graphs that compare how different conditions affected the fish population and its characteristics.
- An explanation of the graphs.
- An explanation of the chosen evidence that uses core ideas from the investigation.

Teaching Tip

It is important to give students time to meet with their original groups so they can discuss what they learned by interacting with individuals from the other groups. At this point in the process, students often realize that their initial problem-solving model was flawed in some way. It is therefore important to encourage students to "redo the science" as needed. At the end of this activity, each group should have a final argument that is much better than their original one.

Cichlids of Lake Malawi Lesson Plan

Stage 6 REFLECT

Total Time for the Stage	25 minutes
Grouping	Whole class, Teams

Learning Goals

Students should make progress on their ability to do the following by the end of this stage of the lesson:

Science

1. Ask questions that arise from careful observation of phenomena, or unexpected results, to clarify and/or seek additional information as needed (SEP1).
2. Describe how natural selection occurs only if there is both variation in the genetic information between organisms in a population and variation in the expression of that genetic information, leading to differences in performance among individuals (LS4.B).
3. Explain how traits that positively affect an organism's survival are more likely to be reproduced, and thus become more common in the population (LS4.B).
4. Describe how natural selection leads to adaptation, or to a population dominated by organisms that are anatomically, behaviorally, and physiologically well suited to survive and reproduce in a specific environment (LS4.C).
5. Identify how changes in an environment have contributed to the expansion of some species, the emergence of new distinct species, and the decline and extinction of other species (LS4.C).
6. Explain how species become extinct because they can no longer survive and reproduce in their altered environment (LS4.C).
7. Use models to simulate systems and interactions within and between systems at different scales (CC4).

English Language Development

1. Paraphrase central ideas in complex evidence, concepts, processes, and information to help explain how or why a phenomenon occurs (ELD-SLC.9-12.Explain.Interpretive).
2. Use scientific explanations about a phenomenon in a different context or situation (ELD-SLC.9-12.Explain.Expressive).

English Language Arts

1. Participate effectively in small-group collaborative discussions with diverse partners on texts by building on others' ideas and expressing their own clearly and persuasively (SL.9-10.1).

Activities

Activity A	Discuss a Core Idea
Time Needed for the Activity	10 minutes
Grouping	Whole class/Teams

What to Do

1. Students should be in small teams for this activity. However, they do not need to be in the same teams as they were during previous stages.
2. Begin this activity by giving out a copy of the Reflect handout to each student.
3. Show the "Discuss a Core Idea" page on a screen using the Presentation mode in the Learning Hub so everyone in the classroom can see it.
4. Say, "Okay. Now that we have all revised our arguments and made them better, let's take some time to discuss some of the core ideas you used during this investigation and to justify the evidence you used in your arguments, so we are all on the same page. Look at the picture on the screen."
5. Pause for about 30 seconds.
6. Say, "Take a few minutes to talk in your teams and see if you can use what you know about speciation, natural selection, and sexual selection to explain what is happening in the picture."
7. Give the students at least three minutes to talk in their small teams.
8. Say, "Okay. Let's share with the whole class. Team A (or whichever team you want to share first), what did you talk about?"
9. Allow the first team to share.
10. Say, "Interesting. Would anyone like to add to, challenge, or refine the ideas shared by Team A?"
11. Allow students to respond.
12. Say, "Okay. Team B (or whichever team you want to share second), what did you talk about?"
13. Allow the second team to share.

(*continued*)

What to Give Out

Reflect handout for **Cichlids of Lake Malawi**

What to Show

The "Discuss a Core Idea" page in the ADI Learning Hub

Explanation of the Image

The image that the students discuss during this activity shows different populations undergoing natural and sexual selection.

What to Look For During the Discussion

Students who understand the content will likely mention some, if not all, of the following ideas when providing an explanation for the image:

- A speciation event can occur due to natural selection, sexual selection, or a combination of the two processes.
- Natural selection is the process by which organisms in a population become more suited for their environment because they inherit traits that are more advantageous for their survival.

(*continued*)

14. Say, "Interesting. Would anyone like to add to, challenge, or refine the ideas shared by Team B?"
15. Allow students to respond.
16. Keep up this pattern of questioning until all teams have shared. Be sure to keep encouraging students to talk about the content until they reach a correct and consensual explanation for what is shown in the image. If they are unable to reach a consensus or reach consensus using an inaccurate idea, be sure to provide an accurate explanation for the image before moving on to the next activity.

- Natural selection is a consequence of the amount of variation within a population, a selective pressure in the environment that results in some individuals surviving to reproduce while others do not, and the inheritability of the traits that support survival.
- Sexual selection describes when traits become more prominent within a population due to how these traits help an organism attract a mate.
- Natural and sexual selection can only occur if there is both variation in the genetic information between organisms in a population and variation in the expression of that genetic information.

Teaching Tip

This activity is a discussion, not a presentation about what the students "should have seen" or "should have learned" during the investigation. The more students talk during this activity, the more meaningful the experience will be for them and the more you can learn about their thinking. Be sure to use talk goals and moves to help facilitate the discussion. Below are some examples of talk goals that you can use to help facilitate the discussion and some questions that you can ask the students to help accomplish each goal. These talk goals and questions are only suggestions.

Goal 1: Encourage students to share, expand on, or clarify an idea.

- What do you see going on here?
- What do you think is happening?
- How might we explain that?
- Can you tell me more about that?
- What do you mean by that?
- Can you give me an example of that?
- Can you say that in another way?

Goal 2: Encourage students to support an idea.

- Why do you think that?
- What is your evidence for that?
- How did you arrive at that conclusion?
- What makes you say that?
- How do you know that?

Goal 3: Encourage students to think with others.

- Would anyone like to add to that idea?
- What should we keep or change about that idea?
- Can anyone else give me some examples of that?
- Can anyone take that idea and push it a little further?
- Does anyone want to respond to that idea?
- Do you agree or disagree with that idea, and why?
- What do we think about that idea?
- Does that fit with what we have figured out so far?
- Is there another way to do it?

Goal 4: Encourage students to think about what they know and don't know yet.

- I'm confused now. What have we figured out so far?
- What do we need to figure out now?
- So, what do we agree on right now, and what are we still trying to figure out?
- What are we sure and not so sure about?
- What questions do you have?

Activity B	Discuss the Practices
Time Needed for the Activity	10 minutes
Grouping	Whole class/Teams

What to Do

1. Show the first page of "Discuss the Practices" on a screen using the Presentation mode in the Learning Hub so everyone in the classroom can see it.
2. Say, "Okay. Now that we have talked about some of the core ideas that we used during our investigation, let's take a few minutes to talk about the science practices that we used during the investigation. An example of a scientific practice used was analyzing and interpreting data. On the screen there is a question that I want you all to think about: What other things did you do during your investigation that you feel were good examples of a scientific practice or attempting to figure out something in a scientific way?"
3. Pause for about 30 seconds.
4. Say, "Take a few minutes to talk about this question in your teams."
5. Give the students at least three minutes to talk in their small teams.
6. Say, "Okay. Let's share with the whole class. Team A (or whichever team you want to share first), what did you talk about?"
7. Allow the first team to share.
8. Say, "Okay. Team B (or whichever team you want to share second), what did you talk about?"
9. Allow the second team to share.
10. Continue asking teams until every team has a chance to share.
11. Show the second page of "Discuss the Practices" on a screen using the Presentation mode in the Learning Hub so everyone in the classroom can see it.
12. Say, "Here is another question that I want you all to think about: What things did you do during your investigation that were not good examples of a scientific practice or not very scientific?"
13. Pause for about 30 seconds.

What to Show First

The first page of "Discuss the Practices" in the ADI Learning Hub

What to Show Second

The second page of "Discuss the Practices" in the ADI Learning Hub

What to Show Third

The third page of "Discuss the Practices" in the ADI Learning Hub

(continued)

14. Say, "Take a few minutes to talk about this question in your team."
15. Give the students at least three minutes to talk in their small teams.
16. Say, "Okay. Let's share with the whole class. Team A (or whichever team you want to share first), what did you talk about?"
17. Allow the first team to share.
18. Say, "Okay. Team B (or whichever team you want to share second), what did you talk about?"
19. Allow the second team to share.
20. Continue asking teams until every team has a chance to share.
21. Show the third page of "Discuss the Practices" on a screen using the Presentation mode in the Learning Hub so everyone in the classroom can see it.
22. Say, "Here is a final question that I want you all to think about: What should we all do during our next investigation to make sure we are using scientific practices and being scientific in the way we try to figure something out?"
23. Pause for about 30 seconds.
24. Say, "Take a few minutes to talk about this question in your team."
25. Give the students at least three minutes to talk in their small teams.
26. Say, "Okay. Let's share with the whole class. Team A (or whichever team you want to share first), what did you talk about?"
27. Allow the first team to share.
28. Say, "Okay. Team B (or whichever team you want to share second), what did you talk about?"
29. Allow the second team to share.
30. Continue asking teams until every team has a chance to share.
31. Once everyone has shared, offer any suggestions or advice about what they can do to improve their next investigation and then move on to the next stage.

Teaching Tip

A great way to wrap up this activity is for the students to contribute to a list of class norms or "gotta-haves" to keep in mind when planning and carrying out an investigation in science. To create this list, each group can suggest for a potential norm or "gotta-have" to add to the list at the end of this activity, then the class votes on the one that they think is most important. The most important one gets added to the list. The list then serves as reminder for students during the Plan and Do stages of subsequent investigations. Over time, the list will grow and each addition to the list will be more nuanced than the last as students develop expertise with scientific processes.

Attending to Equity

Be sure to give all students, especially emerging multilingual students, enough time to talk in small groups before asking anyone to share with the entire class during this activity. Small-group structures offer all students, including emerging multilingual students, a chance to engage in sense-making with their peers as well as space to use their verbal and nonverbal (e.g., writing, drawing pictures, and using gestures) resources to express their ideas. Small-group structures also give students a chance to learn from other students' uses of these resources. When students are tasked with expressing their ideas, be sure that they know it is okay for them to do so in a mode or combination of modes (speaking, writing, drawings, and gestures).

Cichlids of Lake Malawi Lesson Plan

Stage 7 REPORT

Total Time Needed for the Stage	100 minutes (over three days)
Grouping	Individual, Teams

Learning Goals

Students should make progress on their ability to do the following by the end of this stage of the lesson:

Science

1. Apply concepts of statistics and probability to support explanations that organisms with an advantageous heritable trait tend to increase in proportion to organisms lacking this trait (HS-LS4-3).
2. Evaluate the evidence supporting claims that changes in environmental conditions may result in increases in the number of individuals of some species, the emergence of new species over time, and the extinction of other species (HS-LS4-5).
3. Make and defend a claim based on evidence about the natural world that reflects scientific knowledge and student-generated evidence (SEP7).
4. Construct a written argument based on evidence (SEP7).
5. Evaluate the claims, evidence, and/or reasoning to determine the merits of arguments (SEP7).
6. Respectfully give critiques on scientific arguments by and challenging ideas and conclusions (SEP7).
7. Communicate scientific information or ideas in multiple formats including graphically and textually (SEP8).

English Language Development

1. Identify appropriate and sufficient evidence from data, models, and/or information from investigations of a phenomenon (ELD-SLC.9-12.Argue.Interpretive).
2. Compare reasoning and claims based on evidence from competing arguments (ELD-SLC.9-12.Argue.Interpretive).
3. Defend or refute a claim based on data and evidence (ELD-SLC.9-12.Argue.Expressive).
4. Establish and maintain an appropriate tone and stance (ELD-SLC.9-12.Argue.Expressive).
5. Signal logical relationships among reasoning, evidence, data, and/or models when making and defending a claim (ELD-SLC.9-12.Argue.Expressive).

English Language Arts

1. Write routinely over extended time frames (time for reflection and revision) and shorter time frames for a range of discipline-specific tasks, purposes, and audiences (WHST.9-10.10).
2. Write arguments focused on discipline-specific content (WHST.9-10.1).
3. Write informative/explanatory texts, including the narration of scientific procedures/experiments or technical processes (WHST.9-10.2).
4. Produce clear and coherent writing in which the development, organization, and style are appropriate to task, purpose, and audience (WHST.9-10.4).

(continued)

5. Strengthen writing as needed by revising, editing, rewriting, and focusing on addressing what is most significant for a specific purpose and audience (WHST.9-10.5).
6. Determine the central ideas or conclusions of a text; trace the text's explanation or depiction of a complex process, phenomenon, or concept; provide an accurate summary of the text (RST.9-10.2).
7. Determine the meaning of symbols, key terms, and other domain-specific words and phrases as they are used in a specific scientific or technical context (RST.9-10.4).
8. Analyze the structure of the relationships among concepts in a text, including relationships among key terms (RST.9-10.5).
9. Assess the extent to which the reasoning and evidence in a text support the author's claim (RST.9-10.8).
10. Compare and contrast findings presented in a text to those from other sources (including their own experiments), noting when the findings support or contradict other accounts (RST.9-10.9).

Activities

Activity A	Draft Report
Time Needed for the Activity	20 minutes
Grouping	Individual

What to Do

1. Have students work on their own during this activity. Give a copy of the Report handout to each student.
2. Show the "Create a Draft Report" page on a screen using the Presentation mode in the Learning Hub so everyone in the classroom can see it.
3. Say, "You are now going to write an investigation report to share what you figured out. The first part of the report is called the 'Introduction'. In this section, you should explain to the reader what you were investigating, why you were investigating it, and what question you were trying to answer. All this information can be found on your Task handout."
4. Hold up the Task handout.
5. Ask the students, "Do you have any questions about what you need to do?"
6. Answer any questions that come up.
7. Say, "Here are some sentence starters to help you begin writing."
8. Hold up the Report handout and point to the sentence starters.
9. Tell the students, "Okay, let's write."
10. Give the students five minutes to write the "Introduction" section of the report. As they work, move from student to student to check in, ask probing questions, and offer a suggestion if a student gets stuck.

What to Give Out

Report handout for **Cichlids of Lake Malawi**

What to Show

The "Create a Draft Report" page in the ADI Learning Hub

(continued)

11. Say, "Okay, let's move on to the second part of the report. The second part of the report is called the 'Method'. In this section, you should explain to the reader what you did during the investigation to complete the task and answer the guiding question. All this information can be found in the Plan handout. Remember that you all planned and carried out different investigations, so do not assume that the reader will know what you did."
12. Ask the students, "Do you have any questions about what you need to do?"
13. Answer any questions that come up.
14. Say, "Here are some sentence starters to help you begin writing."
15. Hold up the Report handout and point to the sentence starters.
16. Tell the students, "Okay, let's write."
17. Give the students five minutes to write the "Method" section of the report. As they work, move from student to student to check in, ask probing questions, and offer a suggestion if a student gets stuck.
18. Say, "Okay, let's move on to the last part of the report. The third part of the report is called the 'Argument'. In this section, you should explain to the reader what you figured out and how you know you are correct. All this information can be found on your Do handout and on your team's whiteboard."
19. Hold up the Do handout and point to the boards (or remind them that they have a picture of their board).
20. Ask the students, "Do you have any questions about what you need to do?"
21. Answer any questions that come up.
22. Say, "Here are some sentence starters to help you begin writing."
23. Hold up the Report handout and point to the sentence starters.
24. Tell the students, "Okay, let's write."
25. Give the students 10 minutes to write the "Argument" section of the report. As they work, move from student to student to check in, ask probing questions, and offer a suggestion if a student gets stuck.
26. Collect all the draft reports before moving on to the next activity. Tell students to mark their draft reports with a symbol that they can use to identify it later. They should NOT write their names on their draft reports.

Teaching Tip

Writing is an important component of this model. When a student writes the report on their own, it not only helps the student learn to write in the context of science (an important literacy skill) but also allows that student to write to learn (develop a better understanding of scientific concepts by writing about them). There are many writing supports (e.g., sentence starters) already embedded into this activity, so you might not need to provide much extra support for your students.

This activity can be done completely in class or started in class and then finished as homework. Do not assign the entire activity as homework unless you know that every student in your class has dedicated time, space, and resources to write outside of class and everyone understands and is comfortable with the report writing process.

If you are concerned about students feeling overwhelmed, you can develop a class "gotta-have-it" list of information that needs to be included in each section of the report. Students can then refer to this list as they write each section.

Don't worry if students do not produce a perfect report at this point. All the reports will be reviewed and revised before they are turned in to you. The reports should therefore be viewed as a starting point. Be sure to encourage students to "just get something down on paper" so "we can work on it together."

Attending to Equity

Emerging multilingual students' access to instructional tasks that require complex thinking, such as writing a report, is enhanced when linguistic complexity and instructional support match their current level of language proficiency. Therefore, it is important to scaffold the writing of the reports for the emerging multilingual students based on their current English language proficiency level. The use of sentence starters to model written language production skills can help emerging multilingual students develop English language skills because the development of academic language and academic content knowledge are interrelated processes. However, it is important that such scaffolds be used purposefully and removed when no longer needed.

ADI Go Digital

Use the **ADI Learning Hub** to go paperless. When you create a class and add students to it, students can complete all the stages right in the system.

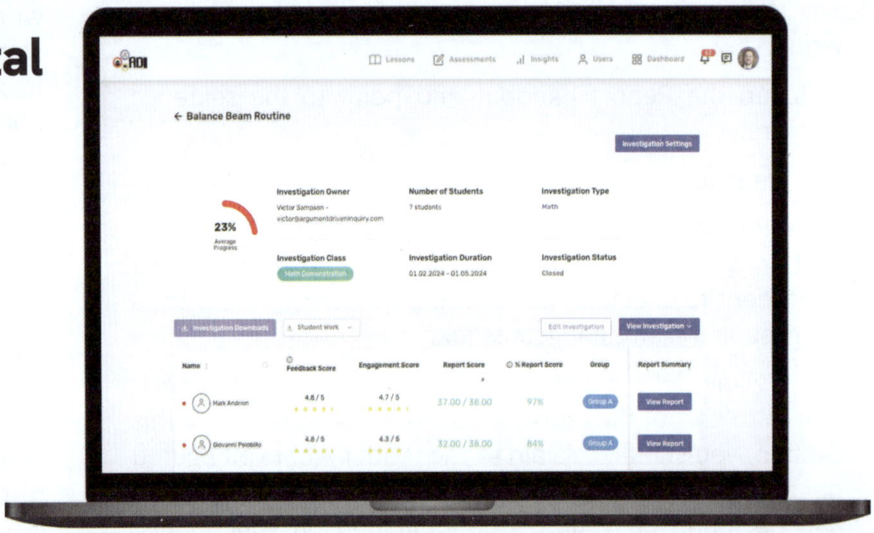

Activity B	Peer Review the Reports
Time Needed for the Activity	45 minutes
Grouping	Teams

What to Do

1. Students should work in teams of three for this activity. These can be completely different teams.
2. Show the "Peer Review Some Draft Reports" page on a screen using the Presentation mode in the Learning Hub so everyone in the classroom can see it.
3. Say, "We are now going to review our reports to find ways to make them better."
4. If possible, use a document camera to project the peer review guide (PRG) on a screen.
5. Say, "We are going to use this peer review guide to give each other feedback." Point to the image.
6. Say, "I'm going to ask you to work in teams of three to do this. I will give each team a draft report to read. You three will then read the report together. Once you are done reading the report, I want you to answer each of the questions on the peer review guide." Point to the review questions on the image of the PRG.
7. Tell the students, "You can check No, Almost, or Yes after each question." Point to the checkboxes on the image of the PRG.
8. Say, "This will be your rating for this part of the report. Make sure you agree on the rating you give the author. If you mark No or Almost, then you need to tell the author what he or she needs to do to get a Yes." Point to the space for the reviewer feedback on the image of the peer review guide.
9. Say, "It is really important for you to give the authors feedback that is helpful. That means you need to tell them exactly what they need to do to make their report better."
10. Ask the students, "Do you have any questions about what you need to do?"
11. Answer any questions that come up.
12. Tell the students, "Please get into your teams of three."
13. Allow the students time to move into their triads.
14. Say, "Okay, I'm now going to give you one report to read." Give out one draft report to each triad. Make sure that the report you give a triad was not written by one of the students in that triad. Give each triad one PRG to fill out.

(continued)

What to Give Out

Peer Review Guide (one per group per review, so for this investigation you need the same number of peer review guides as the number of draft reports that will be reviewed)

What to Show

The "Peer Review Some Draft Reports" page in the ADI Learning Hub

What to Look For

Make sure students are reaching consensus about what rating to give each criterion. Also, be sure that students are giving one another feedback that is specific, detailed, and helpful when they think an aspect of a report needs to be improved.

15. Tell the students, "Okay, I'm going to give you 15 minutes to read the report I gave you and to fill out the peer review guide. Go ahead and get started."
16. Give the students 15 minutes to work. As they work, move around from team to team to check in and see how things are going, answer questions, and offer advice.
17. After 15 minutes pass, tell the students, "Okay, time is up. Please give me the report and the peer review guide that you filled out."
18. Collect the draft reports and the PRGs, or collect the workbooks if students are using them.
19. Say, "Okay, I am now going to give you a different report to read and a new peer review guide to fill out." Give out one more report to each triad. Make sure that the report you give a triad was not written by one of the students in that triad. Give each triad one new peer review guide to fill out as a team.
20. Say, "Okay, I'm going to give you 15 minutes to read this new report and to fill out the peer review guide. Go ahead and get started."
21. Give the students 15 minutes to work. As they work, move around from team to team to check in and see how things are going, answer questions, and offer advice.
22. After 15 minutes pass, tell the students, "Okay, time is up. Please give me the report and the peer review guide that you filled out."
23. Repeat this process for the remaining draft reports. Each triad should one have one more report to review (for a total of three reviews).
24. Collect the last draft reports and PRGs. Set all the draft reports on a table in separate stacks. Each stack should have a draft report on top (with the identification symbol facing up so students can find their reports) and the PRG on the bottom.

Teaching Tip

The peer review process is one of the best ways for students to learn how to write. When students read and review different reports, they have an opportunity to (1) see examples of texts written by others for the same purpose, (2) discuss what counts as quality and why some reports are stronger than others, (3) discuss ways to strengthen a report, and (4) pick up things that they can do to improve their reports. This helps the students learn new ways to organize and present information, which will help them write better on subsequent reports.

Attending to Equity

One helpful scaffold is to give students sentence stems for what to write in their feedback to the author. Ideal feedback is respectful, specific about what needs to be changed, and helpful in that it suggests how the author should go about making that change. You can review the writing of the emerging multilingual students in your class if they have limited English language proficiency, but make sure that all emerging multilingual students participate in the peer review process. You will see tremendous growth in their writing skills as they review their peers' writing and discuss what makes some reports stronger than others.

Activity C	Revise the Reports
Time Needed for the Activity	25 minutes
Grouping	Individuals

What to Do

1. Students should work on their own for this activity.
2. Show the "Reflect on the Feedback" page on a screen using the Presentation mode in the Learning Hub so everyone in the classroom can see it.
3. Say, "Please come pick up your reports from over here on this table." Point to the table.
4. Give the students a few minutes to pick up their reports and return to their seats.
5. Say, "Please take a few minutes to read over the peer review guide. You should use it to figure out what you need to change in your report and how you will change it."
6. Allow the students to read the PRG.
7. Show the "Create a Final Report" page on a screen using the Presentation mode in the Learning Hub so everyone in the classroom can see it.
8. Say, "Okay, let's revise our reports. I would like you to rewrite your report. You can use your draft report as a starting point but use the feedback on the peer review guide to make it better."
9. Ask the students, "Do you have any questions about what you need to do?"
10. Answer any questions that come up.
11. Tell the students, "Okay, let's write." Allow about 20 minutes for the students to revise their reports.
12. After about 20 minutes, give the following directions, "Okay, time's up. I will now come around and collect your first draft, final report, and peer review guide."
13. Collect all the first drafts, PRGs, and final reports.
14. Use the "Teacher Score" columns in the PRG to grade the final report. Be sure to give the students feedback about their writing in the "Teacher Comments" section.

What to Show First

The "Reflect on the Feedback" page in the ADI Learning Hub

What to Show Second

The "Create a Final Report" page in the ADI Learning Hub

Attending to Equity

Students don't always use the feedback they receive from others to improve the quality of their reports. Make sure you move from student to student and encourage them to make changes based on the feedback. If students are not sure how to revise their reports, model how to use feedback. For example, you might say, "Okay, based on the peer review guide, I need to do more to describe what I did to collect data. I bet if I add information about ______, readers will have a better idea of what I did."

Teaching Tip

The peer review guide (PRG) is designed to be used with any ADI investigation, thus allowing you to use the same scoring rubric throughout the entire year. This is beneficial for several reasons. First, the criteria for what counts as a high-quality report do not change from investigation to investigation. Students, therefore, quickly learn what is expected from them when they write a report, and you do not have to spend valuable class time explaining the various components of the PRG each time students write a report. Second, the PRG makes it clear which components of a report need to be improved next time because the grade is not based on a holistic evaluation of the report. The PRG format allows students to see which aspects of their writing are strong enough and which aspects need improvement. Finally, and perhaps most importantly, the PRG provides you with a standardized measure of student performance that can be compared over multiple reports across semesters, thus allowing you to track improvement over time.

Teacher Notes

Cichlids of Lake Malawi

TASK

The Phenomenon

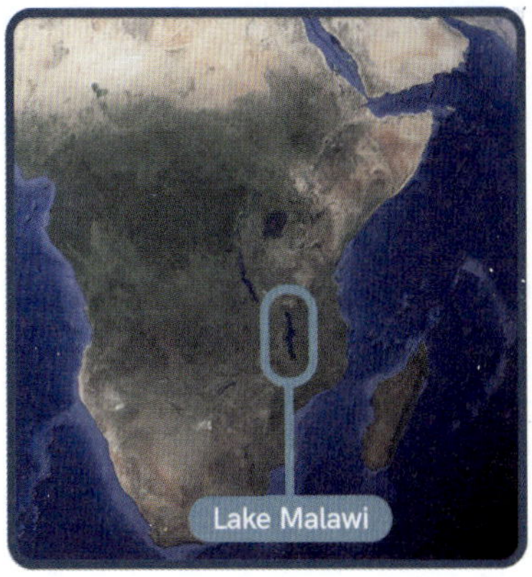

Lake Malawi is one of the African Great Lakes (see picture at right). It is the ninth largest lake in the world and the third largest lake in Africa by surface area. Lake Malawi is known for its biodiversity of fish. In fact, there are least 800 species of cichlids that live only in this lake. Cichlids are often known for their vibrant colors and liveliness. Their behaviors, such as eating and mating, vary from species to species. Take a few minutes to watch a video about the cichlids of Lake Malawi. As you watch the video, write down some things you notice in the space below.

Things I **noticed**...

The Task

The 800 or more species of cichlids in Lake Malawi are the descendants of a single small population of cichlids that entered the lake one or two million years ago. Cichlids in Lake Malawi can live up to eight years and reproduce every two months when it is warm enough and there is plenty of food available. Some species of cichlids in Lake Malawi are herbivores, some are carnivores, and others are omnivores. The cichlids in Lake Malawi have complex mating and breeding behaviors that are unique to each species. Your goal in this investigation is to figure out how a single population of cichlids can result in so many different species. The guiding question of this investigation is:

Why are there so many species of cichlids in Lake Malawi?

Initial Ideas

Before you start this investigation, take a few minutes to think about how biologists define a species and how new species appear over time. Then, draw a model in the space below that shows why you think there are so many species of cichlids in Lake Malawi. Be sure to include labels to help explain your thinking.

Some **ideas** that I have...

Share your model with the rest of your group. Once everyone has had a chance to share their ideas, list some questions that your group has about this phenomenon.

Some **questions** we have now...

Cichlids of Lake Malawi

IDEAS

Some Ideas You Can Use: Species and Speciation

Biologists can define a species in several different ways. One way to define a species is to use the **phenotypic species definition**. According to this definition, a species is a group of organisms that look the same. This definition is often used by biologists who study the history of life on Earth. However, organisms that look very similar may never mate and produce offspring, and the production of offspring is essential for the long-term survival of any species. For example, western meadowlarks and eastern meadowlarks look almost identical to one another (see pictures below), yet do not mate and produce offspring. Also, some organisms that look very different than each other, such as the two mandarin ducks in the picture below, do mate and produce offspring. To address this issue of mating, biologists can also define a species using the **biological species definition**. According to this definition, a species is a group of individuals that interbreed in nature. Ecologists often use this definition of a species.

Similar and different looking birds

The biological species definition, however, does not apply to all life on Earth. For example, many bacteria reproduce asexually. The bacterium in the picture on the next page, for example, does not need to mate with another bacterium to produce an offspring. Many flowering plants can also self-pollinate and produce offspring without mating with another plant. Also, some groups of organisms that look very different and do not usually mate with each other have been observed mating across groups only in a specific area or during a specific time of year. Hooded crows and carrion crows (such as the ones in the pictures on the next page) are examples of organisms that will occasionally mate and produce an offspring. To address this issue of different methods of reproduction, biologists can also define a species using the **phylogenetic species definition**. According to this definition, a species is the smallest set of organisms that share an ancestor and can be distinguished from other such sets. This definition of a species is often used by biologists who study evolution, genetics, or growth and development.

Reproduction in different types of living things

Speciation is the term that biologists use to describe the creation of a new species. A **speciation event** is the point in time and/or location where a new species splits off from an existing one. The new species is created and splits off from an existing one because a small group of individuals within a larger population develops one or more characteristics that change how this small group of individuals interact with the individuals that make up the larger population. The key to a speciation event is the evolution of traits or behavior that makes mating between the two groups unlikely (or at least unsuccessful when it happens). The difference that evolves in the small group of individuals does not need to be substantial for a speciation event to occur. A small change in the timing, location, or rituals of mating could be enough to cause a speciation event; however, a speciation event cannot happen without the change happening first.

Some **useful ideas** from what I read...

Some Ideas You Can Use: Natural Selection and Sexual Selection

A change in timing, location, or rituals of mating that results in a speciation event can evolve by natural selection, sexual selection, or some combination of the two. Evolution by means of **natural selection** is a consequence of the interaction of three factors: (1) the amount of variation within a population, (2) a selective pressure in the environment, such as a predator, lack of food, or a disease, that results in some individuals surviving long enough to reproduce and some dying before they have a chance to produce offspring, and (3) the inheritability of the traits that support survival. These three factors are illustrated in the image below using a hypothetical beetle population.

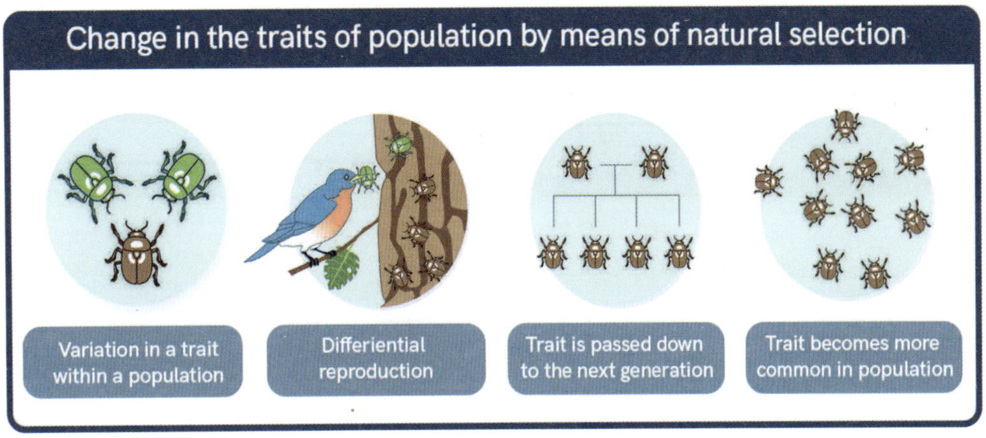

Another mechanism or means of evolution is a process called **sexual selection**. Evolution by means of sexual selection is similar to natural selection in many ways because it requires variation in a trait within a population and a selective pressure that leads to differential reproduction. However, instead of the traits that are selected for or against affecting the survival of an individual, the traits that are selected affect the ability of an individual to attract a mate. These three factors are illustrated in the image below using a hypothetical bird population as an example.

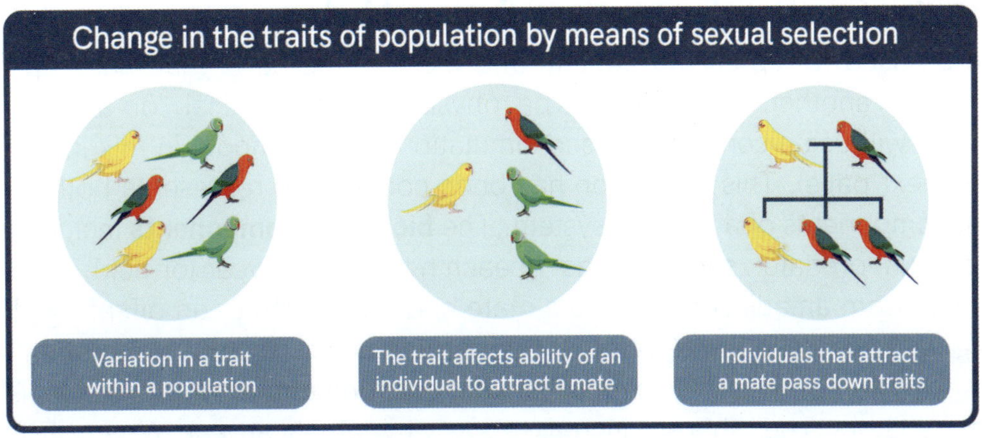

When a population evolves by means of natural selection, traits that support survival in a specific environment become more common; those that do not become less common. Natural selection, in other words, leads to the development of specific adaptations that make a group of organisms well suited to survive within a specific environment. When a population evolves by means of sexual selection, in contrast, traits that support successful mating become more common; those that do not become less common regardless of the environment. Sexual selection, in other words, leads to the development of specific sex characteristics that make it more likely for an individual to obtain a mate regardless of the environment. Natural selection and sexual selection, however, can only occur if there is both (1) variation in the genetic information between organisms in a population and (2) variation in the expression of that genetic information, which leads to a variation in one or more traits. Without **variation**, the processes of natural selection and sexual selection cannot happen.

Some **useful ideas** from what I read...

Some Ideas You Can Use: Systems and System Models

Models can be used to represent a system and the way the parts that make up that system interact with one another. A **model** can be physical, mathematical, or a computer-based simulation. You will use a computer-based simulation as your model in this investigation (see image on the next page). This simulation has been coded so it represents both the biological and physical components of a lake ecosystem. The **biological components** include populations of fish and a visual predator, such as birds. In each hypothetical year (or tick), each female fish chooses a male from an adjacent area as a mate and gives birth to an offspring. The color of the offspring is the average of the color of the parents unless there is a mutation in the color gene. The more different the fish's color is from the environment, the more likely it is to become prey for the predators. The fish's age increases every hypothetical year. When a fish reaches 20 years old, it dies. The **physical components** of the ecosystem in the model include the amount of space and color of the water. This computer-based simulation, as a result, will enable you to figure out how a single population of cichlids can result in so many different species.

A computer-based simulation of a lake

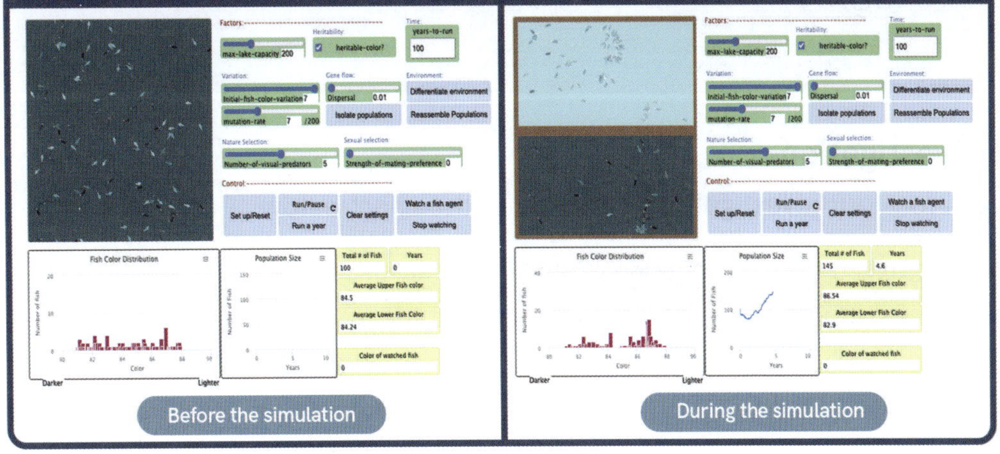

You can change a biological component of the ecosystem, such as the amount of "fish-color variation", if the fish color trait is inheritable or not ("heritable-color?"), how often color mutations occur ("mutation-rate"), the maximum size of the fish population in the lake ("max-lake-capacity"), how much the fish stay together or spread out ("dispersal"), if females prefer to mate with males that are the same color or not ("strength-of-female-mating-preference"), and the "number-of-visual-predators" that feed on the fish in the lake. These biological components of the lake ecosystem will be the **inputs** of the system model. You can also change the physical components of the ecosystem in the model by changing the color of the lake water ("differentiate environment") or creating a barrier that will create two groups of fish ("isolate populations"). These physical components will affect the **interactions** that take place in the system model. How the color of the fish changes over time or the average color of the fish after a period time (such as 100 years) will be the **output of interest** for this investigation. Models like this can be used to predict the behavior of a system under different conditions, such as when the biological and physical components of an ecosystem are altered in some way, but it is important to keep in mind that **these predictions have limited precision and reliability** due to the assumptions and approximation inherent in this, or any, model.

Some **useful ideas** from what I read...

Cichlids of Lake Malawi

PLAN

Plan Your Investigation

Prepare a plan for your investigation by filling out the chart below.

I am trying to answer the following **question**...

I will use the following **observations or measurements**...

I will **collect** these observations or measurements by...

I will **analyze** the observations or measurements that I collect by...

I approve of this investigation plan.

Teacher's signature

Date

Cichlids of Lake Malawi

DO

Collect Data

Keep a record of what you observe or measure during your investigation.

My **observations or measurements...**

Make Sense of Your Data

Use the space below to create your model so you can answer the guiding question.

My **model...**

Cichlids of Lake Malawi

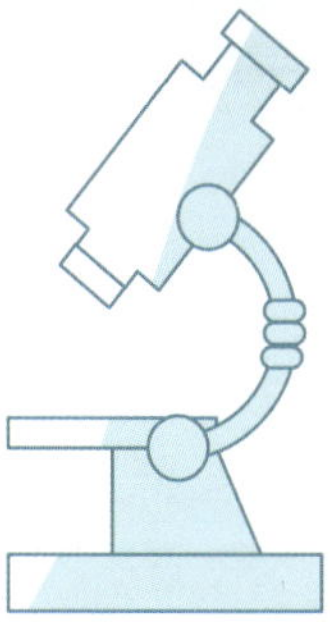

SHARE

Draft Argument

Develop an argument on a whiteboard. It should include:

1. A *claim* that answers the guiding question,
2. *Evidence* in support of the claim that provides an analysis of the data and a description of what the analysis means,
3. A *justification of the evidence* that explains why your group thinks the evidence is important.

Our Guiding Question:

Claim:

Evidence:	Justification of the Evidence:

Argumentation Session

Share your argument with your classmates. In the space below, be sure to keep track of any ideas that you can use to revise your argument and make it better.

Some possible ways to **improve** our argument...

Cichlids of Lake Malawi

REFLECT

Reflective Discussion

In the space below, you can keep track of any ideas from the discussion that you think are important or will be useful in the future.

Some important **ideas**...

Cichlids of Lake Malawi

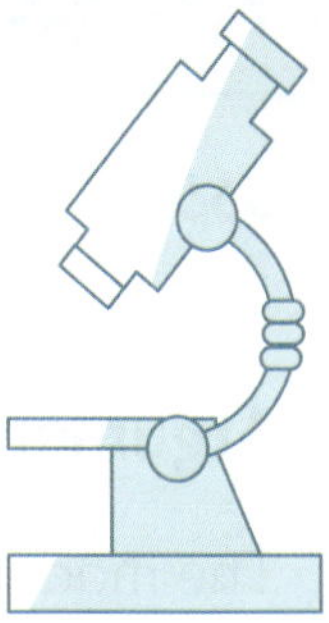

REPORT

Draft Report

Prepare a report to share what you figured out during your investigation.

Introduction

We have been studying ____________________ in class. Before we started this investigation, we explored ____________________

____

____

We noticed ____________________

____

____

Our goal for this investigation was to figure out ____________________

____

____

The guiding question was ____________________

____

Method

To answer this question, ____________________

____

____

____

____

____

____

____

____

____

____

Argument

The model below shows ______________________________

My evidence for this model is ______________________________

This evidence is based on several important scientific ideas. The first one is ______________________________

Other Useful Stuff

This chapter includes the master list of all the materials that will be used by students during the Do stage of each investigation. It also includes three reproducible copy masters. The first reproducible copy master, which is the Safety Rules Acknowledgment form, is intended to be used before you start your first ADI investigation. The other two reproducible copy masters, which include the Peer Review Guide and Teacher Scoring Rubric (PRG-TSR) and exit tickets, are intended to be used during each investigation.

Master List of Hands-on Materials

The master list of hands-on materials includes all the consumables and equipment the students will need to complete all eight investigations in this book. You can purchase all these materials at the ADI Store by purchasing the *Biology Investigations Kit: NGSS Edition* at https://shop.argumentdriveninquiry.com.

Peer Review Guide and Teacher Scoring Rubric (PRG-TSR)

The PRG-TSR is used during the Report stage of an ADI investigation. We recommend that you make one copy of the PRG-TSR for each student. Then, during the peer review stage, have each pair of reviewers fill out one PRG-TSR for each report that they review. The reviewers should rate the report on each criterion and then provide advice to the author about ways to improve it based on their rating. Once the review is complete, the author needs to revise his or her report and respond to the reviewers' rating and comments in the appropriate sections in the PRG-TSR. You should collect the PRG-TSR, the first draft, and the final report for a final evaluation. To score the report, simply fill out the "Teacher Score" column of the PRG/TSR and then total the scores. There is also space at the bottom of the PRG/TSR for teacher feedback.

Exit Tickets

Teachers can require students to fill out an exit ticket after completing each stage of an investigation. The exit ticket consists of two questions asking students to share what they learned and what they are unsure about. The tickets, which take only a few minutes for students to complete, can be used as a formative assessment by teachers because they require students to share their thinking (and how it has or has not changed) during the investigation at key points. Teachers can use these exit tickets to determine what students have picked up and any sources of confusion while there is still time to reteach. As a result, these exit tickets help drive real-time adjustments and decisions about next steps. To use the exit tickets found in the book, simply make one copy for every two students in a class and then cut them in half. Many teachers have also reported that they often use these exit tickets as an academically productive filler activity when that day's lesson runs short. These teachers recommend printing 500+ copies, cutting them in half, and then storing all of them in a deep 4″x 6″ plastic bin at the beginning of the year. They then place the bin near their desk or the door so they can grab a handful and hand them out quickly when needed.

ADI Learning Hub Subscription

The ADI Learning Hub is a browser-based platform designed to make it easier for teachers to facilitate an ADI investigation. To create a free account, follow the directions below:

1. Use a web browser such as Chrome, Firefox, or Edge to visit https://adilearninghub.com.
2. Click on "Create a free teacher account".
3. Fill out the information in the form show below. Be sure to use your school or district email.

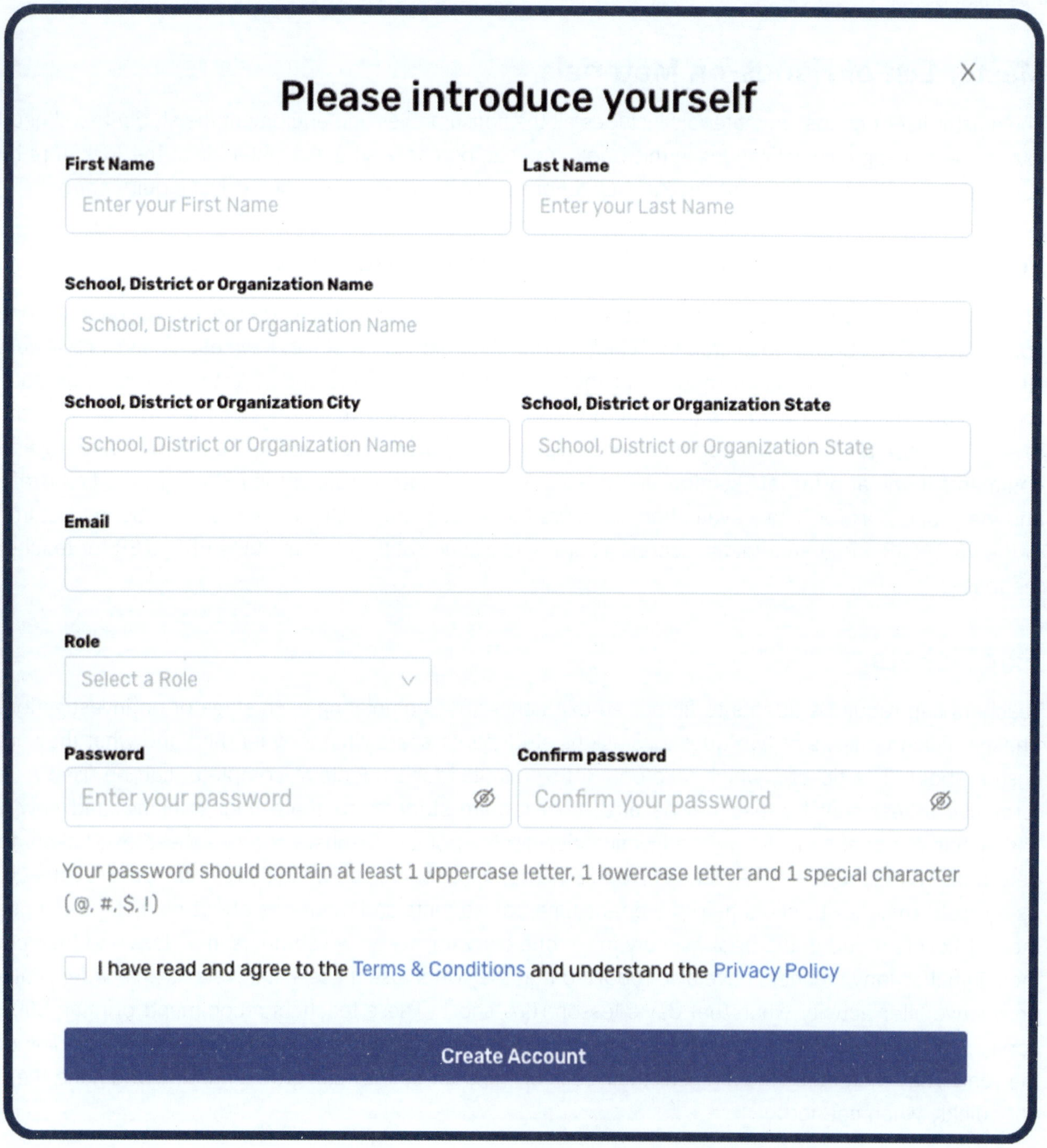

Please introduce yourself

First Name	**Last Name**
Enter your First Name	Enter your Last Name

School, District or Organization Name

School, District or Organization Name

School, District or Organization City	**School, District or Organization State**
School, District or Organization Name	School, District or Organization State

Email

Role

Select a Role

Password	**Confirm password**
Enter your password	Confirm your password

Your password should contain at least 1 uppercase letter, 1 lowercase letter and 1 special character (@, #, $, !)

☐ I have read and agree to the Terms & Conditions and understand the Privacy Policy

Create Account

Complete Hands-on Materials List

Type of Material	Item	Quantity	
		For Entire Class	For Each Group
Consumables	Salt	26 oz	-
	Distilled water*	3 L	-
	Dialysis tubing	20 ft	2 ft
	Saltwater concentration stickers	3	-
Equipment	Pipette	8	1
	Funnel	8	1
	Plastic measuring cups	40	9
	Scale	8	1
	Pitchers	3	-
	Spatula	1	-
	Tissue Repair Card 1: Event A	8	1
	Tissue Repair Card 2: Event B	8	1
	Tissue Repair Card 3: Event C	8	1
	Tissue Repair Card 4: Event D	8	1
	Tissue Repair Card 5: Event E	8	1
	Tissue Repair Card 6: Appearance of Chromosomes	8	1
	Red pop beads	192	24
	Yellow pop beads	384	48
	Hook-and-loop cable ties, 4″	64	8
	Chromosome Reduction in Reproductive Cells Card 1: Meiosis Event A	8	1
	Chromosome Reduction in Reproductive Cells Card 2: Meiosis Event B	8	1
	Chromosome Reduction in Reproductive Cells Card 3: Meiosis Event C	8	1
	Chromosome Reduction in Reproductive Cells Card 4: Meiosis Event D	8	1
	Chromosome Reduction in Reproductive Cells Card 5: Meiosis Event E	8	1
	Chromosome Reduction in Reproductive Cells Card 6: Meiosis Event F	8	1

(*continued*)

Type of Material	Item	Quantity	
		For Entire Class	For Each Group
Equipment	Chromosome Reduction in Reproductive Cells Card 7: Meiosis Event G	8	1
	Chromosome Reduction in Reproductive Cells Card 8: Meiosis Event H	8	1
PPE	Indirectly vented chemical-splash goggles	32	4
	Chemical-resistant apron	32	4
	Chemical-resistant gloves	32	4

Safety Rules Acknowledgment Form

I know that it is very important to be as safe as I can be during an investigation. My teacher has told me how to be safer in science. I agree to follow these 11 safety rules when I am working with my classmates to figure things out in science:

1. I will always act in a responsible manner. I will not run around the classroom, throw things, play jokes on my classmates, or be careless.
2. I will never eat, drink, or chew gum in the class.
3. I will never touch, taste, or smell any materials, tools, or chemicals without permission.
4. I will always wear my safety goggles during the activity setup, hands-on work, and cleanup.
5. I will do my best to take care of the materials and tools that my teacher allows me to use.
6. I will always tell my teacher about any accidents as soon as they happen.
7. I will always dress in a way that will help keep me safer. I will wear closed-toed shoes and pants. My clothes will not be loose, baggy, or bulky. If my hair is long, I will use hair ties to keep it out of the wa while I am working.
8. I will always keep my work area clean and neat. I will put my backpack, books, and other personal items where my teacher tells me to put them, and I will not get them out unless my teacher tells me that it is okay.
9. I will clean my work area and the materials or tools that I use.
10. I will wash my hands with soap and water at the end of the activity.
11. I will always follow my teacher's directions.

______________________	______________________	______________
Print Name	Signature	Date

I have read and reviewed the 11 investigation safety rules with my child. He/She understands how important it is to follow safety rules in science and has agreed to always follow these safety rules. I give my permission for my child to participate in the investigations this year.

______________________	______________________	______________
Parent or Guardian Name	Parent or Guardian Signature	Date

Peer Review Guide and Teacher Scoring Rubric (PRG/TSR)

Report By: ______________________
Name

Author: Did the reviewers do a good job? 1 2 3 4 5
Rate the overall quality of the peer review

Reviewed By: ______________________ ______________________
Name Name

Take a look at the introduction. Do you think:	Reviewer Rating			Teacher Score*
1. There is a good overview of the phenomenon?	☐ No	☐ Somewhat	☐ Yes	0 1 2 3
2. The task and guiding question are clear?	☐ No	☐ Somewhat	☐ Yes	0 1 2 3
3. It is clear why it is important to complete the task?	☐ No	☐ Somewhat	☐ Yes	0 1 2 3
Reviewers: Please explain how the author could improve this part the report.	**Author:** What changes did you make to your report based on feedback from the reviewers?			

Take a look at the method. Do you understand:	Reviewer Rating			Teacher Score*
1. What data or information the author used or collected?	☐ No	☐ Somewhat	☐ Yes	0 1 2 3
2. How the author obtained the data or information?	☐ No	☐ Somewhat	☐ Yes	0 1 2 3
3. How the author made sense of the data or information?	☐ No	☐ Somewhat	☐ Yes	0 1 2 3
Reviewers: Please explain how the author could improve this part of the report.	**Author:** What changes did you make to your report based on feedback from the reviewers?			

* 0: Missing; 1: Approaching Performance Expectation; 2: Meets Performance Expectation; 3: Exceeds Performance Expectation

ARGUMENT-DRIVEN INQUIRY

Take a look at the argument. Do you think:	Reviewer Rating			Teacher Score*
1. The claim answers the guiding question?	☐ No	☐ Somewhat	☐ Yes	0 1 2 3
2. The analysis of the data is correct?	☐ No	☐ Somewhat	☐ Yes	0 1 2 3
3. The interpretation of the analysis makes sense?	☐ No	☐ Somewhat	☐ Yes	0 1 2 3
4. There is enough evidence to support the claim?	☐ No	☐ Somewhat	☐ Yes	0 1 2 3
5. The choice or use of the evidence is well justified?	☐ No	☐ Somewhat	☐ Yes	0 1 2 3
6. All the ideas included in the justification are correct?	☐ No	☐ Somewhat	☐ Yes	0 1 2 3
7. All variables or symbols in the argument are defined?	☐ No	☐ Somewhat	☐ Yes	0 1 2 3
8. The claim is valid or acceptable?	☐ No	☐ Somewhat	☐ Yes	0 1 2 3

Reviewers: Please explain how the author could improve this part of the report.	**Author:** What changes did you make to your report based on feedback from the reviewers?

Look at the entire report. Do you think:	Reviewer Rating			Teacher Score*
1. It is easy to read and well organized?	☐ No	☐ Somewhat	☐ Yes	0 1 2 3
2. The sentences are complete?	☐ No	☐ Somewhat	☐ Yes	0 1 2 3
3. The subject and verb agree in each sentence?	☐ No	☐ Somewhat	☐ Yes	0 1 2 3
4. The voice is active?	☐ No	☐ Somewhat	☐ Yes	0 1 2 3
5. The spelling, punctuation, and capitalization is correct?	☐ No	☐ Somewhat	☐ Yes	0 1 2 3

Additional Review Feedback

What other feedback would you like to give? Is there something that you really liked about their investigation report? Is there anything else that you noticed that could be improved?

Teacher Comments	
	Total: /38

* 0: Missing; 1: Approaching Performance Expectation; 2: Meets Performance Expectation; 3: Exceeds Performance Expectation

Exit Ticket

Name: ______________________ Date: ______________________

My biggest takeaway from today...	I am now wondering...

Exit Ticket

Name: ______________________ Date: ______________________

My biggest takeaway from today...	I am now wondering...